5 STEPS TO A 5™
AP Statistics
2020

5 STEPS TO A 5

AP Statistics

2020

Corey Andreasen
DeAnna Krause McDonald

Mc
Graw
Hill

New York Chicago San Francisco Athens London Madrid
Mexico City Milan New Delhi Singapore Sydney Toronto

1 2 3 4 5 6 7 8 9 LHS 24 23 22 21 20 19 (Cross-Platform Prep Course only)
1 2 3 4 5 6 7 8 9 LHS 24 23 22 21 20 19 (Elite Student Edition)

ISBN   978-1-260-45589-2 (Cross-Platform Prep Course only)
MHID   1-260-45589-0

e-ISBN   978-1-260-45590-8 (Cross-Platform Prep Course only)
e-MHID   1-260-45590-4

ISBN   978-1-260-45591-5 (Elite Student Edition)
MHID   1-260-45591-2

e-ISBN   978-1-260-45592-2 (Elite Student Edition)
e-MHID   1-260-45592-0

The series editor was Grace Freedson, and the project editor was Del Franz. Series design by Jane Tenenbaum.

CONTENTS

PREFACE

Congratulations, you are now an AP Statistics student. AP Statistics is one of the most interesting and useful subjects you will study in school. Sometimes it has the reputation of being easy compared to calculus. However, it can be deceptively difficult, especially in the second half. It is different and challenging in its own way. Unlike calculus, where you are expected to get precise answers, in statistics you are expected to learn to become comfortable with uncertainly. Instead of saying things like, "The answer is ..." you will more often find yourself saying things like, "We are confident that ..." or "We have evidence that ..." It's a new and exciting way of thinking.

How do you do well on the AP exam? By understanding statistical principles. Keep up on the material during your AP Statistics class and read this book to review and prepare for the exam. Note that the questions on the AP exam are generally quite conceptual. In addition to doing computations, you will need to explain why you did them and what they mean. This also means you must communicate your thoughts clearly. You can always use a calculator, so learn to use it well. But success on the AP Statistics exam will require more than pushing buttons.

This book is self-contained in that it covers all of the material required by the course content description published by the College Board. However, it is not designed to substitute for an in-class experience or for your textbook. Use this book as a supplement to your in-class studies, as a reference for a quick refresher on a topic, and as one of your major resources as you prepare for the AP exam.

This edition extends and updates previous editions. It takes into account changes in thinking about AP Statistics since the publication of the first edition in 2004 and includes some topics that should help you understand what is important when answering questions on the exam.

You should begin your preparations by reading through the Introduction and Step 1. However, you shouldn't attempt the Diagnostic Exam in Chapter 3 until you have been through all of the material in the course. Then you can take the exam to help you determine which topics need more of your attention during the course of your review. Note that the Diagnostic Exam simulates the AP exam to a reasonable extent (although the questions are more basic) and the Practice Exams are similar in style and length to the AP exam.

So, how do you get the best possible score on the AP Statistics exam?

- Pick one of the study plans from this book.
- Study the chapters and do the practice problems.
- Take the Diagnostic Exam and the Practice Exams.
- Review as necessary based on your performance on the Diagnostic Exam and the Practice Exams.
- Get a good night's sleep before the exam.

Selected Epigrams about Statistics

All models are wrong, but some of them are useful.
—George Box

Very few people actually use calculus in a conscious, meaningful way in their day-to-day lives. On the other hand, statistics–that's a subject that you could, and should, use on a daily basis.
—Arthur Benjamin

If the statistics are boring, you've got the wrong numbers.
—Edward Tufte

I keep saying the sexy job in the next ten years will be statisticians.
—Hal Varian

The best thing about being a statistician is that you get to play in everyone's backyard.
—John Tukey

The lottery is a tax on people who flunked math.
—Monique Lloyd

So much of the physical world has been explored. But the deluge of data I get to investigate really lets me chart new territory. Genetic data from people living today forms an archaeological record of what happened to their ancestors living 10,000 years ago.
–Pardis Sabeti

As statisticians, we all have important stories to tell. No matter where you work or what you do, I encourage you to TELL OUR STORIES to your employers, to your students, to a high school class, to the media, and to the public. Let's help everyone appreciate statistics!
–Jessica Utts

ACKNOWLEDGMENTS

With gratitude, we acknowledge the following for their assistance and support:

The participants who attended the College Board workshops—I learned as much from them as they did from me.

Our AP Statistics classes at North High School in Sheboygan, Wisconsin, and University High School in Tucson, Arizona, for being willing subjects as we learned to teach AP Statistics.

Grace Feedson, for giving us the opportunity to write this book.

My wife, Jennie, and my children Claire, Charlotte, and Max (C.A.).

The thousands of my former students, especially Katherine, Thomas, and Elizabeth, and to my husband, Michael (D.K.M.).

ABOUT THE AUTHORS

COREY ANDREASEN has been teaching mathematics for 25 years. After many years at North High School in Sheboygan, WI, Corey has begun teaching internationally, currently at American School of the Hague.

Corey is a leader in the mathematics and statistics education communities nationwide. He has been a reader for the AP Statistics exam and has recently consulted on two AP Statistics textbooks and written supplemental classroom materials for statistics teachers. Furthermore, Corey facilitates and cofacilitates workshops for mathematics teachers on a variety of topics, including Common Core workshops on Modeling with Probability and Statistics. He has also served on the Board of Directors of the Wisconsin Mathematics Council and on committees at the state and national levels. Corey has a Bachelor of Mathematics and MEd from the University of Minnesota Twin Cities. He is a National Board Certified Teacher in Adolescent/Young Adult Mathematics and a 2013 winner of the Presidential Award for Excellence in Mathematics and Science Teaching.

DeANNA KRAUSE McDONALD has been teaching for 35 years, and taught Advanced Placement Statistics for 20 years at Palo Verde and University high schools in Tucson, Arizona. She currently teaches statistics courses for the University of Arizona department of Mathematics from which she has a BS and MA in Mathematics and Teacher Education. Her youngest daughter is now a math teacher as well and is also teaching AP Statistics.

INTRODUCTION: THE FIVE-STEP PROGRAM

The Basics

Sometime, probably last spring, you signed up for AP Statistics. Now you are looking through a book that promises to help you achieve the highest grade in AP Statistics: a 5. Your in-class experiences are all-important in helping you toward this goal but are often not sufficient by themselves. In statistical terms, we would say that there is strong evidence that specific preparation for the AP exam beyond the classroom results in significantly improved performance on the exam. If that last sentence makes sense to you, you should probably buy this book. If it didn't make sense, you should definitely buy this book.

Introducing the Five-Step Preparation Program

This book is organized as a five-step program to prepare you for success on the exam. These steps are designed to provide you with the skills and strategies vital to the exam and the practice that can lead you to that perfect 5. Each of the five steps will provide you with the opportunity to get closer and closer to that level of success. Here are the five steps.

Step 1: Set Up Your Study Program

In this step you will read an overview of the AP Statistics exam (Chapter 1). Included in this overview are: an outline of the topics included in the course; the percentage of the exam that you can expect to cover each topic; the format of the exam; how grades on the exam are determined; the calculator policy for Statistics; and what you need to bring to the exam. You will also learn about a process to help determine which type of exam preparation you want to commit yourself to (Chapter 2):

1. Full School Year: September through mid-May
2. One Semester: January through mid-May
3. Basic Training: Six weeks prior to the exam

Step 2: Determine Your Test Readiness

In this step you will take a diagnostic exam in statistics (Chapter 3). This pretest should give you an idea of how prepared you are to take both of the Practice Tests in Step 5 as you prepare for the real exam. The Diagnostic Exam covers the material on the AP exam, but the questions are more basic. Solutions to the exam are given, as well as suggestions for how to use your results to determine your level of readiness. You should go through the Diagnostic Exam and the given solutions step-by-step and question-by-question to build your confidence level.

Step 3: Develop Strategies for Success

In this step, you'll learn strategies that will help you do your best on the exam (Chapter 4). These cover general strategies for success as well as more specific tips and strategies for both the multiple-choice and free-response sections of the exam. Many of these are drawn from our many years of experience as graders for the AP exam; others are the collected wisdom of people involved in the development and grading of the exam.

Step 4: Review the Knowledge You Need to Score High

This step represents the major part, at least in length, of this book. You will review the statistical content you need to know for the exam. Step 4 includes Chapters 5–13 and provides a comprehensive review of statistics as well as sample questions relative to the topics covered in each chapter. If you thoroughly review this material, you will have studied all that is tested on the exam and hence have increased your chances of earning a 5. A combination of good effort all year long in your class and the review provided in these chapters should prepare you to do well.

Step 5: Build Your Test-Taking Confidence

In this step you'll complete your preparation by testing yourself on practice exams. There are two complete sample exams in Step 5 as well as complete solutions to each exam. These exams mirror the AP exam (although they are not reproduced questions from the actual exam) in content and difficulty.

Finally, at the back of this book you'll find additional resources to aid your preparation:

- A summary of formulas related to the AP Statistics exam
- A set of tables needed on the exam
- A brief bibliography
- A short list of websites that might be helpful
- A glossary of terms related to the AP Statistics exam.

The Graphics Used in This Book

To emphasize particular skills and strategies, we use several icons throughout this book. An icon in the margin will alert you to pay particular attention to the accompanying text. We use four icons:

 This icon indicates a very important concept that you should not pass over.

 This icon highlights a strategy that you might want to try.

 This icon alerts you to a tip that you might find useful.

 This icon indicates a tip that will help you with your calculator.

Boldfaced words indicate terms included in the glossary at the end of this book.

5 STEPS TO A 5™

AP Statistics

2020

STEP 1

Set Up Your Study Program

CHAPTER 1

What You Need to Know About the AP Statistics Exam

IN THIS CHAPTER

Summary: Learn what topics are tested, how the test is scored, and basic test-taking information.

Key Ideas

✪ Most colleges will award credit for a score of 4 or 5. Some will award credit for a 3.

✪ Multiple-choice questions account for one-half of your final score.

✪ One point is earned for each correct answer on the multiple-choice section.

✪ Free-response questions account for one-half of your final score.

✪ Your composite score out of a possible 100 on the two test sections is converted to a score on the 1-to-5 scale.

Background Information

The AP Statistics exam that you are taking was first offered by the College Board in 1997. In that year, 7,667 students took the Stats exam (the largest first year exam ever). Since then, the number of students taking the test has grown rapidly. In 2018, about 220,000 students took the AP Statistics exam. Statistics is now one of the 10 largest AP exams.

Some Frequently Asked Questions About the AP Statistics Exam

Why Take the AP Statistics Exam?

Most of you take the AP Statistics exam because you are seeking college credit. Each universtiy makes its own decision about whether and how they recognize AP credit, but the majority of colleges and universities will accept a 4 or 5 as acceptable credit for their noncalculus-based statistics courses. A small number of schools accept a 3 on the exam. This means you are one course closer to graduation before you even begin. Even if you do not score high enough to earn college credit, the fact that you elected to enroll in AP courses tells admission committees that you are a high achiever and serious about your education. You should contact the admissions office of the university or college you plan to attend and ask them their policy for AP Statistics. In recent years, close to 60% of students have scored 3 or higher on the AP Statistics exam.

What Is the Format of the Exam?

AP Statistics

SECTION	NUMBER OF QUESTIONS	TIME LIMIT
I. Multiple-Choice (50% of exam score)	40	90 minutes
II. Free-Response (50% of exam score)	6 (total)	90 minutes
A. Short Answer (75% of free-response score)	5	60 minutes recommended
B. Investigative Task (25% of free-response score)	1	25 minutes recommended

Approved graphing calculators are allowed during all parts of the test. The two sections of the test are completely separate and are administered in separate 90-minute blocks. Please note that you are not expected to be able answer all the questions in order to receive a grade of 5. Specific instructions for each part of the test are given in the Diagnostic Exam and the Practice Exams at the end of this book.

You will be provided with a set of common statistical formulas and necessary tables. Copies of these materials are in the Appendices to this book.

Who Writes the AP Statistics Exam?

Development of each AP exam is a multiyear effort that involves many education and testing professionals and students. At the heart of the effort is the AP Statistics Test Development Committee, a group of college and high school statistics teachers who are typically asked to serve for three years. The committee and other college professors create a large pool of multiple-choice questions. With the help of the testing experts at Educational Testing Service (ETS), these questions are then pretested with college students enrolled in Statistics courses for accuracy, appropriateness, clarity, and assurance that there is only one possible answer. The results of this pretesting allow each question to be categorized by degree of difficulty.

The free-response essay questions that make up Section II go through a similar process of creation, modification, pretesting, and final refinement so that the questions cover the necessary areas of material and are at an appropriate level of difficulty and clarity. The committee also makes a great deal of effort to construct a free-response exam that will allow for clear and equitable grading by the AP readers.

At the conclusion of each AP reading and scoring of exams, the exam itself and the results are thoroughly evaluated by the committee and by ETS. In this way, the College Board can use the results to make suggestions for course development in high schools and to plan future exams.

What Topics Appear on the Exam?

The College Board, after consulting with teachers of statistics, develops a curriculum that covers material that college professors expect to cover in their first-year classes. Based upon this outline of topics, the exams are written such that those topics are covered in proportion to their importance to the expected statistics understanding of the student. There are four major content themes in AP Statistics: exploratory data analysis (20%–30% of the exam); planning and conducting a study (10%–15% of the exam); probability and random variables (20%–30% of the exam); and statistical inference (30%–40% of the exam). Below is an outline of the curriculum topic areas:

SECTION	TOPIC AREA/ PERCENT OF EXAM	TOPICS
I	Exploring Data (20%–30%)	A. Graphical displays of distributions of one-variable data (dotplot, stemplot, histogram, ogive). B. Summarizing distributions of one-variable data (center, spread, position, boxplots, changing units). C. Comparing distributions of one-variable data. D. Exploring two-variable data (scatterplots, linearity, regression, residuals, transformations). E. Exploring categorical data (tables, bar charts, marginal and joint frequencies, conditional relative frequencies, comparing distributions from categorical displays).
II	Sampling and Experimentation (10%–15%)	A. Methods of data collection (census, survey, experiment, observational study). B. Planning and conducting surveys (populations and samples, randomness, sources of bias, sampling methods—SRS, stratified, and cluster sampling). C. Experiments (treatments and control groups, random assignment, replication, sources of bias, confounding, placebo effect, blinding, completely randomized design, randomized block design). D. Generalizability of results and types of conclusions.
III	Anticipating Patterns (Probability and Random Variables and Sampling Distributions) (20%–30%)	A. Probability (long-run relative frequency, law of large numbers, addition and multiplication rules, conditional probability, independence, random variables, simulation, mean and standard deviation of a random variable). B. Combining independent random variables (means and standard deviations).

Continued

SECTION	TOPIC AREA/ PERCENT OF EXAM	TOPICS
		C. The normal distribution. D. Sampling distributions (mean, proportion, differences between two means, difference between two proportions, central limit theorem, simulation, t-distribution, chi-square distribution).
IV	Statistical Inference (30%–40%)	A. Estimation (population parameters, margin of error, point estimators, confidence interval for a proportion, confidence interval for the difference between two proportions, confidence interval for a mean, confidence interval for the difference between two means, confidence interval for the slope of a least-squares regression line). B. Tests of significance (logic of hypothesis testing, Type I and Type II errors, the concept of the power of a test, tests for means and proportions, chi-square test, test for the slope of a least-squares line).

Who Grades My AP Statistics Exam?

Every June a group of statistics teachers (roughly half college professors and half high school teachers of statistics) gather for a week to assign grades to your hard work. Each of these Faculty Consultants spends several hours getting trained on the scoring rubric for each question they grade (an individual reader may read two to three questions during the week). Because each reader becomes an expert on that question, and because each exam book is anonymous, this process provides a very consistent and unbiased scoring of that question. During a typical day of grading, a random sample of each reader's scores is selected and cross-checked by other experienced Table Leaders to ensure that the consistency is maintained throughout the day and the week. Each reader's scores on a given question are also statistically analyzed to make sure that he or she is not giving scores that are significantly higher or lower than the mean scores given by other readers of that question. All measures are taken to maintain consistency and fairness for your benefit.

Will My Exam Remain Anonymous?

Absolutely. Even if your high school teacher happens to randomly read your booklet, there is virtually no way he or she will know it is you. To the reader, each student is a number, and to the computer, each student is a bar code.

What About That Permission Box on the Back?

The College Board uses some responses to help train high school teachers so that they can help the next generation of statistics students avoid common mistakes. If you check this box, you deny permission to use your exam in this way. All responses are used anonymously. There is no way for anyone to find out whose response it is.

How Is My Multiple-Choice Exam Scored?

The multiple-choice section contains 40 questions and is worth one-half of your final score. Your answer sheet is run through the computer, which adds up your correct responses. Then this score is multiplied by 1.25 to scale it to a maximum of 50 points.

How Is My Free-Response Exam Scored?

Your performance on the free-response section is worth one-half of your final score. There are six questions, and each question is given a score from 0–4 (4 = complete response, 3 = substantial response, 2 = developing response, 1 = minimal response, and 0 = insufficient response). Unlike, say, calculus, your response does not have to be perfect to earn the top score. These questions are scored using a carefully prepared rubric to ensure better responses get better scores and that different readers will score the same response the same way.

The raw score on each of questions 1–5 is then multiplied by 1.875 (this forces questions 1–5 to be worth 75% of your free-response score, based on a total of 50) and the raw score on question 6 is multiplied by 3.125 (making question 6 worth 25% of your free-response score). The result is a score based on 50 points for the free-response part of the exam.

How Is My Final Grade Determined and What Does It Mean?

The scores on the multiple-choice and free-response sections of the test are then combined to give a single composite score based on a 100-point scale. As can be seen from the descriptions above, this is not really a percentage score, and it's best not to think of it as one.

In the end, when all of the numbers have been crunched, the Chief Faculty Consultant converts the range of composite scores to the 5-point scale of the AP grades.

The table below gives a typical conversion and, as you complete the practice exams, you may use this to give yourself a hypothetical grade. Keep in mind that the conversion changes every year to adjust for the difficulty of the questions. You should receive your grade in early July.

COMPOSITE SCORING RANGE (OUT OF 100)	AP GRADE	INTERPRETATION OF GRADE
68–100	5	Extremely well qualified
54–67	4	Well qualified
41–53	3	Qualified
32–40	2	Possibly qualified
0–31	1	No recommendation

There is no official passing grade on the AP Exam. However, most people think in terms of 3 or better as passing.

How Do I Register and How Much Does It Cost?

If you are enrolled in AP Statistics in your high school, your teacher is going to provide all of these details, but a quick summary wouldn't hurt. After all, you do not have to enroll in the AP course to register for and complete the AP exam. When in doubt, the best source of information is the College Board's website: www.collegeboard.org.

The fee for taking the exam is currently $94 but tends to go up a little each year. A $30 fee reduction may be available for eligible students with financial need, depending on the student's state. Finally, most states offer exam subsidies to cover all or part of the cost. You can learn more about fee reductions and subsidies from the coordinator of your AP program or by checking specific information on the official website: https://apstudent.collegeboard.org/takingtheexam/exam-fees.

There are also several optional fees that must be paid if you want your scores rushed to you or if you wish to receive multiple grade reports.

The coordinator of the AP program at your school will inform you where and when you will take the exam. If you live in a small community, your exam may not be administered at your school, so be sure to get this information.

What Is the Graphing Calculator Policy?

The following is the policy on graphing calculators as stated on the College Board's AP Central website:

Each student is expected to bring to the exam a graphing calculator with statistical capabilities. The computational capabilities should include standard statistical univariate and bivariate summaries, through linear regression. The graphical capabilities should include common univariate and bivariate displays such as histograms, boxplots, and scatterplots.

- You can bring two calculators to the exam.
- The calculator memory will not be cleared but you may only use the memory to store programs, not notes.
- For the exam, you're not allowed to access any information in your graphing calculators or elsewhere if it's not directly related to upgrading the statistical functionality of older graphing calculators to make them comparable to statistical features found on newer models. The only acceptable upgrades are those that improve the computational functionalities and/or graphical functionalities for data you key into the calculator while taking the examination. Unacceptable enhancements include, but aren't limited to, keying or scanning text or response templates into the calculator.

During the exam, you can't use minicomputers, pocket organizers, electronic writing pads, or calculators with QWERTY (i.e., typewriter) keyboards.

You may use a calculator to do needed computations. However, remember that the person reading your exam needs to see your reasoning in order to score your exam. Your teacher can check for a list of acceptable calculators on AP Central.

What Should I Bring to the Exam?

- Several #2 pencils (and a pencil sharpener) and a good eraser that doesn't leave smudges
- Black or blue colored pens for the free-response section; some students like to use two colors to make their graphs stand out for the reader
- One or two graphing calculators with fresh batteries
- A watch so that you can monitor your time
- Your school code
- A simple snack if the test site permits it

- Your photo identification and social security number
- A light jacket if you know that the test site has strong air conditioning
- Tissues
- Your quiet confidence that you are prepared

What Should I NOT Bring to the Exam?

- A calculator that is not approved for the AP Statistics Exam (for example, anything with a QWERTY keyboard)
- A phone, smartwatch, tablet, or anything else that communicates with the outside world.
- Books, a dictionary, study notes, flash cards, highlighting pens, correction fluid, any other office supplies
- Scrap paper
- Portable music of any kind: no iPods, MP3 players, or CD players

CHAPTER 2

How to Plan Your Time

IN THIS CHAPTER

Summary: The right preparation plan for you depends on your study habits and the amount of time you have before the test.

Key Idea

✪ Choose the study plan that's right for you.

Three Approaches to Preparing for the AP Statistics Exam

No one knows your study habits, likes, and dislikes better than you. So, you are the only one who can decide which approach you want and/or need, to adopt to prepare for the AP Statistics exam. This may help you place yourself in a particular prep mode. This chapter presents three possible study plans, labeled A, B, and C. Look at the brief profiles below and try to determine which of these three plans is right for you.

You're a full-school-year prep student if:

1. You are the kind of person who likes to plan for everything far in advance.
2. You arrive at the airport two hours before your flight.
3. You like detailed planning and everything in its place.
4. You feel that you must be thoroughly prepared.
5. You hate surprises.

If you fit this profile, consider **Plan A.**

You're a one-semester prep student if:

1. You get to the airport one hour before your flight is scheduled to leave.
2. You are willing to plan ahead to feel comfortable in stressful situations, but are okay with skipping some details.
3. You feel more comfortable when you know what to expect, but a surprise or two is okay.
4. You are always on time for appointments.

If you fit this profile, consider **Plan B**.

You're a six-week prep student if:

1. You get to the airport at the last possible moment.
2. You work best under pressure and tight deadlines.
3. You feel very confident with the skills and background you've learned in your AP Statistics class.
4. You decided late in the year to take the exam.
5. You like surprises.
6. You feel okay if you arrive 10–15 minutes late for an appointment.

If you fit this profile, consider **Plan C**.

Summary of the Three Study Plans

MONTH	PLAN A (FULL SCHOOL YEAR)	PLAN B (1 SEMESTER)	PLAN C (6 WEEKS)
September–October	Chapters 5 and 6		
November	Chapter 7 Review Ch. 5		
December	Chapter 8 Review Chs. 5 and 6		
January	Chapter 9 Review Chs. 6–8	Chapters 5–7	
February	Chapter 10 Review Chs. 7–9	Chapters 8 and 9 Review Chs. 5–7	
March	Chapters 11 and 12 Review Chs. 8–10	Chapters 10 and 11 Review Chs. 6–9	
April	Chapter 13 Review Chs. 10–12 Diagnostic Exam	Chapters 12 and 13 Review Chs. 8–11 Diagnostic Exam	Review Chs. 5–13 Rapid Reviews 5–13 Diagnostic Exam
May	Practice Exam 1 Practice Exam 2	Practice Exam 1 Practice Exam 2	Practice Exam 1 Practice Exam 2

Calendar for Each Plan

Plan A: You Have a Full School Year to Prepare

SEPTEMBER–OCTOBER (check off the activities as you complete them)

—— Determine into which student mode you would place yourself.

—— Carefully read Chapters 1–4 of this book.

—— Get on the web and take a look at the AP Central website(s).

—— Skim the Comprehensive Review section (these areas will be part of your yearlong preparation).

—— Buy a few highlighters.

—— Flip through the entire book. Break the book in. Write in it. Toss it around a little bit . . . highlight it.

—— Get a clear picture of your school's AP Statistics curriculum.

—— Begin to use the book as a resource to supplement the classroom learning.

—— Read and study Chapter 5–One-Variable Data Analysis.

—— Read and study Chapter 6–Two-Variable Data Analysis.

NOVEMBER

—— Read and study Chapter 7–Design of a Study: Sampling, Surveys, and Experiments.

—— Review Chapter 5.

DECEMBER

—— Read and study Chapter 8–Probability and Random Variables.

—— Review Chapters 6 and 7.

JANUARY (20 weeks have elapsed)

—— Read and study Chapter 9–Binomial Distributions, Geometric Distributions, and Sampling Distributions.

—— Review Chapters 6–8.

FEBRUARY

—— Read and study Chapter 10–Confidence Intervals and Introduction to Inference.

—— Review Chapters 7–9.

—— Look over the Diagnostic Exam.

MARCH (30 weeks have elapsed)

—— Read and study Chapter 11–Inference for Means and Proportions.

—— Read and study Chapter 12–Inference for Regression.

—— Review Chapters 8–10.

APRIL

—— Read and study Chapter 13–Inference for Categorical Data: Chi-Square.

—— Review Chapters 10–12.

—— Take the Diagnostic Exam.

—— Evaluate your strengths and weaknesses.

—— Study appropriate chapters to correct weaknesses.

MAY (first two weeks)

—— Review Chapters 5–13 (that's everything!).

—— Take and score Practice Exam 1.

—— Study appropriate chapters to correct weaknesses.

—— Take and score Practice Exam 2.

—— Study appropriate chapters to correct weaknesses.

—— Get a good night's sleep the night before the exam.

GOOD LUCK ON THE TEST!

Plan B: You Have One Semester to Prepare

Working under the assumption that you've completed one semester of statistics in the classroom, the following calendar will use those skills you've been practicing to prepare you for the May exam.

JANUARY

—— Carefully read Chapters 1–4 of this book.
—— Read and study Chapter 5–One-Variable Data Analysis.
—— Read and Study Chapter 6–Two-Variable Data Analysis.
—— Read and Study Chapter 7–Design of a Study: Sampling, Surveys, and Experiments.

FEBRUARY

—— Read and study Chapter 8–Probability and Random Variables.
—— Read and study chapter 9–Binomial Distributions, Geometric Distributions, and Sampling Distributions.
—— Review Chapters 5–7.

MARCH

—— Read and study Chapter 10–Confidence Intervals and Introduction to Inference.
—— Read and study Chapter 11–Inference for Means and Proportions.
—— Review Chapters 6–9.

APRIL

—— Read and study Chapter 12–Inference for Regression.
—— Read and study Chapter 13–Inference for Categorical Data: Chi-Square.
—— Review Chapters 8–11.
—— Take Diagnostic Exam.
—— Evaluate your strengths and weaknesses.
—— Study appropriate chapters to correct weaknesses.

MAY (first two weeks)

—— Take and score Practice Exam 1.
—— Study appropriate chapters to correct weaknesses.
—— Take and score Practice Exam 2.
—— Study appropriate chapters to correct weaknesses.
—— Get a good night's sleep the night before the exam.

GOOD LUCK ON THE TEST!

Plan C: You Have Six Weeks to Prepare

At this point, we are going to assume that you have been building your statistics knowledge base for more than six months. You will, therefore, use this book primarily as a specific guide to the AP Statistics Exam.

Given the time constraints, now is not the time to expand your AP Statistics curriculum. Rather, it is the time to limit and refine what you already do know.

APRIL 1 – 15

—— Skim Chapters 1–4 of this book.
—— Skim Chapters 5–9.
—— Carefully go over the "Rapid Review" sections of Chapter 5–9.

APRIL 16 – 30

—— Skim Chapters 10–13.
—— Carefully go over the "Rapid Review" sections of Chapters 10–13.
—— Take the Diagnostic Exam.
—— Evaluate your strengths and weaknesses.
—— Study appropriate chapters to correct weaknesses.

MAY (first two weeks)

—— Take and score Practice Exam 1.
—— Study appropriate chapters to correct weaknesses.
—— Take and score Practice Exam 2.
—— Study appropriate chapters to correct weaknesses.
—— Get a good night's sleep the night before the exam.

GOOD LUCK ON THE TEST!

STEP **2**

Determine Your Test Readiness

CHAPTER 3

Take a Diagnostic Exam

IN THIS CHAPTER

Summary: The following diagnostic exam begins with 40 multiple-choice questions. The diagnostic exam also includes five free-response questions and one investigative task much like those on the actual exam. All of these test questions have been written to approximate the coverage of material that you will see on the AP exam but are intentionally somewhat more basic than actual exam questions (which are more closely approximated by the Practice Exams at the end of the book). Once you are done with the exam, check your work against the given answers, which also indicate where you can find the corresponding material in the book. You will also be given a way to convert your score to a rough AP score.

Key Ideas
- ✪ Practice the kind of questions you will be asked on the real AP Statistics exam.
- ✪ Answer questions that approximate the coverage of topics on the real exam.
- ✪ Check your work against the given answers.
- ✪ Determine your areas of strength and weakness.

AP Statistics Diagnostic Test

ANSWER SHEET FOR SECTION I

1 Ⓐ Ⓑ Ⓒ Ⓓ Ⓔ
2 Ⓐ Ⓑ Ⓒ Ⓓ Ⓔ
3 Ⓐ Ⓑ Ⓒ Ⓓ Ⓔ
4 Ⓐ Ⓑ Ⓒ Ⓓ Ⓔ
5 Ⓐ Ⓑ Ⓒ Ⓓ Ⓔ
6 Ⓐ Ⓑ Ⓒ Ⓓ Ⓔ
7 Ⓐ Ⓑ Ⓒ Ⓓ Ⓔ
8 Ⓐ Ⓑ Ⓒ Ⓓ Ⓔ
9 Ⓐ Ⓑ Ⓒ Ⓓ Ⓔ
10 Ⓐ Ⓑ Ⓒ Ⓓ Ⓔ
11 Ⓐ Ⓑ Ⓒ Ⓓ Ⓔ
12 Ⓐ Ⓑ Ⓒ Ⓓ Ⓔ
13 Ⓐ Ⓑ Ⓒ Ⓓ Ⓔ
14 Ⓐ Ⓑ Ⓒ Ⓓ Ⓔ
15 Ⓐ Ⓑ Ⓒ Ⓓ Ⓔ

16 Ⓐ Ⓑ Ⓒ Ⓓ Ⓔ
17 Ⓐ Ⓑ Ⓒ Ⓓ Ⓔ
18 Ⓐ Ⓑ Ⓒ Ⓓ Ⓔ
19 Ⓐ Ⓑ Ⓒ Ⓓ Ⓔ
20 Ⓐ Ⓑ Ⓒ Ⓓ Ⓔ
21 Ⓐ Ⓑ Ⓒ Ⓓ Ⓔ
22 Ⓐ Ⓑ Ⓒ Ⓓ Ⓔ
23 Ⓐ Ⓑ Ⓒ Ⓓ Ⓔ
24 Ⓐ Ⓑ Ⓒ Ⓓ Ⓔ
25 Ⓐ Ⓑ Ⓒ Ⓓ Ⓔ
26 Ⓐ Ⓑ Ⓒ Ⓓ Ⓔ
27 Ⓐ Ⓑ Ⓒ Ⓓ Ⓔ
28 Ⓐ Ⓑ Ⓒ Ⓓ Ⓔ
29 Ⓐ Ⓑ Ⓒ Ⓓ Ⓔ
30 Ⓐ Ⓑ Ⓒ Ⓓ Ⓔ

31 Ⓐ Ⓑ Ⓒ Ⓓ Ⓔ
32 Ⓐ Ⓑ Ⓒ Ⓓ Ⓔ
33 Ⓐ Ⓑ Ⓒ Ⓓ Ⓔ
34 Ⓐ Ⓑ Ⓒ Ⓓ Ⓔ
35 Ⓐ Ⓑ Ⓒ Ⓓ Ⓔ
36 Ⓐ Ⓑ Ⓒ Ⓓ Ⓔ
37 Ⓐ Ⓑ Ⓒ Ⓓ Ⓔ
38 Ⓐ Ⓑ Ⓒ Ⓓ Ⓔ
39 Ⓐ Ⓑ Ⓒ Ⓓ Ⓔ
40 Ⓐ Ⓑ Ⓒ Ⓓ Ⓔ

AP Statistics Diagnostic Test

SECTION I

Time: 1 hour and 30 minutes

Number of questions: 40

Percent of total grade: 50

Directions: Use the answer sheet provided on the previous page. All questions are given equal weight. There is no penalty for unanswered questions. One point is earned for every correct answer. The use of a calculator is permitted in all parts of this test. You have 90 minutes for this part of the test.

1. Eighteen trials of a binomial random variable X are conducted. If the probability of success for any one trial is 0.4, write the mathematical expression you would need to evaluate to find $P(x = 7)$. Do not evaluate.

   a. $\binom{18}{7}(0.4)^{11}(0.6)^{7}$

   b. $\binom{18}{11}(0.4)^{7}(0.6)^{11}$

   c. $\binom{18}{7}(0.4)^{7}(0.6)^{11}$

   d. $\binom{18}{7}(0.4)^{7}(0.6)^{18}$

   e. $\binom{18}{7}(0.4)^{18}(0.6)^{7}$

2. A survey using a random sample of 1500 households reports that 51% of U.S. households have only mobile phones and no landlines. The survey results have a margin of error of 2.0 percentage points. Which of the following can be said about C, the confidence level?

   a. $C < 90\%$
   b. $90\% < C < 95\%$
   c. $C = 95\%$
   d. $95\% < C < 99\%$
   e. $C = 99\%$

3. You need to construct a 94% confidence interval for a population proportion. What is the upper critical value of z to be used in constructing this interval?

   a. 0.9699
   b. 1.96
   c. 1.555
   d. −1.88
   e. 1.88

GO ON TO THE NEXT PAGE

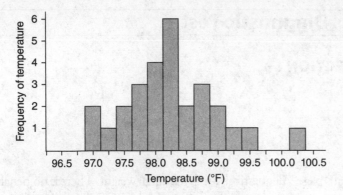

4. For each patient visiting a doctor's office, the nurse records the patient's body temperature. The plot above shows the temperatures of 28 patients for one particular day. Which of the following statements is true about the distribution of body temperatures?

   a. The distribution is skewed to the left.
   b. The median temperature could be 98.2°F.
   c. The median temperature could be 98.4°F.
   d. The minimum temperature is exactly 97.0°F.
   e. The distribution is normal.

5. The student council at a high school surveyed students about a variety of questions. Which of the following displays would be appropriate to display the number of males and females in each grade who plan to attend prom?

   a. Side-by-side bar chart
   b. Scatterplot
   c. Histogram
   d. Dotplot
   e. Boxplot

6. Which of the following are examples of continuous variables?

   I. The speed your car goes
   II. The number of outcomes of a binomial experiment
   III. The average temperature in San Francisco
   IV. The wingspan of a bird
   V. The jersey numbers of a football team
      a. I, III, and IV only
      b. II and V only
      c. I, III, and V only
      d. II, III, and IV only
      e. I, II, and IV only

Use the following computer output for a least-squares regression for Questions 7 and 8.

The regression equation is				
Predictor	Coef	St Dev	*t* ratio	*P*
Constant	22.94	11.79	1.95	0.088
x	−0.6442	0.5466	−1.18	—
s = 2.866	R-sq = 14.8%		R-sq(adj) = 4.1%	

7. What is the equation of the least-squares regression line?

   a. $\hat{y} = -0.6442 + 22.94x$

   b. $\hat{y} = 22.94 + 0.5466x$

   c. $\hat{y} = 22.94 + 2.866x$

   d. $\hat{y} = 22.94 - 0.6442x$

   e. $\hat{y} = -0.6442 + 0.5466x$

8. Assuming that the conditions have been met, which of the following would give you a 95% confidence interval for the true intercept of the regression line, given that the sample size is 10?

   a. $22.94 \pm 2.228(11.79)$
   b. $22.94 \pm 2.262(11.79)$
   c. $22.94 \pm 2.262(11.79/\sqrt{10})$
   d. $22.94 \pm 2.306(11.79)$
   e. $22.94 \pm 2.306(11.79)$

9. "A hypothesis test yields a *P*-value of 0.20." Which of the following best describes what is meant by this statement?

   a. The probability of getting a statistic at least as extreme as that observed by chance alone if the null hypothesis is true is 0.20.
   b. The probability of getting a statistic as extreme as that observed by chance alone from repeated random sampling is 0.20.
   c. The probability is 0.20 that our statistic is significant.
   d. The probability of getting this statistic is 0.20.
   e. The statistic we observed will occur in fewer than 20% of repeated trials of this study.

10. A random sample of 25 adults over 50 years old and a separate random sample of 25 adults under 30 years old are selected to answer questions about attitudes toward abortion. The answers were categorized as "pro-life" or "pro-choice." Which of the following is the proper null hypothesis for this situation?

    a. The variables "age group" and "attitude toward abortion" are related.
    b. The proportion of "pro-life" older adults is the same as the proportion of "pro-life" younger adults.
    c. The proportion of "pro-life" older adults is related to the proportion of "pro-life" younger adults.
    d. The proportion of "pro-choice" older adults is the same as the proportion of "pro-life" younger adults.
    e. The variables "age group" and "attitude toward abortion" are independent.

11. A sports talk show asks people to call in and give their opinion of the officiating in the local basketball team's most recent loss. What will most likely be the typical reaction?

    a. They will most likely feel that the officiating could have been better, but that it was the team's poor play, not the officiating, that was primarily responsible for the loss.
    b. They would most likely call for the team to get some new players to replace the current ones.
    c. The team probably wouldn't have lost if the officials had been doing their job.
    d. Because the team had been foul-plagued all year, the callers would most likely support the officials.
    e. They would support moving the team to another city.

12. Last year, the level of support for a yearly community event was estimated to be 83%. Officials would like to estimate the level of support this year with 95% confidence and a margin of error no larger than 4%. Which of the following is the smallest sample size that would do so?

    a. 30
    b. 300
    c. 350
    d. 610
    e. 1000

GO ON TO THE NEXT PAGE

13. A sample of size 35 is to be drawn from a large population. The sampling technique is such that every possible sample of size 35 that could be drawn from the population is equally likely. What name is given to this type of sample?

    a. Systematic sample
    b. Cluster sample
    c. Voluntary response sample
    d. Random sample
    e. Simple random sample

14. A teachers' union and a school district are negotiating salaries for the coming year. The teachers want more money, and the district, claiming, as always, budget constraints, wants to pay as little as possible. The district, like most, has a large number of moderately paid teachers and a few highly paid administrators. The salaries of all teachers and administrators are included in trying to figure out, on average, how much the professional staff currently earn. Which of the following would the teachers' union be most likely to quote during negotiations?

    a. The mean of all the salaries.
    b. The variance of all the salaries.
    c. The standard deviation of all the salaries.
    d. The interquartile range of all the salaries.
    e. The median of all the salaries.

15. A computerized sewing machine has four independent parts: mechanical, circuit board, touch screen, and power control. The probability that each will fail within the first year is 0.01, 0.02, 0.001, and 0.03, respectively. If one of them fails then the machine does not work. What is the approximate probability that the machine is still working at the end of the first year?

    a. 0.930
    b. 0.939
    c. 0.940
    d. 0.960
    e. greater than 99%

16. A significance test of the hypothesis $H_0: p = 0.3$ against the alternative $H_A: p > 0.3$ found a value of $\hat{p} = 0.35$ for a random sample of size 95. What is the P-value of this test?

    a. 1.06
    b. 0.1446
    c. 0.2275
    d. 0.8554
    e. 0.1535

17. Which of the following is a true statement?
    I. The mean of the sampling distribution of $\bar{x}$ is the same as the mean of the population.
    II. The standard deviation of the sampling distribution of $\bar{x}$ is the same as the standard deviation of $\bar{x}$ divided by the square root of the sample size.
    III. The shape of the sampling distribution of $\bar{x}$ is approximately normal.

    a. I only
    b. I & II only
    c. II only
    d. III only
    e. I, II, and III

GO ON TO THE NEXT PAGE

18. If three fair coins are flipped, P (0 heads) = 0.125, P (exactly 1 head) = 0.375, P (exactly 2 heads) = 0.375, and P (exactly 3 heads) = 0.125. The following results were obtained when three coins were flipped 64 times:

# Heads	Observed
0	10
1	28
2	22
3	4

What is the value of the X^2 statistic used to test if the coins are behaving as expected, and how many degrees of freedom does the determination of the P-value depend on?

a. 3.33, 3
b. 3.33, 4
c. 11.09, 3
d. 3.33, 2
e. 11.09, 4

19.

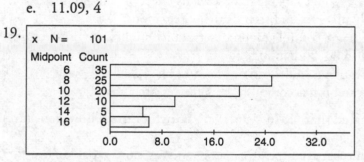

For the histogram pictured above, what is the class interval (boundaries) for the class that contains the median of the data?

a. (5, 7)
b. (9, 11)
c. (11, 13)
d. (15, 17)
e. (7, 9)

20. Thirteen large animals were measured to help determine the relationship between their length and their weight. The natural logarithm of the weight of each animal was taken and a least-squares regression equation for predicting weight from length was determined. The computer output from the analysis is given below:

The regression equation is
$\ln(\text{wt}) = 1.24 + 0.0365 \text{ length}$

Predictor	Coef	St Dev	t ratio	P
Constant	1.2361	0.1378	8.97	0.000
Length	0.036502	0.001517	24.05	0.000

$s = 0.1318$ R-sq = 98.1%   R-sq(adj) = 98.0%

Give a 99% confidence interval for the slope of the regression line. Interpret this interval.

a. (0.032, 0.041); the probability is 0.99 that the true slope of the regression line is between 0.032 and 0.041.
b. (0.032, 0.041); 99% of the time, the true slope will be between 0.032 and 0.041.
c. (0.032, 0.041); we are 99% confident that the true slope of the regression line is between 0.032 and 0.041.
d. (0.81, 1.66); we are 99% confident that the true slope of the regression line is between 0.032 and 0.041.
e. (0.81, 1.66); the probability is 0.99 that the true slope of the regression line is between 0.81 and 1.66.

GO ON TO THE NEXT PAGE

21. A random variable W proves to be binomial with $n = 60$ and $p = 0.10$. Which of the following is a false statement?

    a. There are a fixed number of trials.
    b. The probability of obtaining 4 successes is about 13%.
    c. Because n is greater than 30, you could use the normal approximate to calculate the probability of obtaining more than 4 successes.
    d. The mean of $W = 6$.
    e. The standard deviation of $W = 2.324$.

22. Which of the following is the primary difference between an experiment and an observational study?

    a. Experiments are only conducted on human subjects; observational studies can be conducted on nonhuman subjects.
    b. In an experiment, the researcher manipulates some variable to observe its effect on a response variable; in an observational study, he or she simply observes and records the observations.
    c. Experiments must use randomized treatment and control groups; observational studies also use treatment and control groups, but they do not need to be randomized.
    d. Experiments must be double-blind; observational studies do not need to be.
    e. There is no substantive difference—they can both accomplish the same research goals.

23. The regression analysis of question 20 indicated that "R-sq = 98.1%." Which of the following is (are) true?
      I. There is a strong positive linear relationship between the explanatory and response variables.
     II. There is a strong negative linear relationship between the explanatory and response variables.
    III. About 98% of the variation in the response variable can be explained by the regression on the explanatory variable.

    a. I and III only
    b. I or II only
    c. I or II (but not both) and III
    d. II and III only
    e. I, II, and III

24. A hypothesis test in which $n = 250$ is set up so that $P(\text{rejecting } H_0 \text{ when } H_0 \text{ is true}) = 0.05$. Which of the following is true?

    a. We can increase the power if we let alpha = 0.05.
    b. We can increase the power if we let alpha = 0.01.
    c. We can increase the power if we let $n = 200$.
    d. We can decrease the power if we let $n = 200$.
    e. We can decrease the power if we let $n = 300$.

25. For the following observations collected while doing a chi-square test for independence between the two variables A and B, find the expected value of the cell marked with "**X**."

5	10(**X**)	11
6	9	12
7	8	13

    a. 4.173
    b. 9.00
    c. 11.56
    d. 8.667
    e. 9.33

GO ON TO THE NEXT PAGE

26. The following is a probability histogram for a discrete random variable X.

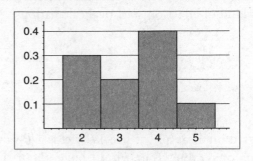

What is μ_x?

a. 3.5
b. 4.0
c. 3.7
d. 3.3
e. 3.0

27. A psychologist believes that positive rewards for proper behavior are more effective than punishment for bad behavior in promoting good behavior in children. A scale of "proper behavior" is developed. μ_1 = the "proper behavior" rating for children receiving positive rewards, and μ_2 = the "proper behavior" rating for children receiving punishment. If H_0: $\mu_1 - \mu_2 = 0$, which of the following is the proper statement of H_A?

a. H_A: $\mu_1 - \mu_2 > 0$
b. H_A: $\mu_1 - \mu_2 < 0$
c. H_A: $\mu_1 - \mu_2 \neq 0$
d. Any of the above is an acceptable alternative to the given null.
e. There isn't enough information given in the problem for us to make a decision.

28. Estrella wants to become a paramedic and takes a screening exam. Scores on the exam have been approximately normally distributed over the years it has been given. The exam is normed with a mean of 80 and a standard deviation of 9. Only those who score in the top 15% on the test are invited back for further evaluation. Estrella received a 90 on the test. What was her percentile rank on the test, and did she qualify for further evaluation?

a. 13.35; she didn't qualify.
b. 54.38; she didn't qualify.
c. 86.65; she qualified.
d. 84.38; she didn't qualify.
e. 88.69; she qualified.

29. Which of the following statements is (are) true?
I. In order to use a χ^2 procedure, the expected value for each cell of a one- or two-way table must be at least 5.
II. In order to use χ^2 procedures, you must have at least 2 degrees of freedom.
III. In a 4×2 two-way table, the number of degrees of freedom is 3.

a. I only
b. I and III only
c. I and II only
d. III only
e. I, II, and III

GO ON TO THE NEXT PAGE

30. In a scatterplot, when the point (15,2) is included, the slope of regression line ($y = a + bx$) is $b = -0.54$. The correlation is $r = -0.82$. When the point is removed, the new slope is -1.04 and the new correlation coefficient is -0.95. What name is given to a point whose removal has this kind of effect on statistical calculations?

    a. Outlier
    b. Statistically significant point
    c. Point of discontinuity
    d. Unusual point
    e. Influential point

31. A one-sided test of a hypothesis about a population mean, based on a sample of size 14, yields a P-value of 0.075. Which of the following best describes the range of t values that would have given this P-value?

    a. $1.345 < t < 1.761$
    b. $1.356 < t < 1.782$
    c. $1.771 < t < 2.160$
    d. $1.350 < t < 1.771$
    e. $1.761 < t < 2.145$

32. Use the following excerpt from a random digits table for assigning six people to treatment and control groups:
    98110 35679 14520 51198 12116 98181 99120 75540 03412 25631
    The subjects are labeled: Arnold: 1; Betty: 2; Clive: 3; Doreen: 4; Ernie: 5; Florence: 6. The first three subjects randomly selected will be in the treatment group; the other three in the control group. Assuming you begin reading the table at the extreme left digit, which three subjects would be in the *control* group?

    a. Arnold, Clive, Ernest
    b. Arnold, Betty, Florence
    c. Betty, Clive, Doreen
    d. Clive, Ernest, Florence
    e. Betty, Doreen, Florence

33. A null hypothesis, H_0: $\mu = \mu_0$ is to be tested against a two-sided hypothesis. A sample is taken; $\bar{x}$ is determined and used as the basis for a C-level confidence interval (e.g., $C = 0.95$) for μ. The researcher notes that μ_0 is not in the interval. Another researcher chooses to do a significance test for μ using the same data. What significance level must the second researcher choose in order to guarantee getting the same conclusion about H_0: $\mu = \mu_0$ (that is, reject or not reject) as the first researcher?

    a. $1 - C$
    b. C
    c. α
    d. $1 - \alpha$
    e. $\alpha = 0.05$

34. Which of the following is *not* required in a binomial setting?

    a. Each trial is considered either a success or a failure.
    b. Each trial is independent.
    c. The value of the random variable of interest is the number of trials until the first success occurs.
    d. There is a fixed number of trials.
    e. Each trial succeeds or fails with the same probability.

GO ON TO THE NEXT PAGE

35. X and Y are independent random variables with mean of $X = 3.5$, variance of $X = 7.29$ and mean of $Y = 1.8$, variance of $Y = 0.4225$. What are μ_{X-Y} and σ_{X-Y}?

   a. $\mu_{X-Y} = 1.7$, $\sigma_{X-Y} = 2.6206$
   b. $\mu_{X-Y} = 1.7$, $\sigma_{X-Y} = 2.7770$
   c. $\mu_{X-Y} = 1.7$, $\sigma_{X-Y} = 7.2777$
   d. $\mu_{X-Y} = 1.7$, $\sigma_{X-Y} = 7.3022$
   e. $\mu_{X-Y} = 1.7$, σ_{X-Y} cannot be determined from the information given.

36. A researcher is hoping to find a predictive linear relationship between the explanatory and response variables in her study. Accordingly, as part of her analysis she plans to generate a 95% confidence interval for the slope of the regression line for the two variables. The interval is determined to be (0.45, 0.80). Which of the following is (are) true? (Assume conditions for inference are met.)

   I. She has good evidence of a linear relationship between the variables.
   II. It is likely that there is a non-zero correlation (r) between the two variables.
   III. It is likely that the true slope of the regression line is 0.

   a. I and II only
   b. I and III only
   c. II and III only
   d. I only
   e. II only

37. A store offers shoppers who make a purchase a scratch card. Each card reveals one number and the probability of each number is listed in the table:

Card	100	50	10	1	0
P(Card)	0.01	0.02	0.03	0.30	0.64

The mean of the distribution of card #'s is 2.6 and the standard deviation is 12.1. Once the number has been revealed, the shopper wins a gift card whose value is given by the formula: Value = 10(Card#) + 5. Which of the following is the mean and standard deviation of Value?

   a. mean = 26, standard deviation = 126
   b. mean = 26, standard deviation = 121
   c. mean = 31, standard deviation = 133
   d. mean = 31, standard deviation = 126
   e. mean = 31, standard deviation = 121

38. You are developing a new strain of strawberries (say, Type X) and are interested in its sweetness as compared to another strain (say, Type Y). You have four plots of land, call them A, B, C, and D, which are roughly four squares in one large plot for your study (see the figure below). A river runs alongside of plots C and D. Because you are worried that the river might influence the sweetness of the berries, you randomly plant type X in either A or B (and Y in the other) and randomly plant type X in either C or D (and Y in the other). Which of the following terms best describes this design?

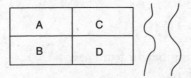

   a. A completely randomized design
   b. A randomized study
   c. A randomized observational study
   d. A block design, controlling for the strain of strawberry
   e. A block design, controlling for the effects of the river

GO ON TO THE NEXT PAGE

39. The scores on two tests in a statistics class were approximately normally distributed. The first test had a mean of 42 and a standard deviation of 5. For the second test, the mean was 45 with a standard deviation of 6. Jemma earned a 38 on the first test. What score on the second test would be at the same percentile as her score on the first?

    a. 38
    b. 39
    c. 40
    d. 41
    e. 42

40. A random sample size of 45 is obtained for the purpose of testing the hypothesis $H_0: p = 0.80$. The sample proportion is determined to be $\hat{p} = 0.75$. What is the value of the standard error of $\hat{p}$ for this test?

    a. 0.0042
    b. 0.0596
    c. 0.0036
    d. 0.0645
    e. 0.0055

SECTION II—PART A, QUESTIONS 1–5

Spend about 65 minutes on this part of the exam. Percentage of Section II grade—75.

Directions: Show all of your work. Indicate clearly the methods you use because you will be graded on the correctness of your methods as well as on the accuracy of your results and explanation.

1. The ages (in years) and heights (in cm) of 10 girls, ages 2 through 11, were recorded. Part of the regression output and the residual plot for the data are given below.

The regression equation is				
Predictor	Coef	St Dev	t ratio	P
Constant	76.641	1.188	64.52	0.000
Age	6.3661	0.1672	38.08	0.000
s = 1.518	R-sq = 99.5%		R-sq(adj) = 99.4%	

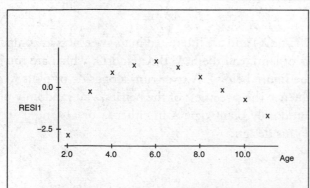

   a. What is the equation of the least-squares regression line for predicting height from age?
   b. Interpret the slope of the regression line in the context of the problem.
   c. Suppose you wanted to predict the height of a girl 5.5 years of age. Would the prediction made by the regression equation you gave in (a) be too small, too large, or is there not enough information to tell?

2. You want to determine whether a greater proportion of men or women purchase vanilla lattes (regular or decaf). To collect data, you hire a person to stand inside the local Scorebucks for 2 hours one morning and

GO ON TO THE NEXT PAGE

tally the number of men and women who purchase the vanilla latte, as well as the total number of men and women customers: 63% of the women and 59% of the men purchase a vanilla latte.

    a. Is this an experiment or an observational study? Explain.

    b. Based on the data collected, you write a short article for the local newspaper claiming that a greater proportion of women than men prefer vanilla latte as their designer coffee of choice. A student in the local high school AP Statistics class writes a letter to the editor criticizing your study. What might the student have pointed out?

    c. Suppose you wanted to conduct a study less open to criticism. How might you redo the study?

3. Melanie battles nervousness when she plays basketball. Over the years she has had a 40% chance of making the first shot she takes in a game. If she makes her first shot, her confidence goes way up, and the probability of her making the second shot she takes rises to 70%. But if she misses her first shot, the probability of her making the second shot she takes doesn't change—it's still 40%.

    a. What is the probability that Melanie makes her second shot?

    b. If Melanie does make her second shot, what is the probability that she missed her first shot?

4. A random sample of 72 likely voters taken 3 weeks before the election for school board identified 60 people who planned to vote for Mr. Little. Then a video circulated of Mr. Little screaming at a server at a restaurant. A second random sample of 80 likely voters taken shortly after the video circulated showed that 56 planned to vote Mr. Little. Does this provide convincing evidence of a drop in support for Mr. Little? Use appropriate statistical reasoning to support your answer.

5. Some researchers believe that education influences IQ. One researcher specifically believes that the more education a person has, the higher, on average, will be his or her IQ. The researcher sets out to investigate this belief by obtaining eight pairs of identical twins reared apart. He identifies the better educated twin as Twin A and the other twin as Twin B for each pair. The data for the study are given in the table below. Do the data give good statistical evidence, at the 0.05 level of significance, that the twin with more education is likely to have the higher IQ? Give appropriate statistical evidence to support your answer.

Pair	1	2	3	4	5	6	7	8
Twin A	103	110	90	97	92	107	115	102
Twin B	97	103	91	93	92	105	111	103

SECTION II—PART B, QUESTION 6

Spend about 25 minutes on this part of the exam. Percentage of Section II grade—25.

Directions: Show all of your work. Indicate clearly the methods you use because you will be graded on the correctness of your methods as well as on the accuracy of your results and explanation.

6. A paint manufacturer claims that the average drying time for its best-selling paint is 2 hours. A random sample of drying times for 20 randomly selected cans of paint are obtained to test the manufacturer's claim. The drying times observed, in minutes, were: 123, 118, 115, 121, 130, 127, 112, 120, 116, 136, 131, 128, 139, 110, 133, 122, 133, 119, 135, 109.

    a. Obtain a 95% confidence interval for the true mean drying time of the paint.

    b. Interpret the confidence interval obtained in part (a) in the context of the problem.

    c. Suppose, instead, that a significance test at the 0.05 level of the hypothesis $H_0: \mu = 120$ was conducted against the alternative $H_A: \mu \neq 120$. What is the P-value of the test?

    d. Are the answers you got in part (a) and part (c) consistent? Explain.

    e. At the 0.05 level, would your conclusion about the mean drying time have been different if the alternative hypothesis had been $H_A: \mu > 120$? Explain.

END OF THE DIAGNOSTIC EXAM

❭ Answers and Explanations

Answers to Diagnostic Test—Section I

1. c	21. c
2. a	22. b
3. a	23. c
4. b	24. d
5. a	25. d
6. a	26. d
7. d	27. a
8. d	28. c
9. a	29. b
10. b	30. e
11. c	31. d
12. c	32. e
13. e	33. a
14. e	34. c
15. c	35. b
16. b	36. a
17. a	37. e
18. a	38. e
19. e	39. c
20. c	40. b

SOLUTIONS TO DIAGNOSTIC TEST—SECTION I

1. From Chapter 9

   The correct answer is (c). If X has $B(n, p)$, then, in general,

   $$P(X = k) = \binom{n}{k}(p)^k(1 - p)^{n-k}.$$

   In this problem, $n = 18$, $p = 0.4$, $x = 7$ so that

   $$P(X = 7) = \binom{18}{7}(0.4)^7(0.6)^{11}.$$

2. From Chapter 10

   The correct answer is (a). You need to solve the equation $0.0175 = z^* \text{sqrt}((0.28)(0.72)/1500)$ for z^*. When you do, you get $z^* = 1.5095$. When you look at table B, you find that 1.5095 is less than 1.645, so z^* is less than 90%.

3. From Chapter 10

   The correct answer is (e). For a 94% z-interval, there will be 6% of the area outside of the interval. That is, there will be 97% of the area less than the upper critical value of z. The nearest entry to 0.97 in the table of standard normal probabilities is 0.9699, which corresponds to a z-score of 1.88. (Using the TI-83/84, we have invNorm(0.97) = 1.8808.)

4. From Chapter 5

   The correct answer is (b). Since there are 28 patients, the median will be between the 14th and 15th data points, which puts it in the bar centered at 98.25°F.

5. From Chapter 8

   The correct answer is (a). These are categorical data. A bar chart is the only one among the choices that is appropriate for categorical data.

6. From Chapter 5

   The correct answer is (a). Discrete data are countable; continuous data correspond to intervals or measured data. Hence, speed, average temperature, and wingspan are examples of continuous data. The number of outcomes of a binomial experiment and the jersey numbers of a football team are countable and, therefore, discrete.

7. From Chapter 6

   The correct answer is (d). The slope of the regression line. −0.6442, can be found under "Coef" to the right of "x." The intercept of the regression line, 22.94, can be found under "Coef" to the right of "Constant."

8. From Chapter 12

   The correct answer is (d). With a sample size of 10, the degrees of freedom for a t-interval are 8. Therefore the t^* value is 2.306. The St Dev in the table has already been divided by the square root of 10. So the margin of error is 2.306 times 11.79.

9. From Chapter 11

   The correct answer is (a). The statement is basically a definition of P-value. It is the likelihood of obtaining, *by chance* alone, a value as extreme or more extreme as that obtained *if* the null hypothesis is true. A very small P-value sheds doubt on the truth of the null hypothesis.

10. From Chapter 13

    The correct answer is (b). Because the samples of men and women represent different populations, this is a chi-square test of homogeneity of proportions: the proportions of each value of the categorical variable (in this case, "pro-choice" or "pro-life") will be the same across the different populations. Had there been only one sample of 50 people drawn, 25 of whom happened to be men and 25 of whom happened to be women, this would have been a test of independence.

11. From Chapter 7

    The correct answer is (c). This is a voluntary response survey and is subject to voluntary response bias. That is, people who feel the most strongly about an issue are those most likely to respond. Because most callers would be fans, they would most likely blame someone besides the team.

12. From Chapter 10

    The correct answer is (c). The formula we use for this is $M > z^*$ times square root $((p)(1 - p)/n)$. z^* for 95% is 1.96 and since we have last year's results we will use $p = 0.83$. Solving for n gives us $n > 338.7811$. The smallest list value greater than this is 350.

13. From Chapter 7

    The correct answer is (e). A random sample from a population is one in which every *member* of the population is equally likely to be selected. A simple random sample is one in which every *sample* of a given size is equally likely to be selected. A sample can be a random sample without being a simple random sample.

14. From Chapter 5

    The correct answer is (e). The teachers are interested in showing that the average teacher salary is low. Because the mean is not resistant, it is pulled in the direction of the few higher salaries and, hence, would be higher than the median, which is not affected by a few extreme values. The teachers would choose the median. The mode, standard deviation, and IQR tell you nothing about the *average* salary.

15. From Chapter 8

    The correct answer is (c). P(still working within one year) means that all four components are still working. You must find the complement for each part of the machine and then, because these components are independent, multiply each. So, P(still working) $= (0.99)(0.98)(0.999)(0.97) = 0.940$

16. From Chapter 11

    The correct answer is (b).

$$P\text{-value} = P\left(z > \frac{0.35 - 0.30}{\sqrt{\dfrac{(0.3)(0.7)}{95}}} = 1.06\right) = 1 - 0.8554 = 0.1446.$$

(Using the TI-83/84, we find `normalcdf(1.06,100) = 0.1446`.)

17. From Chapter 9

    The correct answer is (a). III is only true when either (a) the population is normally distributed or (b) the sample size is large enough—generally greater than 30. II is not true. The standard deviation of the sampling distribution would be the same as the standard deviation of the population divided by the square root of the sample size.

18. From Chapter 13
    The correct answer is (a).

# Heads	Observed	Expected
0	10	$(0.125)(64) = 8$
1	28	$(0.375)(64) = 24$
2	22	$(0.375)(64) = 24$
3	4	$(0.125)(64) = 8$

$$\chi^2 = \frac{(10-8)^2}{8} + \frac{(28-24)^2}{24} + \frac{(22-24)^2}{24} + \frac{(4-8)^2}{8} = 3.33.$$

(This calculation can be done on the TI-83/84 as follows: let L1 = observed values; let L2 = expected values; let L3 = (L2-L1)²/L2; Then χ^2 = LIST MATH sum(L3)=3.33.)
In a chi-square goodness-of-fit test, the number of degrees of freedom equals one less than the number of possible outcomes. In this case, df = $n - 1 = 4 - 1 = 3$.

19. From Chapter 5
    The correct answer is (e). There are 101 terms, so the median is located at the 51st position in an ordered list of terms. From the counts given, the median must be in the interval whose midpoint is 8. Because the intervals are each of width 2, the class interval for the interval whose midpoint is 8 must be (7, 9).

20. From Chapter 12
    The correct answer is (c). df = $13 - 2 = 11 \Rightarrow t^* = 3.106$ (from Table B; if you have a TI-84 with the invT function, $t^* = $ invT(0.995,11)). Thus, a 99% confidence interval for the slope is

    $$0.0365 \pm 3.106(0.0015) = (0.032, 0.041).$$

    We are 99% confident that the true slope of the regression line is between 0.032 units and 0.041 units.

21. From Chapter 9
    The correct answer is (c). We can only use the normal approximation when both np and $n(1 - p)$ are greater than/equal to 10. Since $60(0.1) = 6$, our sample size is not large enough to use the normal approximation.

22. From Chapter 7
    The correct answer is (b). In an experiment, the researcher imposes some sort of treatment on the subjects of the study. Both experiments and observational studies can be conducted on human and nonhuman units; there should be randomization to groups in both to the extent possible; they can both be double blind.

23. From Chapter 6
    The correct answer is (c). III is basically what is meant when we say R-sq = 98.1%. However, R-sq is the square of the correlation coefficient.

    $\sqrt{R^2} = \pm R = \pm 0.99 \Rightarrow r$ could be either positive or negative, but not both. We can't tell direction from R^2.

24. From Chapter 10
    The correct answer is (d). You can increase power by increasing the sample size. Conversely, decreasing the sample size to 200 would decrease power.

25. From Chapter 13
The correct answer is (d). There are 81 observations total, 27 observations in the second column, 26 observations in the first row. The expected number in the first row and second column equals

$$\left(\frac{27}{81}\right)(26) = 8.667.$$

26. From Chapter 8
The correct answer is (d).

$$\mu_X = 2(0.3) + 3(0.2) + 4(0.4) + 5(0.1) = 3.3.$$

27. From Chapter 11
The correct answer is (a). The psychologist's belief implies that, if she's correct, $\mu_1 > \mu_2$. Hence, the proper alternative is H_A: $\mu_1 - \mu_2 > 0$.

28. From Chapter 5
The correct answer is (c).

$$z = \frac{90 - 80}{9} = 1.11 \Rightarrow \text{Percentile rank} = \texttt{normalcdf(-100,1.11) = 0.8665}.$$

Because she had to be in the top 15%, she had to be higher than the 85th percentile, so she was invited back.

29. From Chapter 13
The correct answer is (b). I is true. Another common standard is that there can be no empty cells, and at least 80% of the expected counts are greater than 5. II is not correct because you can have 1 degree of freedom (for example, a 2 × 2 table). III is correct because df = (4 − 1) (2 − 1) = 3.

30. From Chapter 6
The correct answer is (e). An *influential point* is a point whose removal will have a marked effect on a statistical calculation. Because the slope changes from −0.54 to −1.04, it is an influential point.

31. From Chapter 11
The correct answer is (d). df = 14 − 1 = 13. For a one-sided test and 13 degrees of freedom, 0.075 lies between tail probability values of 0.05 and 0.10. These correspond, for a one-sided test, to t^* values of 1.771 and 1.350. (If you have a TI-84 with the `invT` function, $t^* = \texttt{invT(1-0.075,13)} = 1.5299$.)

32. From Chapter 7
The correct answer is (e). Numbers of concern are 1, 2, 3, 4, 5, 6. We ignore the rest. We also ignore repeats. Reading from the left, the first three numbers we encounter for our subjects are 1, 3, and 5. They are in the treatment group, so numbers 2, 4, and 6 are in the control group. That's Betty, Doreen, and Florence. You might be concerned that the three women were selected and that, somehow, that makes the drawing non-random. However, drawing their three numbers had exactly the same probability of occurrence as any other group of three numbers from the six.

33. From Chapter 10
The correct answer is (a). If a significance test at level α rejects a null hypothesis (H_0: $\mu = \mu_0$) against a two-sided alternative, then μ_0 will not be contained in a $C = 1 - \alpha$ level confidence interval constructed using the same value of $\bar{x}$. Thus, $\alpha = 1 - C$.

34. From Chapter 9

The correct answer is (c). The statement in (c) describes the random variable for a geometric setting. In a binomial setting, the random variable of interest is the number count of successes in the fixed number of trials.

35. From Chapter 8

The correct answer is (b).

μ_{X-Y} is correct for any random variables X and Y. Since the variables X and Y are independent, we add the variances even when we are subtracting! However, So, σ_{X-Y} = sqrt of (7.29 + 0.4225).

36. From Chapter 13

The correct answer is (a). Because 0 is not in the interval (0.45, 0.80), it is unlikely that the true slope of the regression line is 0 (III is false). This implies a non-zero correlation coefficient and the existence of a linear relationship between the two variables.

37. From Chapter 8

The correct answer is (e). Here we are transforming a random variable. Since Value = 10(Card#) +5, the mean of Value = 10(2.6) + 5 = 31. But the standard deviation is not affected by addition, so stdev of Value = 10(12.1) = 121

38. From Chapter 7

The correct answer is (e). The choice is made here to treat plots A and B as a block and plots C and D as a block. That way, we are controlling for the possible confounding effects of the river. Hence the answer is (e). If you answered (d), be careful of confusing the treatment variable with the blocking variable.

39. From Chapter 5

The correct answer is (c).

Since both distributions are approximately normally distributed, equal percentiles will have the same $z =$ scores so:

$$z_{\text{first test}} = \frac{38 - 42}{5} = -0.8, \quad z_{\text{2nd test}} = \frac{X - 45}{6} = -0.8.$$

Solving the second z-score equation for X, gives us 40.

40. From Chapter 11

The correct answer is (b).

$$s_{\hat{p}} = \sqrt{\frac{(0.8)(0.2)}{45}} = 0.0596.$$

The standard error of $\hat{p}$ for a test of H_0: $p = p_0$ is

$$s_{\hat{p}} = \sqrt{\frac{p_0(1 - p_0)}{N}}.$$

If you got an answer of 0.0645, it means you used the value of $\hat{p}$ rather than the value of p_0 in the formula for $s_{\hat{p}}$.

SOLUTIONS TO DIAGNOSTIC TEST—SECTION II, PART A

1. a. $\widehat{Height} = 76.641 + 6.3661\ (Age)$
    b. For each additional year of age, the height (in cm) is predicted to increase by 6.36 cm.
    c.

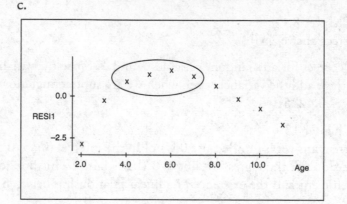

We would expect the residual for 5.5 to be in the same general area as the residuals for 4, 5, 6, and 7 (circled on the graph). The residuals in this area are all positive $\Rightarrow$ actual – predicted > 0 $\Rightarrow$ actual > predicted. The prediction would probably be too small.

2. a. It is an observational study. The researcher made no attempt to impose a treatment on the subjects in the study. The hired person simply observed and recorded behavior.
    b. • The article made no mention of the sample size. Without that you are unable to judge how much sampling variability there might have been. It's possible that the 63–59 split was attributable to sampling variability.
        • The study was done at *one* Scorebucks, on *one* morning, for a *single* 2-hour period. The population at *that* Scorebucks might differ in some significant way from the patrons at other Scorebucks around the city (and there are many, many of them). It might have been different on a different day or during a different time of the day. A single 2-hour period may not have been enough time to collect sufficient data (we don't know because the sample size wasn't given) and, again, a 2-hour period in the afternoon might have yielded different results.
    c. You would conduct the study at multiple Scorebucks, possibly blocking by location if you believe that might make a difference (i.e., would a working-class neighborhood have different preferences than the ritziest neighborhood?). You would observe at different times of the day and on different days. You would make sure that the total sample size was large enough to control for sampling variability (replication).

3. From the information given, we have
    • P(hit the first **and** hit the second) = (0.4) (0.7) = 0.28
    • P(hit the first **and** miss the second) = (0.4) (0.3) = 0.12
    • P(miss the first **and** hit the second) = (0.6) (0.4) = 0.24
    • P(miss the first **and** miss the second) = (0.6) (0.6) = 0.36
    This information can be summarized in the following table:

		Second Shot	
		Hit	Miss
First Shot	**Hit**	0.28	0.12
	Miss	0.24	0.36

    a. P(hit on second shot) = 0.28 + 0.24 = 0.52
    b. P(miss on first | hit on second) = (0.24)/(0.52) = 6/13 = 0.46.

4. Let p_1 be the true proportion who planned to vote for Mr. Little before the video. Let p_2 be the true proportion who plan to vote for him after the video.

$$H_0: p_1 = p_2$$

$$H_A: p_1 > p_2$$

We want to use a 2-proportion z test for this situation. The problem tells us that the samples are random samples.

$$\hat{p}_1 = \frac{60}{72} = 0.83, \quad \hat{p}_2 = \frac{56}{80} = 0.70.$$

Now, 72(0.83), 72(1 − 0.83), 80(0.70), and 80(1 − 0.70) are all greater than 5, so the conditions for the test are present.

$$\hat{p} = \frac{60+56}{72+80} = 0.76, \quad z = \frac{0.83-0.70}{\sqrt{(0.76)(0.24)\left(\frac{1}{72}+\frac{1}{80}\right)}} = \frac{0.13}{0.069} = 1.88 \Rightarrow P\text{-value} = 0.03.$$

Because P is very low, we reject the null hypothesis. We have reason to believe that the level of support for Mr. Little has declined since his "unfortunate" video.

5. The data are paired, so we will use a matched pairs test.
Let μ_d = the true mean difference between Twin A and Twin B for identical twins reared apart.

$$H_0: \mu_d = 0 \text{ (or, } H_0: \mu_d \leq 0)$$

$$H_A: \mu_d > 0$$

We want to use a one-sample t-test for this situation. We need the difference scores:

Pair	1	2	3	4	5	6	7	8
Twin A	103	110	90	97	92	107	115	102
Twin B	97	103	91	93	92	105	111	103
d	6	7	−1	4	0	2	4	−1

A dotplot of the difference scores shows no significant departures from normality:

The conditions needed for the one sample t-test are present.

$$\bar{x}_d = 2.25, \quad s = 2.66.$$

$$t = \frac{2.25-0}{2.66/\sqrt{8}} = 2.39, \quad df = 8 - 1 = 7 \Rightarrow 0.02 < P\text{-value} < 0.025$$

(from Table B; on the TI-83/84, tcdf(2.39,100,7)=0.024).

Because $P < 0.05$, reject H_0. We have evidence that, in identical twins reared apart, the better educated twin is likely to have the higher IQ score.

6. a. $\bar{x} = 123.85$, $s = 9.07$. We are told that the 20 cans of paint have been randomly selected. It is reasonable to assume that a sample of this size is small relative to the total population of such cans. A boxplot of the data shows no significant departures from normality. The conditions necessary to construct a 95% t confidence interval are present.

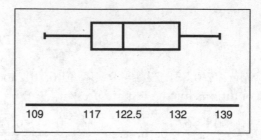

$$n = 20 \Rightarrow df = 19 \Rightarrow t^* = 2.093$$

$$123.85 \pm 2.093\left(\frac{9.07}{\sqrt{20}}\right)$$
$$= 123.85 \pm 2.093(2.03) = (119.60, 128.10).$$

   b.   We are 95% confident that the true mean drying time for the paint is between 119.6 minutes and 128.1 minutes. Because 120 minutes is in this interval, we would not consider an average drying time of 120 minutes for the population from which this sample was drawn to be unusual.

   c. $t = \dfrac{123.85 - 120}{9.07 / \sqrt{20}} = 1.90$ $df = 20 - 1 = 19 \Rightarrow 0.05 < P\text{-value} < 0.10$.

   (On the TI-83/84, we find $P\text{-value} = 2 \times \texttt{tcdf(1.90,100,19)} = 0.073$.)

   d.   We know that if a two-sided α-level significance test rejects (fails to reject) a null hypothesis, then the hypothesized value of μ will not be (will be) in a $C = 1 - \alpha$ confidence interval. In this problem, 120 was in the $C = 0.95$ confidence interval and a significance test at $\alpha = 0.05$ failed to reject the null as expected.

   e.   For the one-sided test, $t = 1.90$, $df = 19 \Rightarrow 0.025 < P\text{-value} < 0.05$
   (On the TI-83/84, we find $P\text{-value} = \texttt{tcdf(1.90,100,19)} = 0.036$.)

For the two-sided test, we concluded that we did not have evidence to reject the claim of the manufacturer. However, for the one-sided test, we have stronger evidence ($P < 0.05$) and would conclude that the average drying time is most likely greater than 120 minutes.

Interpretation: How Ready Are You?

Scoring Sheet for Diagnostic Test

Section I: Multiple-Choice Questions

$$\underbrace{\underline{}}_{\substack{\text{number correct} \\ \text{(out of 40)}}} \times 1.25 = \underbrace{\underline{}}_{\substack{\text{multiple-choice} \\ \text{score}}} = \underbrace{\underline{}}_{\substack{\text{weighted section I} \\ \text{score (do not round)}}}$$

Section II: Free-Response Questions

$$\text{Question 1} \frac{}{\text{(out of 4)}} \times 1.875 = \underbrace{\underline{}}_{\text{(do not round)}}$$

$$\text{Question 2} \frac{}{\text{(out of 4)}} \times 1.875 = \underbrace{\underline{}}_{\text{(do not round)}}$$

$$\text{Question 3} \frac{}{\text{(out of 4)}} \times 1.875 = \underbrace{\underline{}}_{\text{(do not round)}}$$

$$\text{Question 4} \frac{}{\text{(out of 4)}} \times 1.875 = \underbrace{\underline{}}_{\text{(do not round)}}$$

$$\text{Question 5} \frac{}{\text{(out of 4)}} \times 1.875 = \underbrace{\underline{}}_{\text{(do not round)}}$$

$$\text{Question 6} \frac{}{\text{(out of 4)}} \times 3.125 = \underbrace{\underline{}}_{\text{(do not round)}}$$

$$\text{Sum} = \underbrace{\underline{}}_{\substack{\text{weighted section II score} \\ \text{(do not round)}}}$$

Composite Score

$$\underbrace{\underline{}}_{\substack{\text{weighted section I} \\ \text{score}}} + \underbrace{\underline{}}_{\substack{\text{weighted section II} \\ \text{score}}} = \underbrace{\underline{}}_{\substack{\text{composite score} \\ \text{(round to nearest whole number)}}}$$

STEP **3**

Develop Strategies
for Success

CHAPTER **4** Tips for Taking the Exam

Tips for Taking the Exam

IN THIS CHAPTER

Summary: Use these question-answering strategies to raise your AP score.

KEY IDEA

Key Ideas

Before the Test

○ Do as many problems as you can on every topic in preparation for the exam.

○ Know the format of the test in advance.

General Test-Taking Tips

○ Be aware of time.

○ If you get stuck on a question move on. Mark it and come back to it later.

○ Be neat.

Tips for Multiple-Choice Questions

○ Read each question carefully.

○ Try to answer the question yourself before you look at the answers.

○ If you are unsure of the answer, eliminate as many choices as you can and then guess.

○ Answer every question.

○ Drawing a picture can sometimes help.

○ Go back and read the stem of the question again. Make sure the answer you selected matches the question.

Tips for Free-Response Questions

○ Read all the questions before you begin writing.

○ Do not work the problems in order. Answer the easy questions first. Come back to more difficult questions, or parts of questions, later.

○ Question 1 is often, but not always, the easiest. But be thorough and complete when you answer it!

○ Don't wait until the end to start Question 6. Start it early. When you get stuck, go back and work on other questions, then come back to this one.

○ Communicate: make your reasoning clear and in context. Don't ramble.

○ Calculator syntax is not sufficient for communication. Be sure to clearly identify procedures, distributions, and parameters.

✪ Be consistent from one part of your answer to another.
✪ Draw a graph if one is required. If you refer to a graph you made on your calculator, you must sketch it in your response, and it must be labeled clearly and completely.
✪ Always justify your answers. "Bald answers," that is, numbers without calculations, don't receive full credit.

General Test-Taking Tips

Much of being good at test-taking is experience. Your personal test-taking history and these tips should help you demonstrate what you know (and you know a lot) on the exam. The tips in this section are of a general nature—they apply to taking tests in general as well as to both multiple-choice and free-response questions.

1. *Become familiar with the instructions for the different parts of the exam before the day of the exam.* You don't want to have to waste time figuring out *how* to process the exam. You'll have your hands full using the available time figuring out how to do the questions. Look at the Practice Exams at the end of this book so you understand the nature of the test.

2. *Practice working as many exam-like problems as you can in the weeks before the exam.* This will help you know which statistical technique to choose on each question. It's a great feeling to see a problem on the exam and know that you can do it quickly and easily because it's just like a problem you've practiced on.

3. *Make sure your calculator has new batteries or has been fully charged.* There's nothing worse than a "Replace batteries now" warning at the start of the exam. Bring a spare calculator if you have or can borrow one (you are allowed to have two calculators).

4. *Bring a supply of sharpened pencils to the exam.* You don't want to have to waste time walking to the pencil sharpener during the exam. (The other students will be grateful for the quiet, as well.) Also, bring a good-quality eraser to the exam so that any erasures are neat and complete.

5. *Get a good night's sleep before the exam.* You'll do your best if you are relaxed and confident in your knowledge. If you don't know the material by the night before the exam, you aren't going to learn it in one evening. Relax. Maybe watch an early movie. If you know your stuff and aren't overly tired, you should do fine.

6. *Look over the entire exam first,* whichever part you are working on. Find and do the easy questions first.

7. *Don't spend too much time on any one question.* Remember that you have an average of slightly more than two minutes for each multiple-choice question, 12–13 minutes for Questions 1–5 of the free-response section, and 25–30 minutes for the investigative task. Some questions are very short which will give you extra time to spend on the more challenging questions. At the other time extreme, spending 10 minutes on one multiple-choice question (or 30 minutes on one free-response question) is not a good use of time—you won't have time to finish. *Answer the question and move on.*

8. *Be neat!* On the Statistics exam, communication is very important. This means no smudges on the multiple-choice part of the exam and legible responses on the free-response. A machine may score a smudge as incorrect. Readers work hard to figure out what a response says, but you want to make it easy to see that you know what you're doing.

Tips for Multiple-Choice Questions

There are whole industries dedicated to teaching you how to take a test. In reality, no amount of test-taking strategy will replace knowledge of the subject. If you are on top of the subject, you'll most likely do well even if you haven't paid $500 for a test-prep course. The following tips, when combined with your statistics knowledge, should help you do well.

1. *Read the question carefully before beginning.* A lot of mistakes get made because students don't completely understand the question before trying to answer it. The result is that they will often answer a different question than they were asked.

2. *Try to answer multiple choice questions before you look at the answers.* Looking at the choices and trying to figure out which one works best is not a good strategy. You run the risk of being led astray by an incorrect answer. Instead, try to answer the question first, as if there was just a blank for the answer and no choices.

3. *Understand that the incorrect answers* (which are called distractors) *are designed to appear reasonable.* Don't get suckered into choosing an answer just because it sounds good! The question designers try to base the distractors on the most common mistakes and misconceptions. For example, suppose you are asked for the median of the five numbers 3, 4, 6, 7, and 15. The correct answer is 6 (the middle score in the ordered list). But suppose you misread the question and calculated the mean instead. You'd get 7 and, be assured, 7 will appear as one of the distractors.

4. *Drawing a picture can often help* visualize the situation described in the problem. Sometimes, relationships become clearer when a picture is used to display them. For example, using Venn diagrams or two-way tables can often help you "see" the nature of a probability problem. Another example would be using a graph or a scatterplot of some given data as part of doing a regression analysis.

5. *Answer every question.* You will earn one point for each correct answer. Incorrect answers are worth zero points, and no points are earned for blank responses. If you aren't sure of an answer, eliminate as many choices as you can, then guess.

6. *Double check that you have (a) answered the question you are working on*, especially if you've left some questions blank (it's horrible to realize at some point that all of your responses are one question off!) *and (b) that you have filled in the correct bubble* for your answer. If you need to make changes, make sure you erase completely and neatly.

Tips for Free-Response Questions

There are many helpful strategies for maximizing your performance on free-response questions, but actually doing so is a learned skill. Students are often too brief or give incomplete responses. Don't assume the reader will fill in the blanks for you. You have to know the material to do well on the free-response, but knowing the material alone is not sufficient—you must also demonstrate *to the reader* that you understand the statistics in the context of the question. Many of the following tips will help you do just that.

1. *Read all parts of a question first before beginning.* There's been a trend in recent years to have more and more subparts to each question (a, b, c, …). The subparts are usually related, and later parts may rely on earlier parts. Students often make the mistake of

answering, say, part (c) as part of their answer to part (a). Understanding the whole question first can help you answer each part correctly.

2. *Use good English, write complete sentences, and organize your solutions.* Make it easy for the reader to follow your line of reasoning. Remember it is your job to demonstrate that you understand the statistics. The reader will not assume you know something unless you say it. Also, answer questions completely but don't ramble. While some irrelevant ramblings may not hurt you as long as the correct answer is there, you *will* be docked if you say something statistically inaccurate or something that contradicts an otherwise correct answer. Answers should be complete but concise. Don't fill space just because it's there. When you've completely answered a question, move on.

3. *Answers alone* (sometimes called "bald" answers) *may receive some credit but usually not much.* If the correct answer is "I'm 95% confident that the true proportion of voters who favor legalizing statistics is between 75% and 95%" and your answer is "(0.75, 0.95)," you simply won't get full credit. Probability questions generally require that you show some calculation, even if it is a simple one. And be sure to include units when appropriate.

4. *Answers must be in context and justified.* A conclusion to an inference problem that says, "Reject the null hypothesis" is simply not enough. A conclusion in context would be something like, "Because the P-value of 0.012 $\alpha = 0.05$, we reject the null hypothesis and conclude that there is convincing evidence that a majority of people favor legalizing statistics."

5. *Make sure you answer the question you are being asked.* Brilliant answers to questions no one asked will receive no credit. (Seriously, this is very common—some students think they will get credit if they show that they know something, even if it's not what the question asked.) Another too-common mistake is doing all the calculations needed to support an answer, but never giving the answer! Be sure to look back at the question, then at your response, to be sure you answered the question.

6. *Procedures are to be identified by name or by formula.* Copying the correct formula (correctly) will count as identifying your procedure. Correctly substituting the appropriate values into the formula can also save you if you make a calculation error. On the other hand, using an incorrect formula or incorrectly substituting numbers can cost you. Writing the name of the procedure and giving the answer is sufficient, but can not help you if you make a calculation error.

7. If you are using your calculator to do a problem, don't round numbers until you have the final answer. Don't round off at each step of the problem as that creates a cumulative rounding error and can affect the accuracy of your final answer. There is no hard rounding rule, but be reasonable. Don't round means or expected values to whole numbers, but don't keep 8 decimal places, either. A good guideline is to round means to two decimal places beyond the precision of the data, and standard deviations one place beyond that. Make sure that any reference to calculator commands has been accompanied by an appropriate explanation of their usage, and that all parameters are defined.

8. *Try to answer all parts of every question*—you can't get any credit for a blank answer. On the other hand, you can't snow the readers—your response must be reasonable and responsive to the question. Never provide two solutions to a question and expect the reader to pick the better one. In fact, readers have been instructed to pick the *worse* one. Cross out clearly anything you've written that you don't want the reader to look at.

9. *You don't necessarily need to answer a question in paragraph form.* A bulleted list or algebraic demonstration may work well if you are comfortable doing it that way.

10. *Understand that Question #6, the investigative task, may contain questions about material you've never studied.* The goal of such a question is to see how well you think statistically in an unfamiliar situation. The best way to prepare for this question is to practice investigative tasks. Work on applying what you know in a variety of new situations.

Specific Statistics Content Tips

The following set of tips are things that are most worth remembering about specific content issues in statistics. These are things that have been consistent over the years of the reading. This list is *not* exhaustive! The examples, exercises, and solutions that appear in this book are illustrative of the manner in which you are expected to answer free-response problems, but this list is just a sampling of some of the most important things you need to remember.

1. When asked to describe a one-variable data set, always discuss shape, center, and spread in context. That means your answer should mention the variable and include units.
2. If you are asked to compare distributions, use phrases such as *greater than, less than*, and *the same as*. And, again, always answer in context.
3. Understand how skewness can be used to differentiate between the mean and the median.
4. Know how transformations of a data set affect summary statistics.
5. Be careful when using "normal" as an adjective. Normal refers to a specific model, not the general shape of a graph of a data set. It's better to use "mound-shaped and symmetric," etc., instead. You will be docked for saying something like, "The shape of the data set is normal." *No data set is exactly normal.* At least, call it "approximately normal."
6. Remember that a correlation does not necessarily imply a causal relationship between two variables. Conversely, the absence of a strong correlation does not mean there is no relationship (it might not be linear).
7. Be able to use a residual plot to help determine if a linear model for a data set is appropriate. Be able to explain your reasoning.
8. Recognize that the correlation coefficient (r) measures the strength and direction of a relation we have reason to believe is linear. The correlation coefficient does NOT tell us that the linear model is an appropriate model.
9. Be able to interpret, in context, the slope and y-intercept of a least-squares regression line. Be sure to include "predicted" or "tends to" in your description.
10. Be able to read computer regression output.
11. Know the definition of a simple random sample (SRS).
12. Know the definition of, and reasons for, choosing to do a stratified random sample instead of a simple random sample.
13. Be able to design an experiment using a completely randomized design. Understand that an experiment that utilizes blocking cannot, by definition, be a *completely* randomized design.
14. Explain the difference between the purposes of randomization and blocking.
15. Be able to describe what blinding and confounding variables are.
16. Clearly describe how to create a simulation for a probability problem.
17. Be clear on the distinction between independent events and mutually exclusive events (and why mutually exclusive events can't be independent).
18. Be able to find the mean and standard deviation of a discrete random variable.
19. Recognize binomial and geometric situations.
20. Never forget that hypotheses are *always* about parameters, *never* about statistics.

21. Any hypothesis testing procedure involves four steps. Know what they are and that they must always be there. And never forget that your conclusion in context (Step 4) must be linked to your calculations (Step 3) in some way.

22. When doing inference problems, remember that you must *show* that the conditions for the inference procedure are present. It is not sufficient to simply *declare* them present. Realize that you are often not instructed to check the conditions in the question but you must do so anyway.

23. Be clear on the concepts of Type I and Type II errors and the power of a test.

24. If you are required to construct a confidence interval, remember that there are three things you must do to receive full credit: justify that the conditions necessary to construct the interval are present; construct the interval; and interpret the interval in context. You'll need to remember this, because often the only instruction you will see is to construct the interval.

25. If you include graphs as part of your solution, *be sure that axes are labeled and that scales are clearly indicated*. This is part of communication.

STEP → 4

Review the Knowledge You Need to Score High

CHAPTER 5

One-Variable Data Analysis

IN THIS CHAPTER

Summary: We begin our study of statistics by considering distributions of data collected on a single variable. This descriptive aspect of statistics is often referred to as *exploratory data analysis* (EDA). "Seeing" the data can often help us understand it and, to that end, we will look at a variety of plots that display or summarize the data and allow us to describe the first important characteristic of a distribution, its **shape**. Following that, we will consider a range of numerical measures for **center** and **spread** (variability) that help complete our description of the distribution. We will learn to describe distributions in terms of these three characteristics: **shape**, **center**, and **spread**, and always **in context**. We will consider ways to describe the dataset itself as an object, as well as ways to describe individual cases within the dataset. A very important model for many distributions, the normal distribution, will be considered in some detail.

Key Ideas
- ✪ Shape of a Distribution
- ✪ Types of Plots
- ✪ Measures of Center
- ✪ Measures of Spread
- ✪ Five-Number Summary
- ✪ *z*-Score
- ✪ Density Curves
- ✪ Normal Distribution
- ✪ The 68-95-99.7 Rule

Descriptive versus Inferential Statistics

Descriptive statistics, often referred to as **Exploratory Data Analysis (EDA)**, is concerned with only the data at hand, and the story being told by them. **Inferential Statistics** involves using our data to make a stronger statement. We want to say something about the population the data come from, or that a treatment causes a certain effect. We will begin with descriptive statistics, and we will discuss inferential statistics in later chapters.

Parameters versus Statistics

Values that describe a sample are called **statistics**, and values that describe a population or population model are called **parameters**. In inferential statistics, we use statistics to estimate parameters. For example, if we draw a sample of 35 students from a large university and compute their mean GPA (that is, the grade point average, usually on a 4-point scale, for each student), we have a statistic. We might want to estimate the mean GPA for *all* students in the university, which is the *parameter*.

Quantitative versus Categorical (Qualitative) Data

Quantitative data or **numerical data** are measures or counts. They tell us "how much" or "how many."

Categorical data or **qualitative data** are data that can be classified into a group based on a non-quantitative characteristic.

Examples of Quantitative (Numerical) Data: The heights of students in an AP Statistics class; the speeds of cars on a busy expressway; the scores on a final exam; the concentration of lead in water in each house in a city; the daily temperatures in Death Valley; the number of people jailed for marijuana possession in each jurisdiction.

Examples of Categorical Data: Gender; political party preference; eye color; ethnicity; level of education; socioeconomic level; zip codes (Even though they are numbers, these just indicate a location.)

There are times that the distinction between quantitative and categorical data is somewhat less clear than in the examples above. For example, we could view the variable "grade in school" as a categorical variable because it simply groups people into categories. "Students in grade nine are meeting with guidance counselors today to discuss college options." On the other hand, "grade in school" is sort of a measure of how many years of education a student has had. In such cases we need to consider how the variable is being used to determine whether we consider it quantitative or categorical.

Discrete and Continuous Variables

Quantitative variables can be either **discrete** or **continuous. Discrete variables** take on only specific values, such as whole numbers. Counts are discrete variables. **Continuous variables** can be measured, or take on values in an interval. The number of heads we can get on 20 flips of a coin is discrete; the time of day is continuous. The distinction becomes important when we consider the difference between our *data*, which are always discrete (even heights are measured to the nearest inch or centimeter) and *models*, which may be continuous.

If you are given an instruction to "describe" a set of data, be sure you discuss the *shape* of the data (including gaps and clusters in the data), the *center* of the data (mean, median, mode), and the *spread* of the data (range, interquartile range, standard deviation). And be sure to do all of this *in context*. That means to mention the cases and variable and, if there are any, units.

Graphical Analysis

Our purpose in drawing a graph of data is to get a visual sense of it. We are interested in the **shape** of the data as well as **gaps** in the data, **clusters** of datapoints, and **outliers** (which are datapoints that lie well outside of the general pattern of the data).

Shape

When we describe **shape**, what we are primarily interested in is the extent to which the graph appears to be **symmetric** (has symmetry around some axis), **mound-shaped** (**bell-shaped**), **skewed** (data are skewed to the left if the tail is to the left; to the right if the tail is to the right), **bimodal** (has more than one location with many scores), or **uniform** (frequencies of the various values are more-or-less constant).

This graph could be described as roughly *symmetric* and *mound-shaped*. Note that it doesn't have to be *perfectly* symmetrical to be classified as symmetric.

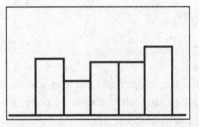

This graph is of a *uniform* distribution. Again, it does not have to be perfectly uniform to be described as *uniform*.

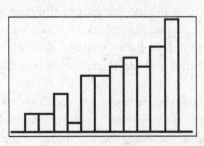

This distribution is **skewed left** because the tail is to the left. If the tail were to the right, the graph would be described at **skewed right**. Note—you might also describe a distribution as **positively skewed** when it is skewed right and **negatively skewed** when the distribution is skewed left.

There are three types of graph we want to look at in order to help us understand the shape of a distribution: dotplot, stemplot, and histogram. We use the following 31 scores from a 50-point quiz given to a community college statistics class to illustrate the first three plots (we will look at a box-and-whiskers plot in a few pages):

28	38	42	33	29	28	41	40	15	36	27	34	22
23	28	50	42	46	28	27	43	29	50	29	32	34
27	26	27	41	18								

Dotplot

A **dotplot** is a very simple type of graph that involves plotting the data values, with dots, above the corresponding values on a number line. A dotplot of the scores on the statistics quiz, drawn by a statistics computer package, looks like this:

[Calculator note: Most calculators do *not* have a built-in function for drawing dotplots. There are work-arounds that will allow you to draw a dotplot on a calculator, but they involve more effort than they are worth.]

Stemplot (Stem-and-Leaf Plot)

A stemplot is a bit more complicated than a dotplot. Each data value has a *stem* and a *leaf.* There are no mathematical rules for what constitutes the *stem* and what constitutes the *leaf.* Rather, the nature of the data will suggest reasonable choices for the stem and leaves. With the given score data, we might choose the first digit to be the *stem* and the second digit to be the *leaf.* So, the number 42 in a stem and leaf plot would show up as 4 | 2. All the leaves for a common stem are often on the same line. Typically, these are listed in increasing order, so the line with stem 4 could be written as: 4 | 0112236. The complete stemplot of the quiz data looks like this:

Score

1	58
2	23677778888999
3	234468
4	0112236
5	00

3 | 5 means 35

Using the 10s digit for the stem and the units digit for the leaf made good sense with this dataset; other choices make sense depending on the type of data. For example, suppose we had a set of gas mileage tests on a particular car (e.g., 28.3, 27.5, 28.1, . . .). In this case, it might make sense to make the stems the integer part of the number and the leaf the decimal part. As another example, consider measurements on a microscopic computer part (0.0018, 0.0023, 0.0021, . . .). Here you'd probably want to ignore the 0.00 (since that doesn't help distinguish between the values) and use the first nonzero digit as the stem and the second nonzero digit as the leaf.

Some data lend themselves to breaking the stem into two or more parts. (Some authors refer to this as *splitting the stems.*) For the statistics quiz data, the stem "4" could be shown with leaves broken up 0–4 and 5–9. Done this way, the stemplot for the scores data would look like this (there is a single "1" because there are no leaves with the values 0–4 for a stem of 1; similarly, there is only one "5" since there are no values in the 55–59 range.):

Score

1	58
2	23
2	677778888999
3	2344
3	68
4	011223
4	6
5	00

3 | 5 means 35

The visual image is of data that are slightly skewed to the right (that is, toward the higher scores). We do notice a *cluster* of scores in the high 20s that was not obvious when we used an increment of 10 rather than 5. There is no hard-and-fast rule about how to break up the stems—it's easy to try different arrangements on most computer packages.

Sometimes plotting more than one stemplot, side-by-side or back-to-back, can provide us with comparative information. The following stemplot shows the results of two quizzes given for the community college statistics class (one of the quizzes is described on the previous page):

Stem-and-Leaf of Scores 1	**Stem-and-Leaf of Scores 2**
1 | 33	1 | 58
2 | 0567799	2 | 23677778888999
3 | 1125	3 | 234468
4 | 0235778889999	4 | 0112236
5 | 00000	5 | 00 KEY: 5 | 0 = score of 50

Or, as a back-to-back stemplot:

Quiz 1		Quiz 2
33	1	58
9977650	2	23677778888999
5211	3	234468
999888775320	4	0112236
00000	5	00

It can be seen from this comparison that the scores on Quiz #1 (on the left) were generally higher than for those on Quiz #2—the center of Quiz #1 scores is higher than the center of Quiz #2 scores. Both distributions are reasonably symmetric. The spreads of the two distributions appear to be similar.

[Note: Most calculators do not have a built-in function for drawing stemplots. However, some computer programs do have this ability, and it's quite easy to experiment with various stem increments.]

Histogram

A **bar chart** is used to illustrate *categorical data*, and a **histogram** is used to illustrate *quantitative data*. The horizontal axis in a *bar chart* contains the categories, and the vertical axis contains the frequencies, or relative frequencies, of each category. The horizontal axis in a *histogram* contains numerical values, and the vertical axis contains the frequencies, or relative frequencies, of the values (often intervals of values). In a bar chart, there is a space between the bars.

example: A teacher asked his 25 students which social media platform they use the most. Their choices are displayed in the bar chart below:

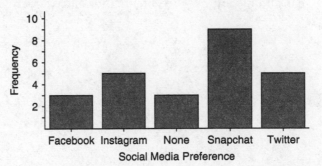

A *histogram* divides the number line into intervals (bins) of equal width. A bar is constructed on each interval, and the height of the bar is the number of cases in that interval. By convention, a value that lies on the boundary between two intervals is included in the interval to the right. So the interval from 25 to 35 contains 25 but not 35.

Consider again the quiz scores we looked at when we discussed dotplots:

28	38	42	33	29	28	41	40	15	36	27	34	22
23	28	50	42	46	28	27	43	29	50	29	32	34
27	26	27	41	18								

Because the data are integral and range from 15 to 50, reasonable intervals might be of size 10 or of size 5. The graphs below show what happens with the two choices:

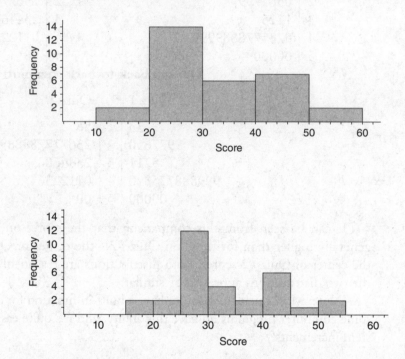

Typically, the interval from 15 to 20 would have class boundaries $15 \leq x < 20$; the interval from 20 to 25 would have class boundaries $20 \leq x < 25$, etc. Notice that a score of 20 would be in the second of those two intervals.

There are no hard-and-fast rules for how wide to make the bars (called "class intervals"). You should use computer or calculator software to help you find a picture to which your eye reacts. In this case, intervals of size 5 give us a better sense of the data.

Calculator Tip: When you are specifically asked to draw a graph for a set of data, you might want to do so without the aid of a calculator. Typing in the raw data is time-consuming and has the potential for entry error.

example: For the histogram below, identify the boundaries for the class intervals.

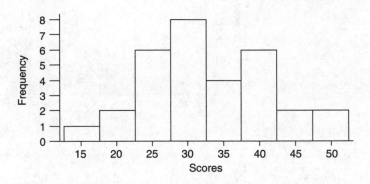

solution: The midpoints of the intervals begin at 15 and go by increments of 5. So the boundaries of each interval are 2.5 above and below the midpoints. Hence, the boundaries for the class intervals are 12.5, 17.5, 22.5, 27.5, 32.5, 37.5, 42.5, 47.5, and 52.5.

example: For the histogram given in the previous example, what proportion of the scores are less than 27.5?

solution: From the graph we see that there is 1 value in the first bar, 2 in the second, 6 in the third, etc., for a total of 31 altogether. Of these, $1 + 2 + 6 = 9$ are less than 27.5. $9/31 = 0.29$.

example: The heights of 100 college age women are displayed in a stem-and-leaf plot and a histogram. Describe the distribution.

Height

This stem-and-leaf plot breaks the heights into increments of 2 inches:

5	9
6	00001111
6	22222333333333333333
6	44444444455555555555555555
6	6666666666666666666677777777777777
6	8888888999
7	00 KEY: 6|2 = 62 inches
7	2

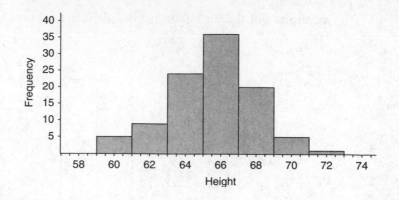

solution: Both the stem-and-leaf plot and the histogram show symmetric, bell-shaped distributions. The graph is symmetric and centered about 66 inches. In the histogram, the boundaries of the bars are $59 \leq x < 61$, $61 \leq x < 63$, $63 \leq x < 65$, . . ., $71 \leq x < 73$. Note that for each interval, the lower bound is contained in the interval, while the upper bound is part of the next larger interval. Also note that the stem-and-leaf plot and the histogram convey the same visual image for the *shape* of the data.

Measures of Center

In the last example of the previous section, we said that the graph appeared to be *centered* at about a height of 66″. In this section, we talk about ways to describe the *center* of a distribution. There are two primary measures of center: the **mean** and the **median**. There is a third measure, the **mode**, but it tells where the most frequent values occur, which may or may not represent the center. In some distributions, the mean, median, and mode will be close in value, but the mode can appear at any point in the distribution and sometimes there is more than one mode.

Mean

Let x_i represent any value in a set of n values ($i = 1, 2, . . ., n$). The **mean** of the set is defined as the sum of the x's divided by n. Symbolically, $\bar{x} = \dfrac{\sum x}{n}$. Because the indices on the summation are almost always 1 to n, they are usually omitted. $\sum x$ means "the sum of x" and is defined as follows: $\sum x = x_1 + x_2 + . . . + x_n$. Think of it as the "add-'em-up" symbol to help remember its definition. $\bar{x}$ is used for the mean of a sample ($\bar{x}$ is a *statistic*). For a population or a theoretical model. (such as in Chapters 8 and 9), the mean is a *paramter* symbolized by the Greek letter μ.

(*Note:* In the previous chapter, we made a distinction between *statistics*, which are values that describe sample data, and *parameters*, which are values that describe populations. Unless we are clear that we have access to an entire population, or that we are discussing a theoretical model, we use the statistics rather than parameters.)

example: During his major league career, Babe Ruth hit the following number of home runs (1914–1935): 0, 4, 3, 2, 11, 29, 54, 59, 35, 41, 46, 25, 47, 60, 54, 46, 49, 46, 41, 34, 22, 6. What was the *mean* number of home runs per year for his major league career?

$$\textbf{Solution:} \quad \bar{x} = \frac{\Sigma x}{n} = \frac{0+4+3+2+\cdots+22+6}{22}$$

$$= \frac{714}{22} = 32.45$$

 Calculator Tip: While each brand of graphing calculator allows you to type raw data into a list and activate a function that will return statistics such as the mean, it is extremely unlikely that an AP test question would require you to use such a function. Very often, those summary statistics are provided for you.

Median

The **median** of a distribution is the value that cuts the dataset in half. At least 50% of the dataset values are greater than or equal to the median and at least 50% are less than or equal to the median. If the dataset has an odd number of values, the *median* is a member of the set and is the middle value. If there are 3 values, the median is the second value. If there are 5, it is the third, etc. If the dataset has an even number of values, the median is the mean of the two middle numbers. If there are 4 values, the median is the mean of the second and third values. In general, if there are n values in the ordered dataset, the median is at the $\frac{n+1}{2}$ position. If you have 28 terms in order, you will find the median at position $\frac{28+1}{2} = 14.5$ (that is, between the 14th and 15th terms). Be careful not to interpret $\frac{n+1}{2}$ as the value of the median rather than as the location of the median.

example: Consider once again the data in the previous example from Babe Ruth's career. What was the median number of home runs per year he hit during his major league career?

solution: First, put the numbers in order from smallest to largest: 0, 2, 3, 4, 6, 11, 22, 25, 29, 34, 35, 41, 41, 46, 46, 46, 47, 49, 54, 54, 59, 60. There are 22 scores, so the *median* is found at the 11.5th position, between the 11th and 12th scores (35 and 41). So the *median* is

$$\frac{35+41}{2} = 38.$$

Resistant

Although the mean and median are both measures of center, the choice of which to use depends on the shape of the distribution. If the distribution is symmetric and mound-shaped, the mean and median will be close. However, if the distribution has outliers or is strongly skewed, the median is probably the better choice to describe the center. This is because it is a **resistant statistic,** one whose numerical value is not dramatically influenced by extreme values, while the mean is not resistant.

> **example:** A group of five teachers in a small school have salaries of $32,700, $32,700, $38,500, $41,600, and $44,500. The mean and median salaries for these teachers are $38,160 and $38,500, respectively. Suppose the highest paid teacher gets sick, and the school superintendent volunteers to substitute for her. The superintendent's salary is $174,300. If you replace the $44,500 salary with the $174,300 one, the median doesn't change at all (it's still $38,500), but the new mean is $64,120—almost everybody is below average if, by "average," you mean *mean*.

> **example:** For the graph given below, would you expect the mean or median to be larger? Why?

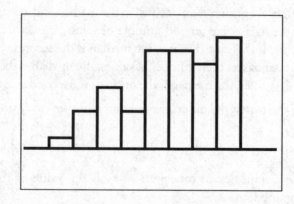

> **solution:** You would expect the median to be larger than the mean. Because the graph is skewed to the left, and the mean is not resistant, you would expect the mean to be pulled to the left (in fact, the dataset from which this graph was drawn from has a mean of 5.4 and a median of 6, as expected, given the skewness).

Estimating Means and Medians From Grouped Data

If you are given data in the form of a histogram or frequency table in which intervals contain more than one value, you cannot calculate the median, mean, or quartiles precisely. But you can estimate them using reasonable techniques.

example: Noor and Jacob were collecting data for a statistics class. They used a precise scale borrowed from the science department and weighed 100 U.S. pennies. They did not provide their raw data, but they are summarized in a frequency table and a histogram. Estimate the mean and median weight of their sample of pennies.

WEIGHT INTERVAL	FREQUENCY
$2.98 \leq x < 3.02$	1
$3.02 \leq x < 3.06$	14
$3.06 \leq x < 3.10$	29
$3.10 \leq x < 3.14$	26
$3.14 \leq x < 3.18$	21
$3.18 \leq x < 3.22$	7
$3.22 \leq x < 2.26$	2

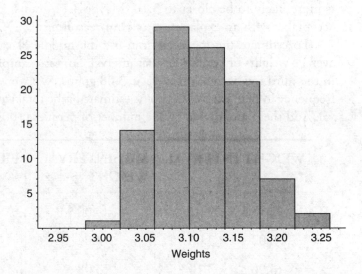

solution: To estimate the median, we recognize that we are looking for the penny in position $\frac{(100+1)}{2} = 50.5$, or between the 50th and 51st pennies. We need to keep track of the cumulative frequency, the number of pennies that weigh a certain amount or less. You can add a column to the frequency table.

WEIGHT INTERVAL	FREQUENCY	CUMULATIVE FREQUENCY
$2.98 \leq x < 3.02$	1	1
$3.02 \leq x < 3.06$	14	15
$3.06 \leq x < 3.10$	29	44
$3.10 \leq x < 3.14$	26	70
$3.14 \leq x < 3.18$	21	91
$3.18 \leq x < 3.22$	7	98
$3.22 \leq x < 2.26$	2	100

The cumulative frequency of 44 in the third row means that 44 pennies weigh less than 3.10 grams. $29 + 14 + 1 = 44$. We know the median is between the 50th and 51st pennies, so that is in the interval $3.10 \leq x < 3.14$. The median is between 3.10 and 3.14 grams. Our estimate could be in the middle of that interval, or 3.12 grams. Or we might say that we are looking between the 6th and 7th pennies out of the 26 in that interval, so the median is more likely to be closer to 3.10. Maybe 3.11 grams. Either of those estimates would be acceptable with an explanation of your reasoning.

To estimate the mean, we can use the middle of each interval as our estimate of the average weights of pennies in that interval. So, for example, we estimate that the 29 pennies in the third row weigh, on average, 3.08 grams. Wo we would multiply that number by the frequency in that interval to get the estimate of the total weight of all the pennies in that interval. Add them and divide by the number of pennies (100) to get the estimate for the mean.

WEIGHT INTERVAL	MID-INTERVAL WEIGHT	FREQUENCY	PRODUCT
$2.98 \leq x < 3.02$	3.00	1	3.00
$3.02 \leq x < 3.06$	3.04	14	42.56
$3.06 \leq x < 3.10$	3.08	29	89.32
$3.10 \leq x < 3.14$	3.12	26	81.12
$3.14 \leq x < 3.18$	3.16	21	66.36
$3.18 \leq x < 3.22$	3.20	7	22.4
$3.22 \leq x < 2.26$	3.24	2	6.48
		Sum =	311.24

$\dfrac{311.24}{100} = 3.1124$, so we estimate the mean weight of the pennies to be about 3.1124 grams.

Measures of Spread

Simply knowing the center of a distribution doesn't tell you all you might want to know about the distribution. One group of 20 people earning $20,000 each will have the same mean and median as a group of 20 where 10 people earn $10,000 and 10 people earn $30,000. These two sets of 20 numbers differ not in terms of their center but in terms of their spread, or variability. Just as there are measures of center based on the mean and the median, we also have measures of spread based on the mean and the median.

Variance and Standard Deviation

One measure of spread based on the mean is the **variance**. By definition, the variance is the average squared deviation from the mean. That is, it is a measure of spread because the more distant a value is from the mean, the larger will be the square of the difference between it and the mean.

Symbolically, the variance is defined by

$$s^2 = \frac{1}{n-1} \sum (x_i - \bar{x})^2.$$

Note that we average by dividing by $n-1$ rather than n as you might expect. This is because there are only $n-1$ independent datapoints, not n, if you know $\bar{x}$. That is, if you know $n-1$ of the values and you also know $\bar{x}$, then the nth datapoint is determined.

One problem using the variance as a measure of spread is that the units for the variance won't match the units of the original data because each difference is squared. For example, if you find the variance of a set of measurements made in inches, the variance will be in square inches. To correct this, we often take the square root of the variance as our measure of spread.

The square root of the variance is known as the **standard deviation**. Symbolically,

$$s = \sqrt{\frac{1}{n-1} \Sigma (x - \bar{x})^2}.$$

It is more important to understand the formula and what standard deviation measures, than to be able to calculate a standard deviation by hand.

Calculator Tip: If you do use the shortcut for summary statistics on the calculator (such as 1-Var Stats on a TI-83/84), the calculator will, in addition to Sx, return σ_x, which is the standard deviation of a distribution. Its formal definition is $\sigma = \sqrt{\frac{1}{n} \sum (x - \mu)^2}$. Note that this assumes you know μ, the population mean, which you rarely do in practice unless you are dealing with a probability distribution (see Chapter 8). Most of the time in statistics, you are dealing with sample data and not a distribution. Thus, with the exception of the type of probability material found in Chapters 8 and 9, you should use *only* s and not σ.

The definition of *standard deviation* has three useful qualities when it comes to describing the spread of a distribution:

- *It is independent of the mean.* Because the standard deviation depends on how far data-points are from the mean, the actual location of the mean is irrelevant.
- *It measures the spread.* The greater the spread, the larger will be the standard deviation. For two datasets with the same mean, the one with the larger standard deviation has more variability.
- *It is independent of n.* Because we are *averaging* squared distances from the mean, the standard deviation will not get larger simply because we add more terms. (The standard deviation *will* get larger if the terms we add are farther from the mean than the current typical distance.)

example: Find the standard deviation of the following 6 numbers: 3, 4, 6, 6, 7, 10.

solution: $\bar{x} = 6 \Rightarrow s = \sqrt{\dfrac{(3-6)^2 + (4-6)^2 + (6-6)^2 + (6-6)^2 + (7-6)^2 + (10-6)^2}{6-1}}$
$= 2.449$

Note that the standard deviation, like the mean, is not *resistant* to extreme values. Because it depends upon distances from the mean, it should be clear that extreme values will have a major impact on the numerical value of the standard deviation. Note also that, in practice, you will never have to do the calculation above by hand—you will rely on your calculator or computer software.

Interquartile Range

Although the standard deviation works well in situations where the mean works well (reasonably symmetric distributions), we need a measure of spread that works well when a mean-based measure is not appropriate. That measure is called the **interquartile range**.

Remember that the median of a distribution divides the distribution in two—it is the middle of the distribution. The medians of the upper and lower halves of the distribution, not including the median itself in either half, are called **quartiles**. The median of the lower half is called the **lower quartile**, or the **first quartile** (which is the 25th percentile—**Q1** on the calculator). The median of the upper half is called the **upper quartile**, or the **third quartile** (which is in the 75th percentile—**Q3** on the calculator). The median itself can be thought of as the second quartile or Q2 (although we usually don't).

The **interquartile range (IQR)** is the difference between Q3 and Q1. That is, IQR = Q3 – Q1. When you do 1-Var Stats, the calculator will return Q1 and Q3 along with a lot of other information. You have to compute the IQR from Q1 and Q3. Note that the IQR comprises the middle 50% of the data.

example: Find Q1, Q3, and the IQR for the following dataset: 5, 5, 6, 7, 8, 9, 11, 13, 17.

solution: Because the data are in order, and there is an odd number of values (9), the median is 8. The bottom half of the data comprises 5, 5, 6, 7. The median of the bottom half is the average of 5 and 6, or 5.5 which is Q1. Similarly, Q3 is the median of the top half, which is the mean of 11 and 13, or 12. The IQR = 12 – 5.5 = 6.5.

example: Find the standard deviation and IQR for the number of home runs hit by Babe Ruth in his major league career. The number of home runs was: 0, 4, 3, 2, 11, 29, 54, 59, 35, 41, 46, 25, 47, 60, 54, 46, 49, 46, 41, 34, 22, 6.

solution: Put these numbers into your calculator's list and do 1-Var Stats on that list. The calculator returns Sx = 20.21, Q1 = 11, and Q3 = 47. Hence the IQR = Q3 − Q1 = 47 − 11 = 36.

The **range** of the distribution is the difference between the maximum and minimum scores in the distribution. For the home run data, the range equals 60 − 0 = 60. Although this is sometimes used as a measure of spread, it is not very useful because we are usually interested in how the data spread out from the center of the distribution, not in just how far it is from the minimum to the maximum values.

Outliers

We have a pretty good intuitive sense of what an *outlier* is: it's a value far removed from the others. There is no rigorous mathematical formula for determining whether or not something is an outlier, but there are a few conventions that people seem to agree on. Not surprisingly, some of them are based on the mean and some are based on the median!

A commonly agreed-upon way to think of outliers based on the mean is to consider how many standard deviations away from the mean a term is. Some texts identify a potential **outlier** as a datapoint that is more than two or three standard deviations from the mean.

In a mound-shaped, symmetric distribution, this is a value that has only about a 5% chance (for two standard deviations) or a 0.3% chance (for three standard deviations) of being as far removed from the center of the distribution as it is. Think of it as a value that is way out in one of the tails of the distribution.

Most texts now use a median-based measure and identify potential outliers in terms of how far a datapoint is above or below the quartiles in a distribution. To find if a distribution has any outliers, do the following (this is known as the "1.5(IQR) rule"):

- Find the IQR.
- Multiply the IQR by 1.5.
- Find Q1 − 1.5(IQR) and Q3 + 1.5(IQR).
- Any value below Q1 − 1.5(IQR) or above Q3 + 1.5(IQR) is a potential **outlier**.

Some texts call an outlier defined as above a *mild* outlier. An *extreme* outlier would then be one that lies more than 3 IQRs beyond Q1 or Q3.

example: The following data represent the amount of money, in British pounds, spent weekly on tobacco for 11 regions in Britain: 4.03, 3.76, 3.77, 3.34, 3.47, 2.92, 3.20, 2.71, 3.53, 4.51, 4.56. Do any of the regions seem to be spending a lot more or less than the other regions? That is, are there any outliers in the data?

solution: Using a calculator, we find $\bar{x} = 3.62$, $Sx = s = .59$, Q1 = 3.2, Q3 = 4.03.

- Using means: 3.62 ± 2(0.59) = (2.44, 4.8). There are no values in the dataset less than 2.44 or greater than 4.8, so there are no outliers by this method. We don't need to check ± 3s since there were no outliers using ± 2s.
- Using the 1.5(IQR) rule: Q1 − 1.5(IQR) = 3.2 − 1.5(4.03 − 3.2) = 1.96, Q3 + 1.5(IQR) = 4.03 + 1.5(4.03 − 3.2) = 5.28. Because there are no values in the data less than 1.96 or greater than 5.28, there are no outliers by this method either.

Outliers are important because they will often tell us that something unusual or unexpected is going on with the data that we need to know about. A manufacturing process that produces products so far out of spec that they are outliers often indicates that something is wrong with the process. Sometimes outliers are just a natural, but rare, variation. Often, however, an outlier can indicate that the process generating the data is out of control in some fashion.

Position of a Term in a Distribution

Up until now, we have concentrated on the nature of a distribution as a whole. We have been concerned with the shape, center, and spread of the entire distribution. Now we look briefly at individual cases in the distribution.

Five-Number Summary

There are positions in a dataset that give us valuable information about the dataset. The **five-number summary** of a dataset is composed of the minimum value, the lower quartile, the median, the upper quartile, and the maximum value.

On the TI-83/84, these are reported on the second screen of data when you do 1-Var Stats as: minX, Q1, Med, Q3, and maxX.

> **example:** The following data are standard of living indices for 20 cities: 2.8, 3.9, 4.6, 5.3, 10.2, 9.8, 7.7, 13, 2.1, 0.3, 9.8, 5.3, 9.8, 2.7, 3.9, 7.7, 7.6, 10.1, 8.4, 8.3. Find the 5-number summary for the data.

> **solution:** Put the 20 values into a list on your calculator and do 1-Var Stats. We find: minX=0.3, Q1=3.9, Med=7.65, Q3=9.8, and maxX=13.

Box-and-Whiskers Plots (Outliers Revisited)

In the first part of this chapter, we discussed three types of graphs: dotplot, stem-and-leaf plot, and histogram. Using the five-number summary, we can add a fourth type of one-variable graph to this group: the **box-and-whiskers plot** (Sometimes called a **box-plot).** A box-and-whiskers plot is simply a graphical version of the five-number summary. A box is drawn that contains the middle 50% of the data (from Q1 to Q3) and "whiskers" extend from the lines at the ends of the box (the lower and upper quartiles) to the minimum and maximum values of the data if there are no outliers. If there are outliers, the "whiskers" extend to the last value before the outlier that is *not* an outlier. The outliers themselves are marked with a special symbol, such as a point, a box, or a plus sign.

> **example:** Consider again the standard of living indices from the previous example: 2.8, 3.9, 4.6, 5.3, 10.2, 9.8, 7.7, 13, 2.1, 0.3, 9.8, 5.3, 9.8, 2.7, 3.9, 7.7, 7.6, 10.1, 8.4, 8.3. A box-and-whiskers plot of this data looks like this (the five-number summary was [0.3, 3.9, 7.65, 9.8, 13]; you would need to add your own horizontal axis and values):

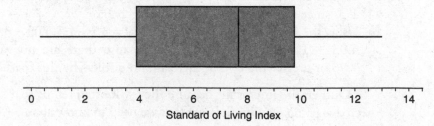

Standard of Living Index

> **Calculator Tip:** Graphing calculators will display a box-and-whiskers plot. Some allow two types: one that displays outliers based on the $1.5 \times$ IQR rule and one that does not. For AP Statistics always use the one that displays the outliers.

example: Using the same dataset as the previous example, but replacing the 10.2 with 20, which would be an outlier in this dataset (the largest possible non-outlier for these data would be $9.8 + 1.5(9.8 - 3.9) = 18.65$), we get the following graph on the calculator (you would need to add your own horizontal axis and values):

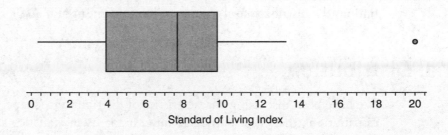

Note that the "whisker" ends at the largest value in the dataset that is **not** an outlier, 13.

Percentile Rank of a Term

The **percentile rank** of a term in a distribution equals the proportion of terms in the distribution less than the term. A term that is *at* the 75th percentile is larger than 75% of the terms in a distribution. If we know the five-number summary for a set of data, then Q1 is at the 25th percentile, the median is at the 50th percentile, and Q3 is at the 75th percentile. Some texts define the percentile rank of a term to be the proportion of terms less than *or equal to* the term. By this definition, being at the 100th percentile is possible.

z-Scores

One way to identify the position of a term in a distribution is to note how many standard deviations the term is above or below the mean. The statistic that does this is the **z-score:**

$$z_{xi} = \frac{x_i - \bar{x}}{s}.$$

The z-score is positive when x is above the mean and negative when it is below the mean.

example: $z_3 = 1.5$ tells us that the value 3 is 1.5 standard deviations above the mean. $z_3 = -2$ tells us that the value 3 is two standard deviations below the mean.

example: For the first test of the year, Harvey got a 68. The class average (mean) was 73, and the standard deviation was 3. What was Harvey's *z*-score on this test?

solution: $z_{68} = \dfrac{68 - 73}{3} = -1.67$.

Thus, Harvey was 1.67 standard deviations *below* the mean.

Suppose we have a set of data with mean $\bar{x}$ and standard deviation s. If we subtract $\bar{x}$ from every term in the distribution, it can be shown that the new distribution will have a mean of $\bar{x} - \bar{x} = 0$. If we divide every term by s, then the new distribution will have a standard deviation of $s/s = 1$. Conclusion: If you compute the *z*-score for every term in a distribution, the distribution of *z* scores will have a mean of 0 and a standard deviation of 1.

Normal Distribution

We have been discussing characteristics of distributions (shape, center, and spread). Many distributions that have particular interest for us in statistics are roughly symmetric and mound-shaped. The following histogram represents the heights of 100 males whose average height is 70″ and whose standard deviation is 3″.

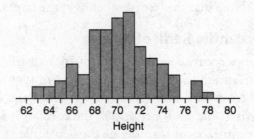

This is clearly approximately symmetric and mound-shaped. We are going to model this with a curve that idealizes what we see in this sample of 100. That is, we will model this with a continuous curve that "describes" the shape of the distribution for very large samples. That curve is the graph of the **normal distribution**. A *normal curve*, when superimposed on the above histogram, looks like this:

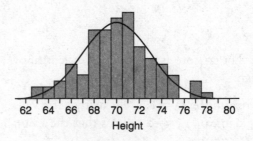

68-95-99.7 Rule

The **68-95-99.7 rule,** or the **empirical rule,** states that *approximately* 68% of the data points in a normal distribution are within one standard deviation of the mean, 95% are within two standard deviations of the mean, and 99.7% are within three standard deviations of the mean. The following three graphs illustrate the 68-95-99.7 rule.

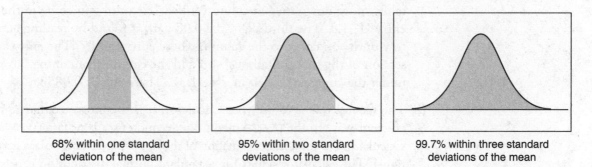

68% within one standard
deviation of the mean

95% within two standard
deviations of the mean

99.7% within three standard
deviations of the mean

Standard Normal Distribution

Because we are dealing with a theoretical distribution, we will use μ and σ, rather than $\bar{x}$ and s, when referring to the normal curve. If X is a variable that has a normal distribution with mean μ and standard deviation σ (we say "X has $N(\mu,\sigma)$"), there is a related distribution we obtain by **standardizing** the data in the distribution to produce the **standard normal distribution**. To do this, we convert the data to a set of z-scores, using the formula

$$z = \frac{x - \mu}{\sigma}.$$

The algebraic effect of this is to produce a distribution of z-scores with mean 0 and standard deviation 1. Computing z-scores is just a linear transformation of the original data, which means that the transformed data will have the same shape as the original distribution. In *this case then*, the distribution of z-scores is normal. We say z has $N(0,1)$.

For the standardized normal curve, the *68-95-99.7 rule* says that approximately 68% of the data points lie between $z = 1$ and $z = -1$, 95% between $z = -2$ and $z = 2$, and 99.7% between $z = -3$ and $z = 3$.

Because many naturally occurring distributions are approximately normal (heights, SAT scores, for example), we are often interested in knowing what proportion of terms lie in a given interval under the normal curve. Problems of this sort can be solved either by use of a calculator or a table of Standard Normal Probabilities (Table A in the appendix to this book). In a typical table, the marginal entries are z-scores, and the table entries are the areas under the curve to the left of a given z-score. All statistics texts have such tables.

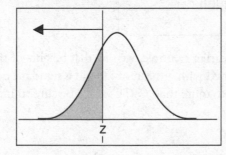

The table entry for z is the area to
the left of z and under the curve.

example: What proportion of the area under a normal curve lies to the left of $z = -1.37$?

solution: There are two ways to do this problem, and you should be able to do it either way.

(i) The first way is to use the table of Standard Normal Probabilities. To read the table, move down the left column (titled "z") until you come to the row whose entry is -1.3. The third digit, the 0.07 part, is found by reading across the top row until you come to the column whose entry is 0.07. The entry at the intersection of the row containing -1.3 and the column containing 0.07 is the area under the curve to the left of $z = -1.37$. That value is 0.0853.

(ii) The second way is to use your calculator. It is the more accurate and more efficient way. In the DISTR menu, the second entry is normalcdf (see the next Calculator Tip for a full explanation of the normalpdf and normalcdf functions). The calculator syntax for a standard normal distribution is normalcdf (lower bound, upper bound). In this example, the lower bound can be any large negative number, say -100. normalcdf(-100,-1.37)= 0.0853435081.

Calculator Tip: Part (ii) of the previous example explained how to use normalcdf for standard normal probabilities. If you are given a nonstandard normal distribution, the full syntax is normalcdf(lower bound, upper bound, mean, standard deviation). If only the first two parameters are given, the calculator assumes a standard normal distribution (that is, $\mu = 0$ and $\sigma = 1$). You will note that there is also a normalpdf function, but it really doesn't do much for you in this course. Normalpdf(X) returns the y-value on the normal curve. You are almost always interested in the area under the curve and between two points on the number line, so normalcdf is what you will use a lot in this course.

example: What proportion of the area under a normal curve lies between $z = -1.2$ and $z = 0.58$?

solution: (i) Reading from Table A, the area to the left of $z = -1.2$ is 0.1151, and the area to the left of $z = 0.58$ is 0.7190. The geometry of the situation (see below) tells us that the area between the two values is $0.7190 - 0.1151 = 0.6039$.

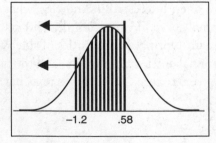

(ii) Using the calculator, we have normalcdf(-1.2, 0.58) = 0.603973005. Round to 0.6040 (difference from the answer in part (i) caused by rounding).

example: In an earlier example, we saw that heights of men are approximately normally distributed with a mean of 70 and a standard deviation of 3. What proportion of men are more than 6′ (72″) tall? Be sure to include a sketch of the situation.

solution: (i) Another way to state this is to ask what proportion of terms in a normal distribution with mean 70 and standard deviation 3 are greater than 72. In order to use the table of Standard Normal Probabilities, we must first convert to z-scores. The z-score corresponding to a height of 72″ is

$$z = \frac{72 - 70}{3} = 0.67.$$

The area to the left of $z = 0.67$ is 0.7486. However, we want the area to the *right* of 0.67, and that is $1 - 0.7486 = 0.2514$.

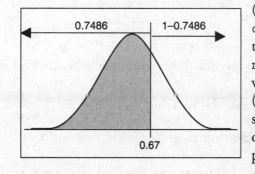

(ii) Using the calculator, we have `normal-cdf(0.67,100) = 0.2514`. We could get the answer from the raw data as follows: `normalcdf(72,1000,70,3) = 0.2525`, with the difference being due to rounding. (As explained in the last Calculator Tip, simply add the mean and standard deviation of a nonstandard normal curve to the list of parameters for `normalcdf`.)

example: For the population of men in the previous example, how tall must a man be to be in the top 10% of all men in terms of height?

solution: This type of problem has a standard approach. The idea is to express z_x in two different ways (which are, of course, equal since they are different ways of writing the z-score for the same point): (i) as a numerical value obtained from Table A or from your calculator, and (ii) in terms of the definition of a z-score.

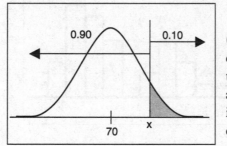

(i) We are looking for the value of x in the drawing. Look at Table A to find the nearest table entry equal to 0.90 (because we know an area, we need to read the table from the inside out to the margins). It is 0.8997 and corresponds to a z-score of 1.28.

So $z_x = 1.28$. But also,

$$z_x = \frac{x - 70}{3}.$$

So,

$$z_x = \frac{x - 70}{3} = 1.28 \Rightarrow x = 73.84.$$

A man would have to be at least 73.84″ tall to be in the top 10% of all men.

(ii) Using the calculator, the z-score corresponding to an area of 90% to the left of x is given by `invNorm(0.90) = 1.28`. Otherwise, the solution is the same as is given in part (i). See the following Calculator Tip for a full explanation of the `invNorm` function.

Calculator Tip: `invNorm` essentially reverses `normalcdf`. That is, rather than reading from the margins in, it reads from the table out (as in the example above). `invNorm(A)` returns the z-score that corresponds to an area equal to A lying to the left of z. `invNorm(A, $\mu$, $\sigma$)` returns the value of x that has area A to the left of x if x has $N(\mu, \sigma)$.

Math Background

The function that yields the *normal curve* is defined *completely* in terms of its mean and standard deviation. Although you are not required to know it, you might be interested to know that the function that defines the normal curve is:

$$f(x) = \frac{1}{\sigma\sqrt{2\pi}} e^{-\frac{1}{2}\left(\frac{x-\mu}{\sigma}\right)^2}.$$

One consequence of this definition is that the total area under the curve, and above the

x-axis, is 1 (for you calculus students, this is because $\int_{-\infty}^{\infty} \frac{1}{\sigma\sqrt{2\pi}} e^{-\frac{1}{2}\left(\frac{x-\mu}{\sigma}\right)^2} dx = 1$).

This fact will be of great use to us later when we consider areas under the normal curve as probabilities.

(**Trivia for calculus students:** one standard deviation from the mean is a *point of inflection*.)

❯ Rapid Review

1. Describe the *shape* of the histogram below:

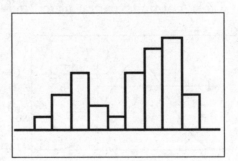

   Answer: Bimodal, somewhat skewed to the left.

2. For the graph of problem #1, would you expect the mean to be larger than the median or the median to be larger than the mean? Why?

   Answer: It is difficult to predict. The general guideline that the mean is lower than the median for distributions that are skewed left applies to smooth unimodal distributions. Bimodal distributions don't necessarily follow that pattern.

3. The first quartile (Q1) of a dataset is 12 and the third quartile (Q3) is 18. What is the largest value above Q3 in the dataset that would not be a potential outlier?

   Answer: Outliers lie more than 1.5 IQRs below Q1 or above Q3. Q3 + 1.5(IQR) = 18 + 1.5(18 − 12) = 27. Any value greater than 27 would be an outlier. 27 is the largest value that would not be a potential outlier.

4. A distribution of quiz scores has $\bar{x} = 35$ and $s = 4$. Sara got 40 points on her quiz. What was her z-score? What information does that give you if the distribution is approximately normal?

   Answer:

   $$z = \frac{40-35}{4} = 1.25.$$

   This means that Sara's score was 1.25 standard deviations above the mean, which puts it at the 89.4th percentile (`normalcdf(-100,1.25)`).

5. In a normal distribution with mean 25 and standard deviation 7, what proportion of terms are less than 20?

   Answer: $z_{20} = \dfrac{20-25}{7} = -0.71 \Rightarrow$ Area $= 0.2389$.

   (By calculator: `normalcdf(-100,20,25,7)=0.2375`.)

6. What are the mean, median, mode, and standard deviation of a *standard normal curve*?

   Answer: Mean = median = mode = 0. Standard deviation = 1.

7. Find the five-number summary and draw the box-and-whiskers plot for the following set of data: 12, 13, 13, 14, 16, 17, 20, 28.

   Answer: The five-number summary is [12, 13, 15, 18.5, 28]. 28 is an outlier (anything larger than $18.5 + 1.5(18.5 - 13) = 26.75$ is an outlier by the 1.5(IQR) rule). Since 20 is the largest nonoutlier in the dataset, it is the end of the upper whisker, as shown in the following diagram:

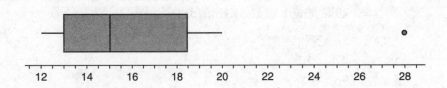

8. A distribution is strongly skewed to the right. Would you prefer to use the mean and standard deviation, or the median and interquartile range, to describe the center and spread of the distribution?

   Answer: Because the mean is not resistant and is pulled toward the tail of the skewed distribution, you would prefer to use the median and IQR.

Practice Problems

Multiple-Choice

1. The following list is ordered from smallest to largest: 25, 26, 26, 30, y, y, y, 33, 150. Which of the following statements is (are) true?

   I. The mean is greater than the median
   II. The mode is 26

III. There are no outliers in the data
   a. I only
   b. I and II only
   c. III only
   d. I and III only
   e. II and III only

2. Megan wonders how the size of her beagle Herbie compares with other beagles. Herbie is 40.6 cm tall. Megan learned on the internet that beagles heights are approximately normally distributed with a mean of 38.5 cm and a standard deviation of 1.25 cm. What is the percentile rank of Herbie's height?

   a. 59
   b. 65
   c. 74
   d. 92
   e. 95

3. The mean and standard deviation of a normally distributed dataset are 19 and 4, respectively. 19 is subtracted from every term in the dataset and then the result is divided by 4. Which of the following *best* describes the resulting distribution?

   a. It has a mean of 0 and a standard deviation of 1.
   b. It has a mean of 0, a standard deviation of 4, and its shape is normal.
   c. It has a mean of 1 and a standard deviation of 0.
   d. It has a mean of 0, a standard deviation of 1, and its shape is normal.
   e. It has a mean of 0, a standard deviation of 4, and its shape is unknown.

4. The five-number summary for a one-variable dataset is {5, 18, 20, 40, 75}. If you wanted to construct a box-and-whiskers plot for the dataset, what would be the maximum possible length of the right side "whisker"?

   a. 35
   b. 33
   c. 5
   d. 55
   e. 53

5. A set of 5000 scores on a college readiness exam are known to be approximately normally distributed with mean 72 and standard deviation 6. To the nearest integer value, how many scores are there between 63 and 75?

   a. 0.6247
   b. 4,115
   c. 3,650
   d. 3,123
   e. 3,227

6. For the data described in #5 above, which score is closest to the 63rd percentile?

   a. 45
   b. 63
   c. 72
   d. 74
   e. 81

7. The following histogram shows the number of students who visited the Career Center each week during the school year.

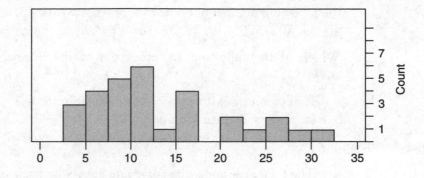

The shape of this graph could best be described as

a. Mound-shaped and symmetric
b. Bimodal
c. Skewed to the left
d. Uniform
e. Skewed to the right

8. Which of the following statements is (are) true?
    I. The median is resistant to extreme values.
    II. The mean is resistant to extreme values.
    III. The standard deviation is resistant to extreme values.
        a. I only
        b. II only
        c. III only
        d. II and III only
        e. I and III only

9. One of the values in a normal distribution is 43 and its z-score is 1.65. If the mean of the distribution is 40, what is the standard deviation of the distribution?
    a. 3
    b. −1.82
    c. 0.55
    d. 1.82
    e. −0.55

10. Free-response questions on the AP Statistics Exam are graded 4, 3, 2, 1, or 0. Question #2 on the exam was of moderate difficulty. The average score on question #2 was 2.05 with a standard deviation of 1. To the nearest tenth, what score was achieved by a student who was at the 90th percentile of all students on the test? You may assume that the scores on the question were approximately normally distributed.
    a. 3.5
    b. 3.3
    c. 2.9
    d. 3.7
    e. 3.1

Free-Response

1. Mickey Mantle played with the New York Yankees from 1951 through 1968. He had the following number of home runs for those years: 13, 23, 21, 27, 37, 52, 34, 42, 31, 40, 54, 30, 15, 35, 19, 23, 22, 18. Were any of these years outliers? Explain.

2. Which of the following are properties of the normal distribution? Explain your answers.

    a. It has a mean of 0 and a standard deviation of 1.
    b. Its mean = median = mode.
    c. All terms in the distribution lie within four standard deviations of the mean.
    d. It is bell-shaped.
    e. The total area under the curve and above the horizontal axis is 1.

3. Make a stem-and-leaf plot for the number of home runs hit by Mickey Mantle during his career (from question #1, the numbers are: 13, 23, 21, 27, 37, 52, 34, 42, 31, 40, 54, 30, 15, 35, 19, 23, 22, 18). Do it first using an increment of 10, then do it again using an increment of 5. What can you see in the second graph that was not obvious in the first?

4. A group of 15 students were identified as needing supplemental help in basic arithmetic skills. Two of the students were put through a pilot program and achieved scores of 84 and 89 on a test of basic skills after the program was finished. The other 13 students received scores of 66, 82, 76, 79, 72, 98, 75, 80, 76, 55, 77, 68, and 69. Find the z-scores for the students in the pilot program and comment on the success of the program.

5. For the 15 students whose scores were given in question #4, find the five-number summary and construct a boxplot of the data. What are the distinguishing features of the graph?

6. Assuming that the batting averages in major league baseball over the years have been approximately normally distributed with a mean of 0.265 and a standard deviation of 0.032, what would be the percentile rank of a player who bats 0.370 (as Barry Bonds did in the 2002 season)?

7. In problem #1, we considered the home runs hit by Mickey Mantle during his career. The following is a stem-and-leaf plot of the number of doubles hit by Mantle during his career. What is the interquartile range (IQR) of this data? (Hint: $n = 18$.)

1	0	8
5	1	1224
(5)	1	56777
8	2	1234
4	2	558
1	3	
1	3	7

Note: The column of numbers to the left of the stemplot gives the cumulative frequencies from each end of the stemplot (e.g., there are 5 values, reading from the top, when you finish the second row). The (5) identifies the location of the row that contains the median of the distribution. It is standard for computer packages to draw stemplots in this manner.

8. For the histogram pictured below, what proportion of the terms are less than 3.5?

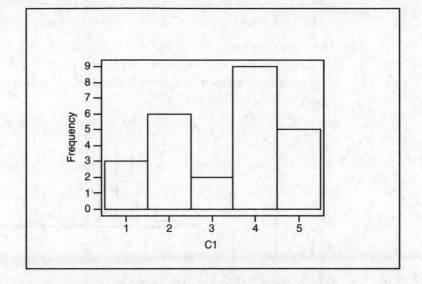

9. The following graph shows boxplots for the number of career home runs for Hank Aaron and Barry Bonds. Comment on the graphs. Which player would you rather have on your team *most* seasons? A season in which you needed a *lot* of home runs?

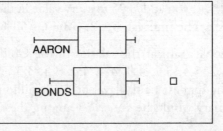

10. Suppose that being in the top 20% of people with high blood cholesterol level is considered dangerous. Assume that cholesterol levels are approximately normally distributed with mean 185 and standard deviation 25. What is the maximum cholesterol level you can have and not be in the top 20%?

11. The following are the salaries, in millions of dollars, for members of the 2001–2002 Golden State Warriors: 6.2, 5.4, 5.4, 4.9, 4.4, 4.4, 3.4, 3.3, 3.0, 2.4, 2.3, 1.3, 0.3, 0.3. Which gives a better "picture" of these salaries, mean-based or median-based statistics? Explain.

12. The following table gives the results of an experiment in which the ages of 525 pennies from current change were recorded. "0" represents the current year, "1" represents pennies one year old, etc.

Age	0	1	2	3	4	5	6	7	8	9	10	11
Count	163	87	52	75	44	24	36	14	11	5	12	2

Describe the distribution of ages of pennies (remember that the instruction "describe" means to discuss center, spread, and shape). Justify your answer.

13. The mean of a set of 150 values is 35, its median is 33, its standard deviation is 6, and its IQR is 12. A new set is created by first subtracting 10 from every term and then multiplying by 5. What are the mean, median, variance, standard deviation, and IQR of the new set?

14. The following graph shows the distribution of the heights of 300 women whose average height is 65" and whose standard deviation is 2.5". Assume that the heights of women are approximately normally distributed. How many of the women would you expect to be less than 5'2" tall?

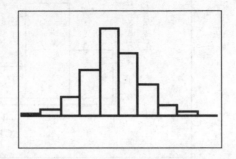

15. Which of the following are properties of the *standard deviation*? Explain your answer.

    a. It's the square root of the average squared deviation from the mean.
    b. It's resistant to extreme values.
    c. It's independent of the number of terms in the distribution.
    d. If you added 25 to every value in the dataset, the standard deviation wouldn't change.
    e. The interval $\bar{x} \pm 2s$ contains 50% of the data in the distribution.

16. Look again at the salaries of the Golden State Warriors in problem 11 (in millions, 6.2, 5.4, 5.4, 4.9, 4.4, 4.4, 3.4, 3.3, 3.0, 2.4, 2.3, 1.3, 0.3, 0.3). Erick Dampier was the highest paid player at $6.2 million. What sort of raise would he need so that his salary would be an *outlier* among these salaries?

17. Given the histogram below, draw, as best you can, the box-and-whiskers plot for the same data.

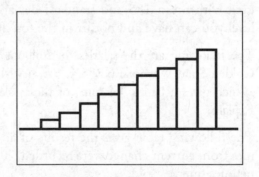

18. On the first test of the semester, the class average was 72 with a standard deviation of 6. On the second test, the class average was 65 with a standard deviation of 8. Nathan scored 80 on the first test and 76 on the second. Compared to the rest of the class, on which test did Nathan do better?

19. What is the mean of a set of data where $s = 20$, $\Sigma x = 245$, and $\Sigma(x-\bar{x})^2 = 13,600$?

Cumulative Review Problems

1. Which of the following are examples of quantitative data?

   a. The number of years each of your teachers has taught
   b. Classifying a statistic as quantitative or categorical
   c. The length of time spent by the typical teenager watching television in a month
   d. The daily amount of money lost by the airlines in the 15 months after the 9/11 attacks
   e. The colors of the rainbow

2. Which of the following are *discrete* and which are *continuous*?

   a. The weights of a sample of dieters from a weight-loss program
   b. The SAT scores for students who have taken the test over the past 10 years
   c. The AP Statistics exam scores for the almost 50,000 students who took the exam in 2009
   d. The number of square miles in each of the 20 largest states
   e. The distance between any two points on the number line

3. Just exactly what is *statistics* and what are its two main divisions?

4. What are the main differences between the goals of a *survey* and an *experiment*?

5. Why do we need to understand the concept of a *random variable* in order to do inferential statistics?

Solutions to Practice Problems

Multiple-Choice

1. The correct answer is (a). I is correct since the mean is pulled in the direction of the large maximum value, 150 (well, large compared to the rest of the numbers in the set). II is not correct because the mode is *y*—there are three *y*s and only two 26s. III is not correct because 150 is an outlier (you can't actually compute the upper boundary for an outlier since the third quartile is *y*, but even if you use a larger value, 33, in place of *y*, 150 is still an outlier).

2. The correct answer is (e).
   $$z = \frac{40.6 - 38.5}{1.25} = 1.68 \rightarrow \text{percentile} = 0.9535 \text{ (see drawing below):}$$

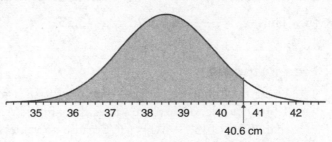

3. The correct answer is (d). The effect on the mean of a dataset of subtracting the same value is to reduce the old mean by that amount (that is, $\mu_{x-k} = \mu_x - k$). Because the original mean was 19, and 19 has been subtracted from every term, the new mean is 0. The effect on the standard deviation of a dataset of dividing each term by the same value is to divide the standard deviation by that value, that is,

$$\sigma_{x/k} = \frac{\sigma_x}{k}.$$

Because the old standard deviation was 4, dividing every term by 4 yields a new standard deviation of 1. Note that the process of subtracting the mean from each term and dividing by the standard deviation creates a set of z-scores

$$z_x = \frac{x - \bar{x}}{s}$$

so that any complete set of z-scores has a mean of 0 and a standard deviation of 1. The shape is normal since any linear transformation of a normal distribution will still be normal.

4. The correct answer is (b). The maximum length of a "whisker" in a box-and-whiskers plot is $1.5(IQR) = 1.5(40 - 18) = 33$.

5. The correct (best) answer is (d). Using Table A, the area under a normal curve between 63 and 75 is 0.6247 ($z_{63} = -1.5 \Rightarrow A_1 = 0.0668$, $z_{75} = 0.5 \Rightarrow A_2 = 0.6915 \Rightarrow A_2 - A_1 = 0.6247$). Then $(0.6247)(5,000) = 3123.5$. Using the TI-83/84, `normal-cdf(63,75,72,6)` $\times$ `5000` $=$ `3123.3`.

6. The correct answer is (b). Using Table A, a percentile of 63 corresponds to an area 0.63. This gives $z = 0.34$. Solving $0.34 = (x - 72)/6$ yields 74.04. Using the TI-83/84, `invNorm(.63,72,6)` $=$ `73.99`.

7. The correct answer is (e). The graph is clearly not symmetric, bimodal, or uniform. It is skewed to the right since that's the direction of the "tail" of the graph.

8. The correct answer is (a). The median is resistant to extreme values, and the mean is not (that is, extreme values will exert a strong influence on the numerical value of the mean but not on the median). II and III involve statistics equal to or dependent upon the mean, so neither of them is resistant.

9. The correct answer is (d). $z = 1.65 = \dfrac{43 - 40}{\sigma} \Rightarrow \sigma = \dfrac{3}{1.65} = 1.82$.

10. The correct answer is (b). A score at the 90th percentile has a z-score of 1.28. Thus, $z_x = \dfrac{x - 2.05}{1} = 1.28 \Rightarrow x = 3.33$.

Free-Response

1. Using the calculator, we find that $\bar{x} = 29.78$, $s = 11.94$, $Q1 = 21$, $Q3 = 37$. Using the $1.5(IQR)$ rule, outliers are values that are less than $21 - 1.5(37 - 21) = -3$ or greater than $37 + 1.5(37 - 21) = 61$. Because no values lie outside of those boundaries, there are no outliers by this rule.

Using the $\bar{x} \pm 2s$ rule, we have $\bar{x} \pm 2s = 29.78 \pm 2(11.94) = (5.9, 53.66)$. By this standard, the year he hit 54 home runs would be considered an outlier.

2. (a) is a property of the *standard normal distribution,* not a property of normal distributions in general. (b) is a property of the normal distribution. (c) is not a property of the normal distribution—*almost* all of the terms are within four standard deviations of the mean but, at least in theory, there are terms at any given distance from the mean. (d) is a property of the normal distribution—the normal curve is the perfect bell-shaped curve. (e) is a property of the normal distribution and is the property that makes this curve useful as a probability density curve.

Home Runs		Home Runs	
1	3589	1	3
2	12337	1	589
3	01457	2	1233
4	02	2	7
5	24	3	014
		3	57
		4	02
		4	
		5	24

3.

What shows up when done by 5 rather than 10 is the gap between 42 and 52. In 16 out of 18 years, Mantle hit 42 or fewer home runs. He hit more than 50 only twice.

4. $\bar{x} = 76.4$ and $s = 10.17$.

$$z_{84} = \frac{84 - 76.4}{10.17} = 0.75. \quad z_{89} = \frac{89 - 76.4}{10.17} = 1.24.$$

Using the Standard Normal Probability table, a score of 84 corresponds to the 77.34th percentile, and a score of 89 corresponds to the 89.25th percentile. Both students were in the top quartile of scores after the program and performed better than all but one of the other students. We don't know that there is a cause-and-effect relationship between the pilot program and the high scores (that would require comparisons with a pretest), but it's reasonable to assume that the program had a positive impact. You might wonder how the student who got the 98 did so well!

5.

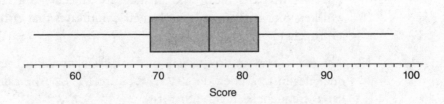

The most distinguishing thing about the graph is that it is quite symmetric. There are no outliers.

6. $z_{0.370} = \dfrac{0.370 - 0.265}{0.032} = 3.28. \Rightarrow$ Area to the left of 3.28 is 0.9995.

That is, Bond's average in 2002 would have placed him in the 99.95th percentile of batters.

7. There are 18 values in the stemplot. The median is 17 (actually between the last two 7s in the row marked by the (5) in the count column of the plot —it's still 17). Because there are 9 values in each half of the stemplot, the median of the lower half of the data, Q1, is the 5th score from the top. So, Q1 = 14. Q3 = the 5th score counting from the bottom = 24. Thus, IQR = 24 − 14 = 10.

8. There are 3 values in the first bar, 6 in the second, 2 in the third, 9 in the fourth, and 5 in the fifth for a total of 25 values in the dataset. Of these, 3 + 6 + 2 = 11 are less than 3.5. There are 25 terms altogether, so the proportion of terms less than 3.5 is 11/25 = 0.44.

9. With the exception of the one outlier for Bonds, the most obvious thing about these two is just how similar the two are. The medians of the two are almost identical and the IQRs are very similar. The data do not show it, but with the exception of 2001, the year Bonds hit 73 home runs, neither batter ever hit 50 or more home runs in a season. So, for any given season, you should be overjoyed to have either on your team, but there is no good reason to choose one over the other. However, if you based your decision on who had the most home runs in a single season, you would certainly choose Bonds.

10. Let x be the value in question. Because we do not want to be in the top 20%, the area to the left of x is 0.8. Hence $z_x = 0.84$ (found by locating the nearest table entry to 0.8, which is 0.7995 and reading the corresponding z-score as 0.84). Then

$$z_x = 0.84 = \frac{x - 185}{25} \Rightarrow x = 206.$$

[Using the calculator, the solution to this problem is given by `invNorm` `(0.8,185,25)`.]

11. $\bar{x}$ = \$3.36 million, s = \$1.88 million, *Med* = \$3.35 million, IQR = \$2.6 million. A box-and-whiskers plot of the data looks like this:

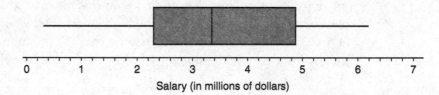

Salary (in millions of dollars)

The fact that the mean and median are virtually the same, and that the box-and-whiskers plot is more or less symmetric, indicates that either set of measures would be appropriate.

12. The easiest way to do this is to use the calculator. Put the age data in `L1` and the frequencies in `L2`. Then do `1-Var Stats L1,L2` (the calculator will read the second list as frequencies for the first list).

    • The mean is 2.48 years, and the median is 2 years. This indicates that the mean is being pulled to the right—and that the distribution is skewed to the right or has outliers in the direction of the larger values.

- The standard deviation is 2.61 years. Because one standard deviation to left would yield a negative value, this also indicates that the distribution extends farther to the right than the left.
- A histogram of the data, drawn on the TI-83/84, is shown below. This definitely indicates that the ages of these pennies is skewed to the right.

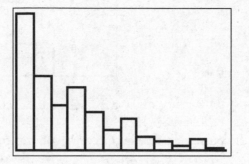

13. The new mean is $5(35 - 10) = 125$.
   The new median is $5(33 - 10) = 115$.
   The new variance is $5^2(6^2) = 900$.
   The new standard deviation is $5(6) = 30$.
   The new IQR is $5(12) = 60$.

14. First we need to find the *proportion* of women who would be less than 62″ tall:

$$z_{55} = \frac{62 - 65}{2.5} = -1.2 \Rightarrow \text{Area} = 0.1151.$$

So 0.1151 of the terms in the distribution would be less than 62″. This means that $0.1151(300) = 34.53$, so you would expect that 34 or 35 of the women would be less than 62″ tall.

15. (a), (c), and (d) are properties of the standard deviation. (a) serves as a definition of the standard deviation. It is independent of the number of terms in the distribution in the sense that simply adding more terms will not necessarily increase or decrease s. (d) is another way of saying that the standard deviation is independent of the mean—it's a measure of spread, not a measure of center.

   The standard deviation is *not* resistant to extreme values (b) because it is based on the mean, not the median. (e) is a statement about the interquartile range. In general, unless we know something about the curve, we don't know what proportion of terms are within 2 standard deviations of the mean.

16. For these data, Q1 = $2.3 million, Q3 = $4.9 million. To be an outlier, Erick would need to make at least $4.9 + 1.5(4.9 - 2.3) = 8.8$ million. In other words, he would need a $2.6 million raise in order to have his salary be an outlier.

17. You need to estimate the median and the quartiles. Note that the histogram is skewed to the left, so that the scores tend to pack to the right. This means that the median is

to the right of center and that the box-and-whiskers plot would have a long whisker to the left. The box-and-whiskers plot looks like this:

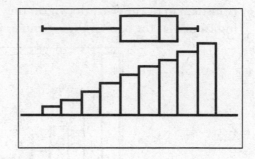

18. If you standardize both scores, you can compare them on the same scale. Accordingly,

$$z_{80} = \frac{80 - 72}{6} = 1.333, \; z_{76} = \frac{76 - 65}{8} = 1.375.$$

Nathan did slightly, but only slightly, better on the second test.

19.

$$s = 20 = \sqrt{\frac{\sum (x - \bar{x})^2}{n-1}} = \sqrt{\frac{13,600}{n-1}}$$

$$20^2 = \frac{13,600}{n-1} \Rightarrow n = 35$$

$$\bar{x} = \frac{\sum x}{n} = \frac{245}{35} = 7$$

Solutions to Cumulative Review Problems

1. (a), (c), and (d) are quantitative.

2. (a), (d), and (e) are continuous; (b) and (c) are discrete. Note that (d) could be considered discrete if what we meant by "number of square miles" was the integer number of square miles.

3. Statistics is the science of data. Its two main divisions are *data analysis* and *inference*. Exploratory data analysis (EDA) utilizes graphical and analytical methods to try to see what the data "say." That is, EDA looks at data in a variety of ways in order to understand them. Inference involves using information from samples to make statements or predictions about the population from which the sample was drawn.

4. A survey, based on a sample from some population, is usually given in order to be able to make statements or predictions about the population. An experiment, on the other hand, usually has as its goal studying the differential effects of some treatment on two or more samples, which are often composed of volunteers.

5. Statistical inference is based on being able to determine the probability of getting a particular sample statistic from a population with a hypothesized parameter. For example, we might ask how likely it is to get 55 heads on 100 flips of a fair coin. If it seems unlikely, we might reject the notion that the coin we actually flipped is fair. The probabilistic underpinnings of inference can be understood through the language of random variables. In other words, we need random variables to bridge the gap between simple data analysis and inference.

CHAPTER 6

Two-Variable Data Analysis

IN THIS CHAPTER

Summary: In the previous chapter we used *exploratory data analysis* to help us understand what a one-variable dataset was saying to us. In this chapter we extend those ideas to consider the relationships between two variables that might, or might not, be related. In this chapter, and in this course, we are primarily concerned with variables that have a *linear* relationship and, for the most part, leave other types of relationships to more advanced courses. We will spend some time considering nonlinear relationships that, through some sort of transformation, can be analyzed as though the relationship was linear. Finally, we'll consider a statistic that tells the proportion of variability in one variable that can be attributed to the linear relationship with another variable.

Key Ideas
✪ Scatterplots
✪ Linear Models
✪ The Correlation Coefficient
✪ Least Squares Regression Line
✪ Coefficient of Determination
✪ Residuals
✪ Outliers and Influential Points
✪ Transformations to Achieve Linearity

Scatterplots

In the previous chapter, we looked at several different ways to graph one-variable data. By choosing from dotplots, stemplots, histograms, or boxplots, we were able to examine visually patterns in the data. In this chapter, we consider techniques of data analysis for two-variable **(bivariate)** quantitative data. Specifically, our interest is whether or not two variables have a linear relationship and how changes in one variable can allow us to predict changes in the other variable.

> **example:** For an AP Statistics class project, a statistics teacher had her students keep diaries of how many hours they studied before their midterm exam. The following are the data for 15 of the students.
>
> The teacher wanted to know if additional studying is associated with higher grades and drew the following graph, called a **scatterplot**. It seemed pretty obvious to the teacher that students who studied more tended to have higher grades on the exam.

STUDENT	HOURS STUDIED	SCORE ON EXAM
A	0.5	65
B	2.5	80
C	3.0	77
D	1.5	60
E	1.25	68
F	0.75	70
G	4.0	83
H	2.25	85
I	1.5	70
J	6.0	96
K	3.25	84
L	2.5	84
M	0.0	51
N	1.75	63
O	2.0	71

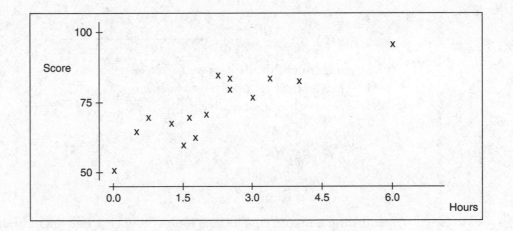

In the previous example, we were interested in seeing whether studying is associated with test performance. To do this we drew a **scatterplot**, which is just a two-dimensional graph of ordered pairs. We put one variable on the horizontal axis and the other on the vertical axis. In the example, the horizontal axis is for "hours studied" and the vertical axis is for "score on test." Each point on the graph represents the ordered pair for one student. If we have an **explanatory variable**, it should be on the horizontal axis, and the **response variable** should be on the vertical axis.

We observed that students who study more tend to have higher scores. We say that two variables are **positively associated** if higher than average values for one variable are generally paired with higher than average values of the other variable. We say they are negatively associated if higher than average values for one variable tend to be paired with lower than average values of the other variable.

Calculator Tip: In order to draw a scatterplot on your calculator, first enter the data in two lists, say the horizontal-axis variable in L1 and the vertical-axis variable in L2. Then go to STAT PLOT and choose the scatterplot icon from Type. Enter L1 for Xlist and L2 for Ylist. Choose whichever Mark pleases you. Be sure there are no equations active in the Y = list. Then do ZOOM ZoomStat (Zoom-9) and the calculator will draw the scatterplot for you. The calculator seems to do a much better job with scatterplots than it does with histograms but, if you wish, you can still go back and adjust the WINDOW in any way you want.

The scatterplot of the data in the example, drawn on the calculator, looks like this (the window used was [0, 6.5, 1, 40, 105, 5, 1]):

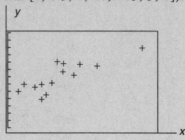

On the AP exam, it is much more efficient to quickly make a sketch of a scatterplot yourself, rather than typing in your data and trying to copy what you see on your calculator screen. Regardless, for any graph you draw, axes must be labeled and must include a scale.

example: Which of the following statements best describes the scatterplot pictured?

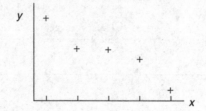

   I. A line might fit the data well.
  II. The variables are positively associated.
 III. The variables are negatively associated.
     a. I only
     b. II only
     c. III only
     d. I and II only
     e. I and III only

solution: e is correct. The data look as though a line might be a good model, and the y-variable decreases as the x-variable increases so that they are negatively associated.

Correlation

We have seen how to graph two-variable data in a scatterplot. Following the pattern we set in the previous chapter, we now want to do some numerical analysis of the data in an attempt to understand the relationship between the variables better.

In AP Statistics, we are primarily interested in determining the extent to which two variables are *linearly* associated. Two variables are *linearly related* to the extent that their relationship can be modeled by a line. Sometimes, and we will look at this more closely later in this chapter, variables may not be linearly related but can be transformed in such a way that the transformed data are linear. Sometimes the data are related but not linearly (e.g., the height of a thrown ball, t seconds after it is thrown, is a quadratic function of the number of seconds elapsed since it was released).

The first statistic we have to quantify a linear relationship is the Pearson product moment correlation, or more simply, the **correlation coefficient**, denoted by the letter r. The correlation coefficient is a measure of the *strength* of the linear relationship between two variables as well as an indicator of the *direction* of the linear relationship (whether the variables are positively or negatively associated).

If we have a sample of size n of paired data, say (x,y), and assuming that we have computed summary statistics for x and y (means and standard deviations), the **correlation coefficient r** is defined as follows:

$$r = \frac{1}{n-1} \sum \left(\frac{x_i - \bar{x}}{s_x} \right) \left(\frac{y_i - \bar{y}}{s_y} \right).$$

Because the terms after the summation symbol are nothing more than the z-scores of the individual x and y values, a different way to look at this definition is

$$r = \frac{1}{n-1} \sum z_x z_y.$$

Fortunately, the original formula for the correlation coefficient is found on the formula sheet provided on both the multiple-choice and free-response portions of the statistics exam.

> **example:** Earlier in the section, we saw some data for hours studied and the corresponding scores on an exam. It can be shown that, for these data, $r = 0.864$ and the scatterplot appears roughly linear. Together, this indicates a strong positive linear relationship between hours studied and exam score. That is, the more hours studied, the higher the exam score. (We will need to confirm that a linear model is appropriate by using a residual plot. More to come on that note.)

The correlation coefficient r has a number of properties you should be familiar with:

- $-1 \leq r \leq 1$. If $r = -1$ or $r = 1$, the points all lie on a line.
- Although there are no hard-and-fast rules about how strong a correlation is based on its numerical value, the following guidelines might help you categorize r:

VALUE OF r	STRENGTH OF LINEAR RELATIONSHIP
$-1 \leq r \leq -0.8$	strong
$0.8 \leq r \leq 1$	
$-0.8 \leq r \leq -0.5$	moderate
$0.5 \leq r \leq 0.8$	
$-0.5 \leq r \leq 0.5$	weak

- If $r > 0$, it indicates that the variables are positively associated. If $r < 0$, it indicates that the variables are negatively associated.
- If $r = 0$, it indicates that there is no linear association that would allow us to predict y from x. It *does not* mean that there is no relationship—just not a linear one.
- It does not matter which variable you call x and which variable you call y. r will be the same. In other words, r depends only on the paired points, not the *ordered* pairs.
- r does not depend on the units of measurement. In the previous example, convert "hours studied" to "minutes studied" and r would still equal 0.864.
- r is not resistant to extreme values because it is based on the mean. A single extreme value can have a powerful impact on r and may cause us to overinterpret the relationship. You must look at the scatterplot of the data as well as r.

> **example:** To illustrate that r is not resistant, consider the following two graphs. The graph on the left, with 12 points, has a marked negative linear association between x and y. The graph on the right has the same basic visual pattern but, as you can see, the addition of the new point has a dramatic effect on r—making what is generally a negative association between two variables appear to have a moderate, positive association.

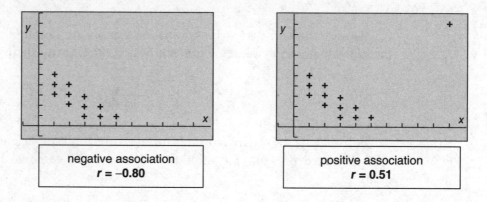

> **example:** The following computer output, again for the hours studied versus exam score data, indicates R-sq, which is the square of r. Accordingly, $r = \pm\sqrt{0.747} = \pm 0.864$. We will learn how to tell if r is positive or negative using other information. There is a lot of other stuff in the box that doesn't concern us just yet. We will learn about other important parts of the output as we proceed through the rest of this book. Note that we cannot determine the sign of r from R-sq. We need additional information.

The regression equation is
Score = 59.0 + 6.77 hours

Predictor	Coef	St Dev	t ratio	P
Constant	59.026	2.863	20.62	.000
Hours	6.767	1.092	6.20	.000

s = 6.135   R-sq = 74.7%   R-sq(adj) = 72.8%

("R-sq" is called the "coefficient of determination" and has a lot of meaning in its own right in regression. We will consider the coefficient of determination later in this chapter.)

Calculator Tip: In order to find r on your TI calculator, you will first need to change a setting from the factory. Enter CATALOG and scroll down to "Diagnostic On." Press ENTER twice. Now you are ready to find r.

Assuming you have entered the x- and y-values in L1 and L2, enter STAT CALC LinReg(a+bx) [that's STAT CALC 8 on the TI-83/84] and press ENTER. Then enter L1, L2 and press ENTER. You should get a screen that looks like this (using the data from the Study Time vs. Score on Test study):

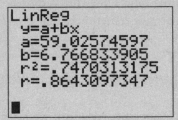

(Note that reversing L1 and L2 in this operation—entering STAT CALC LinReg(a+bx) L2, L1—will change a and b but will not change r since order doesn't matter in correlation.) If you compare this with the computer output above, you will see that it contains some of the same data, including both r and r^2. At the moment, all you care about in this printout is the value of r.

Correlation and Causation

Two variables, x and y, may have a strong correlation, but you need to take care *not* to interpret that as causation. That is, just because two things seem to go together does not mean that one caused the other—some third variable may be influencing them both. Seeing a fire truck at almost every fire doesn't mean that fire trucks cause fires!

example: The table below shows, for the given years, the per capita consumption of cheese in the U.S. and the number of doctorates in civil engineering awarded in the U.S.

YEAR	U.S. PER CAPITA CHEESE CONSUMPTION (LBS)	NUMBER OF CIVIL ENGINEERING PHDS AWARDED
2000	9.3	480
2001	9.7	501
2002	9.7	540
2003	9.7	552
2004	9.9	547
2005	10.2	622
2006	10.5	655
2007	11	701
2008	10.6	712
2009	10.6	708

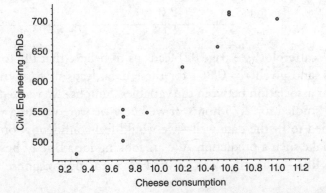

For these data, it turns out that $r = 0.959$.

Is the increase in the amount of cheese consumed in the U.S. responsible for the increase in civil engineering PhDs? Of course not! Nor the reverse. But you must be careful in situations where a causal relationship might sound more plausible. We still cannot conclude that one causes the other just because we see a correlation between the variables. Even a very strong correlation.

In some cases, there may be a third variable that causes both the variables to change in the same direction. This will be discussed more in the next chapter, but in the meantime remember, always remember, that *correlation does not imply causation*.

Linear Models

When we discussed correlation, we learned that it didn't matter which variable we called x and which variable we called y—the correlation r is the same. That is, there is no explanatory and response variable, just two variables that may or may not vary linearly. In this section we will be more interested in predicting, once we've determined the strength of the linear relationship between the two variables, the value of one variable (the response) based on the value of the other variable (the explanatory). In this situation, called linear regression, it matters greatly which variable we call x and which one we call y.

Least-Squares Regression Line

Recall again the data from the study that looked at hours studied versus score on test:

HOURS STUDIED	SCORE ON EXAM
0.5	65
2.5	80
3.0	77
1.5	60
1.25	68
0.75	70
4.0	83
2.25	85
1.5	70
6.0	96
3.25	84
2.5	84
0.0	51
1.75	63
2.0	71

The scatterplot (see page 92) leads us to believe that the form of this relationship is linear. This, and given $r = 0.864$ for these data, leads us to say that we have a strong, positive, linear association between the variables. Suppose we wanted to predict the score of a person who studied for 2.75 hours. If we knew we were working with a linear model—a line that seemed to fit the data well—we would feel confident about using the equation of the line to make such a prediction. We are looking for a **line of best fit**. We want to find a **regression line**—a line that can be used for predicting response values from explanatory values.

In this situation, we would use the regression line to predict the exam score for a person who studied 2.75 hours.

The line we are looking for is called the **least-squares regression line**. We could draw a variety of lines on our scatterplot trying to determine which has the best fit. Let $\hat{y}$ be the predicted value of y for a given value of x. Then $y - \hat{y}$ represents the error in prediction. We want our line to minimize errors in prediction, so we might first think that $\Sigma\,(y - \hat{y})$ would be a good measure ($y - \hat{y}$ is the *actual value* minus the *predicted value*). However, because our line is going to average out the errors in some fashion, we find that $\Sigma\,(y - \hat{y}) = 0$. To get around this problem, we use $\Sigma\,(y - \hat{y})^2$. This expression will vary with different lines and is sensitive to the fit of the line. That is, $\Sigma\,(y - \hat{y})^2$ is small when the linear fit is good and large when it is not.

The **least-squares regression line** (LSRL) is the line that minimizes the sum of squared errors. If $\hat{y} = a + bx$ is the LSRL, then $\hat{y}$ minimizes $\Sigma\,(y - \hat{y})^2$.

Digression for Calculus Students Only: It should be clear that trying to find a and b for the line $\hat{y} = a + bx$ that minimizes $\Sigma\,(y - \hat{y})^2$ is a typical calculus problem. The difference is that, since $\hat{y}$ is a function of two variables, it requires multivariable calculus to derive it. That is, you need to be beyond first-year calculus to derive the results that follow.

For n ordered pairs (x, y), we calculate: $\bar{x}$, $\bar{y}$, s_x, s_y, and r. Then we have:

If $\hat{y} = a + bx$ is the LSRL, $b = r\,\dfrac{s_y}{s_x}$, and $a = \bar{y} - b\bar{x}$.

---

example: For the hours studied (x) versus score (y) study, the LSRL is $\hat{y} = 59.03 + 6.77x$. We asked earlier what score would we predict for someone who studied 2.75 hours. Plugging this value into the LSRL, we have $\hat{y} = 59.03 + 6.77(2.75) = 77.63$. It's important to understand that this is the *predicted* value, not the exact value such a person will necessarily get.

example: Consider once again the computer printout for the data of the preceding example:

The regression equation is
Score = 59.0 + 6.77 Hours

Predictor	Coef	St Dev	t ratio	P
Constant	59.026	2.863	20.62	.000
Hours	6.767	1.092	6.20	.000

s = 6.135   R-sq = 74.7%   R-sq(adj) = 72.8%

Exam Tip: An AP Exam question in which you are asked to determine the regression equation from the printout has been common. Be sure you know where the intercept and slope of the regression line are located in the printout (they are under "Coef").

The regression equation is given as "$\widehat{\text{Score}} = 59 + 6.77$ Hours." The y-intercept, which is the predicted score when the number of hours studied is zero, and the slope of the regression line are listed in the table under the column "Coef."

example: We saw earlier that the calculator output for these data was

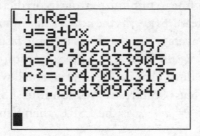

```
LinReg
 y=a+bx
 a=59.02574597
 b=6.766833905
 r²=.7470313175
 r=.8643097347
```

The values of a and b are given as part of the output. Remember that these values were obtained by putting the "Hours Studied" data in L1, the "Test Score" data in L2, and doing LinReg(a+bx)L1,L2. When using LinReg(a+bx), the explanatory variable *must* come first and the response variable second.

example: An experiment is conducted on the effects of having convicted criminals provide restitution to their victims rather than serving time. The following table gives the data for 10 criminals. The monthly salaries (X) and monthly restitution payments (Y) were as follows:

X	300	880	1000	1540	1560	1600	1600	2200	3200	6000
Y	200	380	400	200	800	600	800	1000	1600	2700

(a) Find the correlation between X and Y and the regression equation that can be used to predict monthly restitution payments from monthly salaries.
(b) Draw a scatterplot of the data and put the LSRL on the graph.
(c) Interpret the slope of the regression line in the context of the problem.
(d) How much would a criminal earning \$1400 per month be expected to pay in restitution?

solution: Put the monthly salaries (x) in L1 and the monthly restitution payments (y) in L2. Then enter STAT CALC LinReg(a+bx)L1,L2,Y1.

(a) $r = 0.97$, $\widehat{\text{Payments}} = -56.22 + 0.46$ (Salary). (If you answered $\hat{y} = 56.22 + 0.46x$, you must define x and y so that the regression equation can be understood in the context of the problem. An algebra equation, without a contextual definition of the variables, will not receive full credit.)

(b)

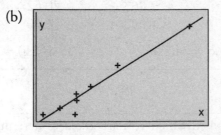

(c) The slope of the regression line is 0.46. This tells us that, for each \$1 increase in the criminal's salary, the amount of restitution is predicted to increase by \$0.46.
(d) *Payment* = $-56.22 + 0.46$ (1400) = \$587.78.

Calculator Tip: The fastest, and most accurate, way to perform the computation above, assuming you have stored the LSRL in Y1 (or some "Y=" location), is to do Y1(1400) on the home screen. To paste Y1 to the home screen, remember that you enter VARS Y-VARS Function Y1. If you do this, you will get an answer of $594.64 with the difference caused by rounding due to the more accurate 12-place accuracy of the TI-83/84.

Residuals

When we developed the LSRL, we referred to $y - \hat{y}$ (the *actual value* – the *predicted value*) as an error in prediction. The formal name for $y - \hat{y}$ is the **residual**. Note that the order is always "actual" – "predicted" so that a positive residual means that the prediction was too small and a negative residual means that the prediction was too large.

> **example:** In the previous example, a criminal earning $1560/month paid restitution of $800/month. The predicted restitution for this amount would be $\hat{y} = -56.22 + 0.46(1560) = \$661.38$. Thus, the residual for this case is $800 - \$ 661.38 = \$138.62$.

Calculator Tip: The TI-83/84 will generate a complete set of residuals when you perform a LinReg. They are stored in a list called RESID which can be found in the LIST menu. RESID stores only the current set of residuals. That is, a new set of residuals is stored in RESID each time you perform a new regression.

Residuals can be useful to us in determining the extent to which a linear model is appropriate for a dataset. If a line is an appropriate model, we would expect to find the residuals more or less randomly scattered about the average residual (which is, of course, 0). In fact, we expect to find them approximately normally distributed about 0. A pattern of residuals that does not appear to be more or less randomly distributed about 0 (that is, there is a systematic nature to the graph of the residuals) is evidence that a line is not a good model for the data. If the residuals are small, the line may predict well even though it isn't a good theoretical model for the data. The usual method of determining if a line is a good model is to examine visually a plot of the residuals plotted against the explanatory variable.

Calculator Tip: In order to draw a residual plot on the TI-83/84, and assuming that your *x*-data are in L1 and your *y*-data are in L2, first do LinReg(a+bx)L1,L2. Next, you create a STAT PLOT scatterplot, where Xlist is set to L1 and Ylist is set to RESID. RESID can be retrieved from the LIST menu (remember that only the residuals for the most recent regression are stored in RESID). ZOOM ZoomStat will then draw the residual plot for the current list of residuals. It's a good idea to turn off any equations you may have in the Y= list before doing a residual plot or you may get an unwanted line on your plot.

> **example:** The data given below show the height (in cm) at various ages (in months) for a group of children.
> (a) Does a line seem to be a good model for the data? Explain.
> (b) What is the value of the residual for a child of 19 months?

Age	18	19	20	21	22	23	24	25	26	27	28	29
Height	76	77.1	78.1	78.3	78.8	79.4	79.9	81.3	81.1	82.0	82.6	83.5

solution:

(a) Using the calculator (`LinReg(a+bx) L1, L2, Y1`), we find $\widehat{height} = 64.94 + 0.634(age)$, $r = 0.993$. The large value of r tells us that the points are close to a line. The scatterplot and LSLR are shown below on the graph at the left.

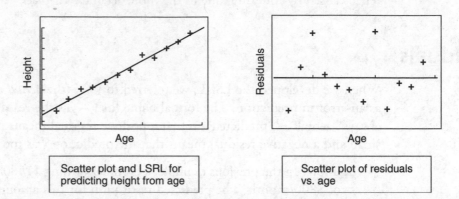

| Scatter plot and LSRL for predicting height from age | Scatter plot of residuals vs. age |

From the graph on the left, a line appears to be a good fit for the data (the points lie close to the line). The residual plot on the right shows no readily obvious pattern, so we have good evidence that a line is a good model for the data and we can feel good about using the LSRL to predict height from age.

(b) The residual (actual minus predicted) for $age = 19$ months is $77.1 - (64.94 + 0.634 \cdot 19) = 0.114$. Note that `77.1-Y1(19)=0.112`.

Note that you can generate a complete set of residuals, which will match what is stored in `RESID`, in a list. Assuming your data are in `L1` and `L2` and that you have found the LSRL and stored it in `Y1`, let `L3 = L2-Y1(L1)`. The residuals for each value will then appear in `L3`. You might want to let `L4 = RESID` (by pasting `RESID` from the `LIST` menu) and observe that `L3` and `L4` are the same.

Digression: Whenever we have graphed a residual plot in this section, the vertical axis has been the residuals and the horizontal axis has been the x-variable. On some computer printouts, you may see the horizontal axis labeled "Fits" (as in the graph below) or "Predicted Value."

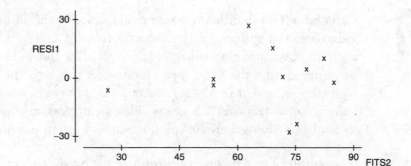

What you are interested in is the visual image given by the residual plot, and it doesn't matter if the residuals are plotted against the x-variable or something else, like "FITS2"—the scatter of the points above and below 0 stays the same. All that changes are the horizontal distances between points. This is the way it must be done in multiple regression, since there is more than one independent variable and, as you can see, it can be done in simple linear regression.

If we are trying to predict a value of *y* from a value of *x*, it is called **interpolation** if we are predicting from an *x*-value within the range of *x*-values. It is called **extrapolation** if we are predicting from a value of *x* outside of the *x*-values.

Age	18	19	20	21	22	23	24	25	26	27	28	29
Height	76	77.1	78.1	78.3	78.8	79.4	79.9	81.3	81.1	82.0	82.6	83.5

> **example:** Using the age/height data from the previous example, we are *interpolating* if we attempt to predict height from an age between 18 and 29 months. It is interpolation if we try to predict the height of a 20.5-month-old baby. We are *extrapolating* if we try to predict the height of a child less than 18 months old or more than 29 months old.

If a line has been shown to be a good model for the data and if it fits the line well (i.e., we have a strong *r* and a more or less random distribution of residuals), we can have confidence in interpolated predictions. We can rarely have confidence in extrapolated values. In the example above, we might be willing to go slightly beyond the ages given because of the high correlation and the good linear model, but it's good practice *not* to extrapolate beyond the data given. If we were to extrapolate the data in the example to a child of 12 years of age (144 months), we would predict the child to be 156.2 inches, or more than 13 feet tall!

Coefficient of Determination

In the absence of a better way to predict *y*-values from *x*-values, our best guess for any given *x* might well be $\bar{y}$, the mean value of *y*.

> **example:** Suppose you had access to the heights and weights of each of the students in your statistics class. You compute the average weight of all the students. You write the heights of each student on a slip of paper, put the slips in a hat, and then draw out one slip. You are asked to predict the weight of the student whose height is on the slip of paper you have drawn. What is your best guess as to the weight of the student?

> **solution:** In the absence of any known relationship between height and weight, your best guess would have to be the average weight of all the students. You know the weights vary about the average and that is about the best you could do.

If we guessed at the weight of each student using the average, we would be wrong most of the time. If we took each of those errors and squared them, we would have what is called the *sum of squares total* (SST). It's the total squared error of our guesses when our best guess is simply the mean of the weights of all students, and represents the total variability of *y*.

Now suppose we have a least-squares regression line that we want to use as a model for predicting weight from height. It is, of course, the LSRL we discussed in detail earlier in this chapter, and our hope is that there will be less error in prediction than by using $\bar{y}$. Now, we still have errors from the regression line (called *residuals*, remember?). We call the sum of *those* errors the **sum of squared errors** (SSE). So, SST represents the total error from using $\bar{y}$ as the basis for predicting weight from height, and SSE represents the total error from using the LSRL. SST − SSE represents the benefit of using the regression line rather than $\bar{y}$ for prediction. That is, by using the LSRL rather than $\bar{y}$, we have *explained* a certain proportion of the total variability by regression.

The proportion of the total variability in y that is explained by the regression of y on x is called the **coefficient of determination**. The *coefficient of determination* is symbolized by r^2. Based on the above discussion, we note that

$$r^2 = \frac{SST - SSE}{SST}.$$

It can be shown algebraically that this r^2 is actually the square of the familiar r, the correlation coefficient. Many computer programs will report the value of r^2 only (usually as "R-sq"), which means that we must take the square root of r^2 if we only want to know r (remember that r and b, the slope of the regression line, are either both positive or negative so that you can check the sign of b to determine the sign of r if all you are given is r^2). The TI-83/84 calculator will report both r and r^2, as well as the regression coefficient, when you do `LinReg(a+bx)`.

example: Consider the following output for a linear regression:

Predictor	Coef	St Dev	t ratio	P
Constant	−1.95	21.97	−0.09	.931
x	0.8863	0.2772	3.20	.011
s = 16.57		R-sq = 53.2%		R-sq(adj) = 48.0%

We can see that the LSRL for these data is $\hat{y} = -1.95 + 0.8863x$. $r^2 = 53.2\% = 0.532$. This means that 53.2% of the total variability in y can be explained by the regression of y on x. Further, $r = \sqrt{0.532} = 0.729$ (r is positive since $b = 0.8863$ is positive). We learn more about the other items in the printout later.

You might note that there are two standard errors (estimates of population standard deviations) in the computer printout above. The first is the "St Dev of x" (0.2772 in this example). This is the standard error of the slope of the regression line, s_b, the estimate of the standard deviation of the slope (for information, although you don't need to know this, $s_b = \dfrac{s}{\sqrt{\sum(x-\bar{x})^2}}$). The second standard error given is the standard error of the residuals,

the "s" ($s = 16.57$) at the lower left corner of the table. This is the estimate of the standard deviation of the residuals (again, although you don't need to know this, $s = \sqrt{\dfrac{\sum(y-\hat{y})^2}{n-2}}$).

Outliers and Influential Observations

Some observations have an impact on correlation and regression. We defined an outlier when we were dealing with one-variable data (remember the 1.5 [IQR] rule?). There is no analogous numerical guideline when dealing with two-variable data, but it is the same basic idea: an **outlier** lies outside of the general pattern of the data. Specifically, an outlier has a large residual. An outlier can certainly influence the correlation and, depending on where it is located, may also exert an influence on the slope of the regression line.

An **influential observation** is one that has a strong influence on the regression model. It may influence the slope, the correlation, the y-intercept, or more than one of these. Its influence, if it doesn't line up with the rest of the data, is on the slope of the regression line. More generally, an influential observation is a datapoint that exerts a strong influence on a measure.

example: Graphs I, II, and III are the same except for the point symbolized by the box in graphs II and III. Graph I below has no outliers or influential points. Graph II has an outlier that is an influential point that has an effect on the correlation. Graph III has an outlier that is an influential point that has an effect on the regression slope. Compare the correlation coefficients and regression lines for each graph. Note that the outlier in Graph II has some effect on the slope and a significant effect on the correlation coefficient. The influential point in Graph III has about the same effect on the correlation coefficient as the outlier in Graph II, but a major influence on the slope of the regression line.

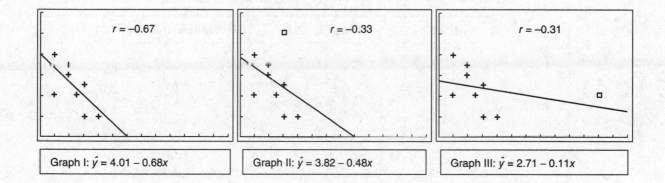

| Graph I: $\hat{y} = 4.01 - 0.68x$ | Graph II: $\hat{y} = 3.82 - 0.48x$ | Graph III: $\hat{y} = 2.71 - 0.11x$ |

Transformations to Achieve Linearity

Until now, we have been concerned with data that can be modeled with a line. Of course, there are many two-variable relationships that are nonlinear. The path of an object thrown in the air is parabolic (quadratic). Population tends to grow exponentially, at least for a while. Even though you could find an LSRL for nonlinear data, it makes no sense to do so. The AP Statistics course deals only with two-variable data that can be modeled by a line *OR* nonlinear two-variable data that can be *transformed* in such a way that the transformed data can be modeled by a line.

example: The number of a certain type of bacteria present (in thousands) after a certain number of hours is given in the following chart:

HOURS	NUMBER
1.0	1.8
1.5	2.4
2.0	3.1
2.5	4.3
3.0	5.8
3.5	8.0
4.0	10.6
4.5	14.0
5.0	18.0

What would be the predicted quantity of bacteria after 3.75 hours?

solution: A scatterplot of the data and a residual plot [for $\widehat{Number} = a + b(Hour)$] shows that a line is not a good model for this data:

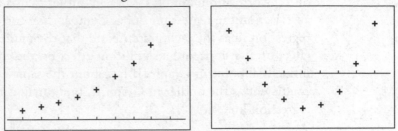

Now, take $ln(Number)$ to produce the following data:

HOURS (L1)	NUMBER (L2)	LN(NUMBER) (L3 = LN (L2))
1.0	1.8	0.59
1.5	2.4	0.88
2.0	3.1	1.13
2.5	4.3	1.46
3.0	5.8	1.76
3.5	8.0	2.08
4.0	10.6	2.36
4.5	14.0	2.64
5.0	18.0	2.89

The scatterplot of *Hours* versus $ln(Number)$ and the residual plot for $ln(\widehat{Number}) = -0.0047 + 0.586(Hours)$ are as follows:

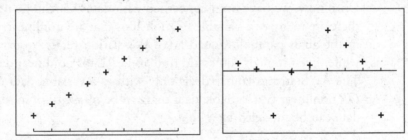

The scatterplot looks much more linear, and the residual plot no longer has the distinctive pattern of the raw data. We have transformed the original data in such a way that the transformed data is well modeled by a line. The regression equation for the transformed data is: $ln(\widehat{Number}) = -0.047 + 0.586(Hours)$.

The question asked for how many bacteria are predicted to be present after 3.75 hours. Plugging 3.75 into the regression equation, we have $ln(\widehat{Number}) = -0.0048 + 0.586(3.75) = 2.19$. But that is $ln(\widehat{Number})$, not $\widehat{Number}$. We must back-transform this answer to the original units. Doing so, we have $\widehat{Number} = e^{2.19} = 8.94$ thousand bacteria.

Calculator Tip: You do not need to take logarithms by hand in the above example—your calculator is happy to do it for you. Simply put the Hours data in L1 and the Number data in L2. Then let L3 = LN(L2). The LSRL for the transformed data is then found by `LinReg(a+bx) L1,L3,Y1`.

Remember that the easiest way to find the value of a number substituted into the regression equation is to simply find Y1(#). Y1 is found by entering VARS Y-VARS Function Y1.

Interesting Diversion: You will find a number of different regression expressions in the STAT CALC menu: LinReg(ax+b), QuadReg, CubicReg, QuartReg, LinReg(a+bx), LnReg, ExpReg, PwrReg, Logistic, and SinReg. While each of these has its use, only LinReg(a+bx) needs to be used in this course (well, LinReg(ax+b) gives the same equation—with the *a* and *b* values reversed, just in standard algebraic form rather than in the usual statistical form).

Exam Tip: Also remember, when taking the AP exam, NO calculatorspeak. If you do a linear regression on your calculator, simply report the result. The person reading your exam will know that you used a calculator and is NOT interested in seeing something like LinReg L1, L2, Y1 written on your exam.

On the AP exam, it is highly unlikely that you would be asked to try transformations on raw data. Most likely, you will be given a scatterplot, residual plot, and/or a regression output for transformed data, and you will be asked to comment on it. You may be asked to evaluate the effectiveness of the transformation, or to compare the effectiveness of a couple models.

› Rapid Review

1. The correlation between two variables *x* and *y* is 0.85. Interpret this statement.

   Answer: There is a strong, positive, linear association between *x* and *y*. That is, as one of the variables increases, the other variable increases as well.

2. The following is a residual plot of a least-squares regression. Does it appear that a line is a good model for the data? Explain.

   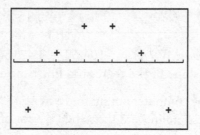

   Answer: The residual plot shows a definite pattern. If a line was a good model, we would expect to see a more or less random pattern of points about 0. A line is unlikely to be a good model for this data.

3. Consider the following scatterplot. Is the point A an outlier, an influential observation, or both? What effect would its removal have on the slope of the regression line?

   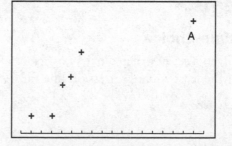

   Answer: A is not an *outlier* because it would not have a large residual. It is an *influential observation* since its removal would have an effect on a calculation, specifically the slope of the regression line. Removing A would increase the slope of the LSRL.

4. A researcher finds that the LSRL for predicting GPA based on average hours studied per week is $\widehat{GPA} = 1.75 + 0.11 \, (hours \, studied)$. Interpret the slope of the regression line in the context of the problem.

   Answer: A student that studies an hour more than another is predicted to have a GPA 0.11 points higher than the other student.

5. One of the variables that is related to college success (as measured by GPA) is socioeconomic status. In one study of the relationship, $r^2 = 0.45$. Explain what this means in the context of the problem.

   Answer: $r^2 = 0.45$ means that 45% of the variability in college GPA is explained by the regression of GPA on socioeconomic status.

6. Each year of Governor Jones's tenure, the crime rate has decreased in a linear fashion. In fact, $r = -0.8$. It appears that the governor has been effective in reducing the crime rate. Comment.

   Answer: Correlation does not necessarily imply causation. The crime rate could have gone down for a number of reasons besides Governor Jones's efforts.

7. What is the regression equation for predicting weight from height in the following computer printout, and what is the correlation between height and weight?

   The regression equation is
   weight = _____ + _____ height

Predictor	Coef	St Dev	t-ratio	P
Constant	−104.64	39.19	−2.67	.037
Height	3.4715	0.5990	5.80	.001

   s = 7.936   R-sq = 84.8%   R-sq(adj) = 82.3%

   Answer: $\widehat{Weight} = -104.64 + 3.4715(Height)$; $r = \sqrt{0.848} = 0.921$. r is positive since the slope of the regression line is positive and both must have the same sign.

8. In the computer output for Exercise #7 above, identify the standard error of the slope of the regression line and the standard error of the residuals. Briefly explain the meaning of each.

   Answer: The standard error of the slope of the regression line is 0.5990. It is an estimate of the change in the mean response y as the independent variable x changes. The standard error of the residuals is $s = 7.936$ and is an estimate of the variability of the response variable about the LSRL.

Practice Problems

Multiple-Choice

1. Given a set of ordered pairs (x, y) so that $s_x = 1.6$, $s_y = 0.75$, $r = 0.55$. What is the slope of the least-square regression line for these data?

   a. 1.82
   b. 1.17
   c. 2.18
   d. 0.26
   e. 0.78

2.

x	23	15	26	24	22	29	32	40	41	46
y	19	18	22	20	27	25	32	38	35	45

The regression line for the two-variable dataset given above is $\hat{y} = 2.35 + 0.86x$. What is the value of the residual for the point whose x-value is 29?

a. 1.71
b. −1.71
c. 2.29
d. 5.15
e. −2.29

3. A study found a correlation of $r = -0.58$ between hours per week spent watching television and hours per week spent exercising. That is, the more hours spent watching television, the less hours spent exercising per week. Which of the following statements is most accurate?

a. About one-third of the variation in hours spent exercising can be explained by hours spent watching television.
b. A person who watches less television will exercise more.
c. For each hour spent watching television, the predicted decrease in hours spent exercising is 0.58 hrs.
d. There is a cause-and-effect relationship between hours spent watching television and a decline in hours spent exercising.
e. 58% of the hours spent exercising can be explained by the number of hours watching television.

4. A response variable appears to be exponentially related to the explanatory variable. The natural logarithm of each y-value is taken and the least-squares regression line is found to be $ln(\hat{y}) = 1.64 - 0.88x$. Rounded to two decimal places, what is the predicted value of y when $x = 3.1$?

a. −1.09
b. −0.34
c. 0.34
d. 0.082
e. 1.09

5. Consider the following residual plot:

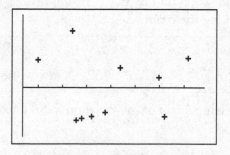

Which of the following statements is (are) true?
I. The residual plot indicates that a line is a reasonable model for the data.
II. The residual plot indicates that there is no relationship between the data.
III. The correlation between the variables is probably non-zero.

    a. I only
    b. II only
    c. I and III only
    d. II and III only
    e. I and II only

6. Suppose the LSRL for predicting weight (in pounds) from height (in inches) is given by $\widehat{Weight} = -115 + 3.6\,(Height)$. Which of the following statements is correct?
    I. A person who is 61 inches tall will weigh 104.6 pounds.
    II. For each additional inch of height, weight will increase on average by 3.6 pounds.
    III. There is a strong positive linear relationship between height and weight.

    a. I only
    b. II only
    c. III only
    d. II and III only
    e. I and II only

7. A least-squares regression line for predicting performance on a college entrance exam based on high school grade point average (GPA) is determined to be $\widehat{Score} = 273.5 + 91.2\,(GPA)$. One student in the study had a high school GPA of 3.0 and an exam score of 510. What is the residual for this student?

    a. 26.2
    b. 43.9
    c. −37.1
    d. −26.2
    e. 37.1

8. The correlation between two variables x and y is −0.26. A new set of scores, x^* and y^*, is constructed by letting $x^* = -x$ and $y^* = y + 12$. The correlation between x^* and y^* is

    a. −0.26
    b. 0.26
    c. 0
    d. 0.52
    e. −0.52

9. A study was done on the relationship between high school grade point average (GPA) and scores on the SAT. The following 8 scores were from a random sample of students taking the exam:

GPA	3.2	3.8	3.9	3.3	3.6	2.8	2.9	3.5
SAT	725	752	745	680	700	562	595	730

What percent of the variation in SAT scores is explained by the regression of SAT score on GPA?

    a. 62.1%
    b. 72.3%
    c. 88.8%
    d. 94.2%
    e. 78.8%

10. A study of mileage found that the least squares regression line for predicting mileage (in miles per gallon) from the weight of the vehicle (in hundreds of pounds) was $\widehat{mpg} = 32.50 - 0.45(weight)$. The mean weight for the vehicles in the study was 2980 pounds. What was the mean miles per gallon in the study?

    a. 19.09
    b. 15.27
    c. −1308.5
    d. 18.65
    e. 20.33

Free-Response

1. Given a two-variable dataset such that $\bar{x} = 14.5$, $\bar{y} = 20$, $s_x = 4$, $s_y = 11$, $r = .80$, find the least-squares regression line of y on x.

2. The data below give the first and second exam scores of 10 students in a calculus class.

Test 1	63	32	87	73	60	63	83	80	98	85
Test 2	51	21	52	90	83	54	73	85	83	46

   (a) Draw a scatterplot of these data.
   (b) To what extent do the scores on the two tests seem related?

3. The following is a residual plot of a linear regression. A line would not be a good fit for these data. Why not? Is the regression equation likely to underestimate or overestimate the y-value of the point in the graph marked with the square?

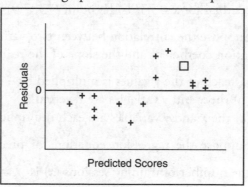

4. The regional champion in 10 and under 100 m backstroke has had the following winning times (in seconds) over the past 8 years:

Year	1	2	3	4	5	6	7	8
Time	77.3	80.2	77.1	76.4	75.5	75.9	75.1	74.3

   How many years until you expect the winning time to be one minute or less? What's wrong with this estimate?

5. Measurements are made of the number of cockroaches present, on average, every 3 days, beginning on the second day, after apartments in one part of town are vacated. The data are as follows:

Days	2	5	8	11	14
# Roaches	3	4.5	6	7.9	11.5

   How many cockroaches would you expect to be present after 9 days?

6. A study found a strongly positive relationship between number of miles walked per week and overall health. A local news commentator, after reporting on the results of the study, advised everyone to walk more during the coming year because walking more results in better health. Comment on the reporter's advice.

7. Carla, a young sociologist, is excitedly reporting on the results of her first professional study. The finding she is reporting is that 72% of the variation in math grades for girls can be explained by the girls' socioeconomic status. What does this mean, and is it indicative of a strong linear relationship between math grades and socioeconomic status for girls?

8. Which of the following statements are true of a least-squares regression equation?

   (a) It is the unique line that minimizes the sum of the residuals.
   (b) The average residual is 0.
   (c) It minimizes the sum of the squared residuals.
   (d) The slope of the regression line is a constant multiple of the correlation coefficient.
   (e) The slope of the regression line tells you how much the response variable will change for each unit change in the explanatory variable.

9. Consider the following dataset:

x	45	73	82	91
y	15	7.9	5.8	3.5

   Given that the LSRL for these data is $\hat{y} = 26.211 - 0.25x$, what is the value of the residual for $x = 73$? Is the point $(73, 7.9)$ above or below the regression line?

10. Suppose the correlation between two variables is $r = -0.75$. What is true of the correlation coefficient and the slope of the regression line if

    (a) each of the y values is multiplied by -1?
    (b) the x and y variables are reversed?
    (c) the x and y variables are each multiplied by -1?

11. Suppose the regression equation for predicting success on a dexterity task (y) from the number of training sessions (x) is $\hat{y} = 45 + 2.7x$ and that $\frac{s_y}{s_x} = 3.33$.

    What percentage of the variation in y is not explained by the regression on x?

12. Consider the following scatterplot. The highlighted point is both an outlier and an influential point. Describe what will happen to the correlation and the slope of the regression line if that point is removed.

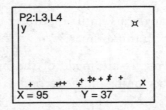

13. The computer printout below gives the regression output for predicting *crime rate* (in crimes per 1000 population) from the *number of casino employees* (in 1000s).

The regression equation is

Rate = _____ + _____ Number

Predictor	Coef	St Dev	*t* ratio	*P*
Constant	−0.3980	0.1884	−2.11	.068
Number	0.118320	0.006804	17.39	.000

s = 0.1499   R-sq = 97.4%   R-sq(adj) = 97.1%

Based on the output,

(a) give the equation of the LSRL for predicting *crime rate* from *number*.
(b) give the value of *r*, the correlation coefficient.
(c) give the predicted *crime rate* for 20,000 casino employees.

14. A study was conducted in a mid-size U.S. city to investigate the relationship between the number of homes built in a year and the mean percentage appreciation for that year. The data for a 5-year period are as follows:

Number	110	80	95	70	55
Percent appreciation	15.7	10	12.7	7.8	10.4

(a) Obtain the LSRL for predicting appreciation from number of new homes built in a year.
(b) The following year, 85 new homes are built. What is the predicted appreciation?
(c) How strong is the linear relationship between number of new homes built and percentage appreciation? Explain.
(d) Suppose you didn't know the number of new homes built in a given year. How would you predict appreciation?

15. A set of bivariate data has $r^2 = 0.81$.

(a) *x* and *y* are both standardized, and a regression line is fitted to the standardized data. What is the slope of the regression line for the standardized data?
(b) Describe the scatterplot of the original data.

16. Estimate *r*, the correlation coefficient, for each of the following graphs:

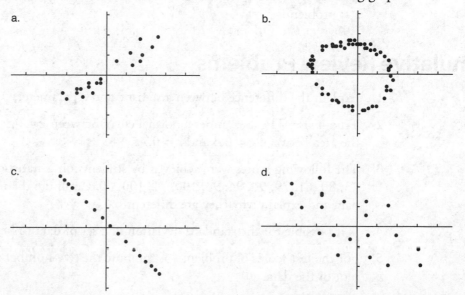

a.

b.

c.

d.

17. The least-squares regression equation for the given data is $\hat{y} = 3 + x$. Calculate the sum of the squared residuals for the LSRL.

x	7	8	11	12	15
y	10	11	14	15	18

18. Many schools require teachers to have evaluations done by students. A study investigated the extent to which student evaluations are related to grades. Teacher evaluations and grades are both given on a scale of 100. The results for Prof. Socrates (y) for 10 of his students are given below together with the average for each student (x).

x	40	60	70	73	75	68	65	85	98	90
y	10	50	60	65	75	73	78	80	90	95

   (a) Do you think student grades and the evaluations students give their teachers are related? Explain.

   (b) What evaluation score do you think a student who averaged 80 would give Prof. Socrates?

19. Which of the following statements are true?

   (a) The correlation coefficient, r, and the slope of the regression line, b, always have the same sign.

   (b) The correlation coefficient is the same no matter which variable is considered to be the explanatory variable and which is considered to be the response variable.

   (c) The correlation coefficient is resistant to outliers.

   (d) x and y are measured in inches, and r is computed. Now, x and y are converted to feet, and a new r is computed. The two computed values of r depend on the units of measurement and will be different.

   (e) The idea of a correlation between height and gender is not meaningful because gender is not numerical.

20. A study of right-handed people found that the regression equation for predicting left-hand strength (measured in kg) from right-hand strength is *left-hand strength* = 7.1 + 0.35 (*right-hand strength*).

   (a) What is the predicted left-hand strength for a right-handed person whose right-hand strength is 12 kg?

   (b) Interpret the intercept and the slope of the regression line in the context of the problem.

Cumulative Review Problems

1. Explain the difference between a statistic and a parameter.

2. True–False. The area under a normal curve between $z = 0.1$ and $z = 0.5$ is the same as the area between $z = 0.3$ and $z = 0.7$.

3. The following scores were achieved by students on a statistics test: 82, 93, 26, 56, 75, 73, 80, 61, 79, 90, 94, 93, 100, 71, 100, 60. Compute the mean and median for these data and explain why they are different.

4. Is it possible for the standard deviation of a set of data to be negative? Zero? Explain.

5. For the test scores of problem #3, compute the five-number summary and draw a box-plot of the data.

Solutions to Practice Problems

Multiple-Choice

1. The correct answer is (d).

    $$b = r \cdot \frac{s_y}{s_x} = (0.55)\left(\frac{0.75}{1.6}\right) = 0.26$$

2. The correct answer is (e). The value of a residual = actual value − predicted value = $25 - [2.35 + 0.86(29)] = -2.29$.

3. The correct answer is (a). $r^2 = (-0.58)^2 = 0.3364$. This is the *coefficient of determination*, which is the proportion of the variation in the response variable that is explained by the regression on the independent variable. Thus, about one-third (33.3%) of the variation in hours spent exercising can be explained by hours spent watching television. (b) is incorrect since correlation does not imply causation. (c) would be correct if $b = -0.58$, but there is no obvious way to predict the response value from the explanatory value just by knowing r. (d) is incorrect for the same reason (b) is incorrect. (e) is incorrect since r, not r^2, is given. In this case $r^2 = 0.3364$, which makes (a) correct.

4. The correct answer is (c). $ln(y) = 1.64 - 0.88(3.1) = -1.088 \Rightarrow y = e^{-1.088} = 0.337$.

5. The correct answer is (c). The pattern is more or less random about 0, which indicates that a line would be a good model for the data. If the data are linearly related, we would expect them to have a non-zero correlation.

6. The correct answer is (b). I is incorrect—the *predicted* weight of a person 61 inches tall is 104.6 pounds. II is a correct interpretation of the slope of the regression line (you could also say that "For each additional inch of height, weight *is predicted to* increase by 3.6 pounds)." III is incorrect. It may well be true, but we have no way of knowing that from the information given.

7. The correct answer is (c). The predicted score for the student is $273.5 + (91.2)(3) = 547.1$. The residual is the actual score minus the predicted score, which equals $510 - 547.1 = -37.1$.

8. The correct answer is (b). Consider the expression for r. $r = \frac{1}{n-1} \sum \left(\frac{x - \bar{x}}{s_x}\right)\left(\frac{y - \bar{y}}{s_y}\right)$.

    Adding 12 to each Y-value would not change s_y. Although the average would be 12 larger, the differences $y - \bar{y}$ would stay the same since each y-value is also 12 larger. By taking the negative of each x-value, each term $\frac{x - \bar{x}}{s_x}$ would reverse sign (the mean also reverses sign) but the absolute value of each term would be the same. The net effect is to leave unchanged the absolute value of r but to reverse the sign.

9. The correct answer is (e). The question is asking for the coefficient of determination, r^2 (R-sq on many computer printouts). In this case, $r = 0.8877$ and $r^2 = 0.7881$, or 78.8%. This can be found on your calculator by entering the GPA scores in L1, the SAT scores in L2, and doing STAT CALC 1-Var Stats L1, L2.

10. The correct answer is (a). The point $(\bar{x}, \bar{y})$ always lies on the LSRL. Hence, $\bar{y}$ can be found by simply substituting $\bar{x}$ into the LSRL and solving for $\bar{y}$. Thus, $\bar{y} = 32.5 - 0.45(29.8) = 19.09$ miles per gallon. Be careful: you are told that the equation uses the weights in hundreds of pounds. You must then substitute 29.8 into the regression equation, not 2980, which would get you answer (c).

Free-Response

1. $b = r\dfrac{s_y}{s_x} = (0.80)\left(\dfrac{11}{4}\right) = 2.2$, $a = \bar{y} - b\bar{x} = 20 - (2.2)(14.5) = -11.9$.

   Thus, $\hat{y} = -11.9 + 2.2x$.

2. (a)

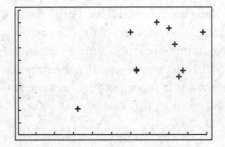

   (b) There seems to be a moderate positive relationship between the scores: students who did better on the first test tend to do better on the second, but the relationship isn't very strong; $r = 0.55$.

3. A line is not a good model for the data because the residual plot shows a definite pattern: the first 8 points have negative residuals and the last 8 points have positive residuals. The box is in a cluster of points with positive residuals. We know that, for any given point, the residual equals actual value minus predicted value. Because actual − predicted > 0, we have actual > predicted, so that the regression equation is likely to underestimate the actual value.

4. The regression equation for predicting time from year is $\widehat{time} = 79.21 - 0.61(year)$. We need $time = 60$. Solving $60 = 79.1 - 0.61(year)$, we get $year = 31.3$. So, we would predict that times will drop under one minute in about 31 or 32 years. The problem with this is that we are extrapolating far beyond the data. Extrapolation is dangerous in any circumstance, and especially so 24 years beyond the last known time. It's likely that the rate of improvement will decrease over time.

5. A scatterplot of the data (graph on the left) appears to be exponential. Taking the natural logarithm of each y-value, the scatterplot (graph on the right) appears to be more linear.

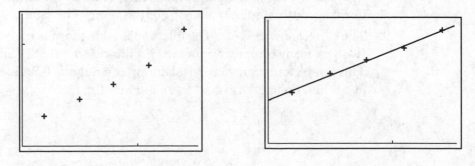

Taking the natural logarithm of each y-value and finding the LSRL, we have $\widehat{ln(\#roaches)} = 0.914 + 0.108\,(days) = 0.914 + 0.108(9) = 1.89$. Then $\widehat{\#roaches} = e^{1.89} = 6.62$.

6. The correlation between walking more and better health may or may not be causal. It may be that people who are healthier walk more. It may be that some other variable, such as general health consciousness, results in walking more and in better health. There may be a causal association, but in general, correlation does not imply causation.

7. Carla has reported the value of r^2, the coefficient of determination. If she had predicted each girl's grade based on the average grade only, there would have been a large amount of variability. But, by considering the regression of grades on socioeconomic status, she has reduced the total amount of variability by 72%. Because $r^2 = 0.72$, $r = 0.85$, which is indicative of a strong positive linear relationship between grades and socioeconomic status. Carla has reason to be happy.

8. (a) is false. $\Sigma(y - \hat{y}) = 0$ for the LSRL, but there is no unique line for which this is true.
   (b) is true.
   (c) is true. In fact, this is the definition of the LSRL—it is the line that minimizes the sum of the squared residuals.
   (d) is true since $b = r\dfrac{s_y}{s_x}$ and $\dfrac{s_y}{s_x}$ is constant.
   (e) is false. The slope of the regression lines tell you by how much the response variable changes *on average* for each unit change in the explanatory variable.

9. $\hat{y} = 26.211 - 0.25x = 26.211 - 0.25(73) = 7.961$. The residual for $x = 73$ is the actual value at 73 minus the predicted value at 73, or $y - \hat{y} = 7.9 - 7.961 = -0.061$. $(73, 7.9)$ is below the LSRL since $y - \hat{y} < 0 \Rightarrow y < \hat{y}$.

10. (a) $r = +0.75$; the slope is positive and is the opposite of the original slope.
    (b) $r = -0.75$. It doesn't matter which variable is called x and which is called y.
    (c) $r = -0.75$; the slope is the same as the original slope.

11. We know that $b = r\dfrac{s_y}{s_x}$, so that $2.7 = r(3.33) \rightarrow r = \dfrac{2.7}{3.33} = 0.81 \rightarrow r^2 = 0.66$. The proportion of the variability that is *not* explained by the regression of y on x is $1 - r^2 = 1 - 0.66 = 0.34$.

12. Because the linear pattern will be stronger, the correlation coefficient will increase. The influential point pulls up on the regression line so that its removal would cause the slope of the regression line to decrease.

13. (a) $\widehat{rate} = -0.3980 + 0.1183\,(number)$.
    (b) $r = \sqrt{0.974} = 0.987$ (r is positive since the slope is positive).
    (c) $\widehat{rate} = -0.3980 + 0.1183(20) = 1.97$ crimes per thousand employees. Be sure to use 20, not 200.

14. (a) $\widehat{Percentage\ appreciation} = 1.897 + 0.115\,(number)$
    (b) $\widehat{Percentage\ appreciation} = 1.897 + 0.115(85) = 11.67\%$.
    (c) $r = 0.82$, which indicates a strong linear relationship between the number of new homes built and percent appreciation.
    (d) If the number of new homes built was unknown, your best estimate would be the average percentage appreciation for the 5 years. In this case, the average percentage appreciation is 11.3%. [For what it's worth, the average error (absolute value) using the mean to estimate appreciation is 2.3; for the regression line, it's 1.3.]

15. (a) If $r^2 = 0.81$, then $r = \pm 0.9$. The slope of the regression line for the standardized data is either 0.9 or −0.9.

(b) If $r = +0.9$, the scatterplot shows a strong positive linear pattern between the variables. Values above the mean on one variable tend to be above the mean on the other, and values below the mean on one variable tend to be below the mean on the other. If $r = -0.9$, there is a strong negative linear pattern to the data. Values above the mean on one variable are associated with values below the mean on the other.

16. (a) $r = 0.8$
    (b) $r = 0.0$
    (c) $r = -1.0$
    (d) $r = -0.5$

17. Each of the points lies on the regression line → every residual is 0 → the sum of the squared residuals is 0.

18. (a) $r = 0.90$ for these data, indicating that there is a strong positive linear relationship between student averages and evaluations of Prof. Socrates. Furthermore, $r^2 = 0.82$, which means that most of the variability in student evaluations can be explained by the regression of student evaluations on student average.
    (b) If y is the evaluation score of Prof. Socrates and x is the corresponding average for the student who gave the evaluation, then $\hat{y} = -29.3 + 1.34x$. If $x = 80$, then $\hat{y} = -29.3 + 1.34(80) = 77.9$, or 78.

19. (a) True, because
    $b = r\dfrac{s_y}{s_x}$ and $\dfrac{s_y}{s_x}$ is positive.
    (b) True. r is the same if explanatory and response variables are reversed. This is not true, however, for the slope of the regression line.
    (c) False. Because r is defined in terms of the means of the x and y variables, it is not resistant.
    (d) False. r does not depend on the units of measurement.
    (e) True. The definition of r, $r = \dfrac{1}{n-1}\sum_{i=1}^{n}\left(\dfrac{x_i - \overline{x}}{s_x}\right)\left(\dfrac{y_i - \overline{y}}{s_y}\right)$,

    necessitates that the variables be numerical, not categorical.

20. (a) $\overline{Left\text{-}hand\ strength} = 7.1 + 0.35(12) = 11.3$ kg.
    (b) **Intercept:** The predicted left-hand strength of a person who has zero right-hand strength is 7.1 kg.
    Slope: On average, left-hand strength increases by 0.35 kg for each 1 kg increase in right-hand strength. Or left-hand strength is predicted to increase by 0.35 kg for each 1 kg increase in right-hand strength.

Solutions to Cumulative Review Problems

1. A *statistic* is a measurement that describes a sample. A *parameter* is a value that describes a population.

2. False. For an interval of fixed length, there will be a greater proportion of the area under the normal curve if the interval is closer to the center than if it is removed from the center. This is because the normal distribution is mound shaped, which implies that the terms tend to group more in the center of the distribution than away from the center.

3. The mean is 77.1, and the median is 79.5. The mean is lower than the median because the mean is not resistant to extreme values, while the median is resistant.

4. By definition, $s = \sqrt{\dfrac{\sum(x-\overline{x})^2}{n-1}}$, which is a positive square root. Since $n > 1$ and $\sum(x-\overline{x})^2 \geq 0$, s cannot be negative. It *can* be zero, but only if $x = \overline{x}$ for all values of x.

5. The 5-number summary is: [26, 66, 79.5, 93, 100].

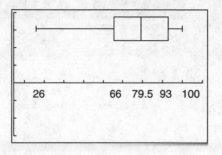

26　　　　　　66　79.5　93　100

CHAPTER 7

Design of a Study: Sampling, Surveys, and Experiments

IN THIS CHAPTER

Summary: Most chapters in this course are concerned with how to analyze and make inferences from data that have been collected. In this chapter, we discuss how to collect the data in a way that allows us to draw meaningful conclusions. We will discuss surveys and sampling, observational studies, experiments, and the types of problems that can creep into each of them, as well as how to avoid these problems. Once you understand all of this, you will also have the background to understand when and why our inference procedures apply.

Key Ideas
- ✪ Samples and Sampling
- ✪ Surveys
- ✪ Sampling Bias
- ✪ Experiments and Observational Studies
- ✪ Statistical Significance
- ✪ Completely Randomized Design
- ✪ Matched-Pairs Design
- ✪ Blocking

Samples

In the previous two chapters we concentrated on the analysis of data at hand—we didn't worry much about how the data came into our possession. The last part of this book deals with statistical inference—making statements about a population based on samples drawn from the population. In both data analysis and inference, we would like to believe that our analyses, or inferences, are meaningful. If we make a claim about a population based on a sample, we want that claim to be true. Our ability to do meaningful analyses and make reliable inferences is a function of the data we collect. To the extent that the sample data we deal with are representative of the population of interest, we are on solid ground. No interpretation of data that are poorly collected or systematically biased will be meaningful. We need to understand how to gather quality data before proceeding on to inference. In this chapter, we study techniques for gathering data so that we have reasonable confidence that they are representative of our population of interest.

Census

We usually want to know something about the entire population of interest. The way to find that out for sure is to conduct a **census**, a procedure by which every member of a population is selected for study. Doing a census, especially when the population of interest is quite large, is often impractical, too time consuming, or too expensive. Interestingly enough, relatively small samples can give quite good estimates of population values if the samples are selected properly. For example, it can be shown that approximately 1500 randomly selected voters can give reliable information about the entire voting population of the United States.

The goal of sampling is to produce a **representative sample**, one that has the essential characteristics of the population being studied and is free of any type of bias. We can never be certain that our sample has the characteristics of the population from which it was drawn. Our best chance of making a sample representative is to use some sort of random process in its selection. It is important to note that "bias" does not mean the same thing as "nonrepresentative." Bias refers to a *method* that produces estimates that are either too high on average, or too low on average. Nonrepresentative refers to a particular sample that differs from the population.

Probability Sample

A list of all members of the population from which we can draw a sample is called a **sampling frame**. We would like the sampling frame to be the same set of individuals we are studying. Unfortunately, this is often not the case. (Think, for example, about how selecting individuals from a phonebook is not the same as all adult residents of a city!) A **probability sample** is one in which each member of the population has a known probability of being in the sample. Each member of the population may or may not have an equal chance of being selected. Probability samples are used to avoid the bias that can arise in a nonprobability sample (such as when a researcher selects the subjects she will use). Probability samples use some sort of random mechanism to choose the members of the sample. The following list includes some types of probability samples:

- **random sample**: Each member of the population is equally likely to be included.
- **simple random sample (SRS)**: A sample of a given size is chosen in such a way that every possible sample of that size is equally likely to be chosen. Note that a sample can be a *random sample* and not be a *simple random sample* (SRS). For example, suppose you want a sample of 64 NFL football players. One way to produce a random sample would be to randomly select two players from each of the 32 teams. This is a random sample but not a simple random sample because not all possible samples of size 64 are possible.

- **systematic sample**: The first member of the sample is chosen according to some random procedure, and then the rest are chosen according to some well-defined pattern. For example, if you wanted 100 people in your sample to be chosen from a list of 10,000 people, you could randomly select one of the first 100 people and then select every 100th name on the list after that. A systematic sample is random because of the random start, but not a *simple* random sample because not every sample of size n is equally likely. For example, two people right next to each other in the list would never be chosen for the sample.
- **stratified random sample**: This is a sample in which the population is first divided into distinct homogenous subgroups called *strata* and then a random sample is chosen from each subgroup. For example, you might divide the population of voters into groups by political party and then select an SRS of 250 from each group.
- **cluster sample:** The population is first divided into sections or "clusters." Then we randomly select one or more clusters and include *all* of the members of the selected cluster(s) in the sample.

  example: You are going to conduct a survey of your senior class concerning plans for graduation. You want to sample 10% of the class. Describe a procedure by which you could use a systematic sample to obtain your sample and explain why this sample isn't a simple random sample. Is this a random sample?

  solution: One way would be to obtain an alphabetical list of all the seniors. Use a random number generator (such as a table of random digits or a scientific calculator with a random digits function) to select one of the first 10 names on the list. Then proceed to select every 10th name on the list after the first.

  Note that this is not an SRS because not every possible sample of 10% of the senior class is equally likely. For example, people next to each other in the list can't both be in the sample. Theoretically, the first 10% of the list could be the sample if it were an SRS. This clearly isn't possible with this technique.
  Before the first name has been randomly selected, every member of the population has an equal chance to be selected for the sample. Hence, this is a random sample, although it is not a simple random sample.

  example: A large urban school district wants to determine the opinions of its elementary schools teachers concerning a proposed curriculum change. The district administration randomly selects three schools from all the elementary schools in the district and surveys each teacher in that school. What kind of sample is this?

  solution: This is a cluster sample. The individual schools represent previously defined groups (clusters) from which we have randomly selected one (it could have been more) for inclusion in our sample.

  example: You are sampling from a population with mixed ethnicity. The population is 45% Caucasian, 25% Asian American, 15% Latinx, and 15% African American. How would a *stratified random sample* of 200 people be constructed?

  solution: You want your sample to mirror the population in terms of its ethnic distribution. Accordingly, from the Caucasians, you would draw an SRS of 90 (that's 45%), an SRS of 50 (25%) from the Asian Americans, an SRS of 30(15%) from the Latinos, and an SRS of 30 (15%) from the African Americans.

Of course, not all samples are probability samples. At times, people try to obtain samples by processes that are nonrandom but still hope, through design or faith, that the resulting sample is representative. The danger in all nonprobability samples is that some (unknown) bias may affect the degree to which the sample is representative. That isn't to say that random samples can't be nonrepresentative, just that we have a better chance of avoiding bias. (Remember the difference!) Some types of nonrandom sampling techniques that tend to be biased are:

- **self-selected sample** or **voluntary response sample:** People choose whether or not to participate in the survey. A radio call-in show is a typical voluntary response sample.
- **convenience sampling:** The pollster obtains the sample any way he can, usually with the ease of obtaining the sample in mind. For example, handing out questionnaires to every member of a given class at school would be a convenience sample. The key issue here is that the surveyor makes the decision whom to include in the sample.
- **quota sampling:** The pollster attempts to generate a representative sample by choosing sample members based on matching individual characteristics to known characteristics of the population. This is similar to a stratified random sample, only the process for selecting the sample is nonrandom.

Sampling Bias

We are trying to avoid **bias** in our sampling techniques, which would mean our method chooses samples that produce estimates that are, on average, either too high or too low.

Undercoverage

One type of sampling bias results from **undercoverage**. This happens when some part of the population being sampled is somehow excluded. This can happen when the sampling frame (the list from which the sample will be drawn) isn't the same as the target population.

> **example:** A pollster conducts a telephone survey to gather opinions of the general population about welfare. Persons too poor to be able to afford a telephone are certainly interested in this issue, but will be systematically excluded from the sample. The resulting sample will be biased because of the exclusion of this group.

Voluntary Response Bias

Voluntary response bias occurs with self-selected samples. Persons who feel most strongly about an issue are most likely to respond. **Nonresponse bias**, the possible biases of those who choose not to respond, is a related issue.

> **example:** You decide to find out how your neighbors feel about the neighbor who seems to be running a car repair shop on his front lawn. You place a questionnaire in every mailbox within sight of the offending home and ask the people to fill it out and return it to you. About 1/2 of the neighbors return the survey, and 95% of those who do say that they find the situation intolerable. We have no way of knowing the feelings of the 50% of those who didn't return the survey—they may be perfectly happy with the "bad" neighbor. Those who have the strongest opinions are those most likely to return your survey—and they may not represent the opinions of all. Most likely they do not.

> **example:** In response to a question once posed in Ann Landers's advice column, some 70% of respondents (almost 10,000 readers) wrote that they would choose not to have children if they had the choice to do it over again. This is most likely representative only of those parents who were

having a *really* bad day with their children when they decided to respond to the question. In fact, a properly designed opinion poll a few months later found that more than 90% of parents said they *would* have children if they had the chance to do it all over again.

Wording Bias

Wording bias occurs when the wording of the question itself influences the response in a systematic way. A number of studies have demonstrated that welfare gathers more support from a random sample of the public when it is described as "helping people until they can better help themselves" than when it is described as "allowing people to stay on the dole."

> **example:** Compare the probable responses to the following ways of phrasing a question.

(i) "Do you support a woman's right to make medical decisions concerning her own body?"

(ii) "Do you support a woman's right to kill an unborn child?"

It's likely that (i) is designed to show that people are in favor of the right to choose abortion and that (ii) is designed to show that people are opposed to the right to choose abortion. The authors of both questions would probably argue that both responses reflect society's attitudes toward abortion.

> **example:** Two different Gallup Polls were conducted in Dec. 2003. Both involved people's opinion about the U.S. space program. Here is one part of each poll.

Poll A: Would you favor or oppose a new U.S. space program that would send astronauts to the moon? Favor—53%; oppose—45%.

Poll B: Would you favor or oppose U.S. government spending billions of dollars to send astronauts to the moon? Favor—31%; oppose—67%.

(source: http://www.stat.ucdavis.edu/~jie/stat13.winter2007/lec18.pdf)

Response Bias

Response bias arises in a variety of ways. The respondent may not give truthful responses to a question (perhaps she or he is ashamed of the truth); the respondent may fail to understand the question (you ask if a person is educated but fail to distinguish between levels of education); the respondent desires to please the interviewer (questions concerning race relations may well solicit different answers depending on the race of the interviewer); the ordering of the question may influence the response ("Do you prefer A to B?" may get different responses than "Do you prefer B to A?").

> **example:** What form of bias do you suspect in the following situation? You are a school principal and want to know students' level of satisfaction with the counseling services at your school. You direct one of the school counselors to ask her next 25 counselees how favorably they view the counseling services at the school.

> **solution:** A number of things would be wrong with the data you get from such a survey. First, the sample is nonrandom—it is a sample of convenience obtained by selecting 25 consecutive counselees. They may or may not be representative of students who use the counseling service. You don't know.

> Second, you are asking people who are seeing their counselor about their opinion of counseling. You will probably get a more favorable view of the

counseling services than you would if you surveyed the general population of the school (would students really unhappy with the counseling services voluntarily be seeing their counselor?). Also, because the counselor is administering the questionnaire, the respondents would have a tendency to want to please the interviewer. The sample certainly suffers from undercoverage—only a small subset of the general population is actually being interviewed. What do those *not* being interviewed think of the counseling?

Experiments and Observational Studies

Statistical Significance

One of the desirable outcomes of a study is to help us determine cause and effect. We do this by looking for differences between groups in an experiment that are so large that we cannot reasonably attribute the difference to chance. We say that a difference between what we would expect to find if there were no treatment and what we actually found is **statistically significant** if the difference is too large to attribute to chance. We discuss numerical methods of determining *significance* in Chapters 10–13.

An **experiment** is a study in which the researcher imposes some sort of treatment on the **experimental units** (which can be human—usually called **subjects** in that case). In an experiment, the idea is to determine the extent to which treatments, the explanatory variable(s), affect outcomes, the response variable (s). For example, a researcher might vary the rewards to different work group members to see how that affects the group's ability to perform a particular task.

An **observational study**, on the other hand, simply observes and records behavior but does not attempt to impose a treatment in order to manipulate the response. Therefore in an observational study, we cannot address cause and effect because we have not imposed a treatment. Sample surveys are actually one type of observational study because no treatment is imposed on the subjects of the study.

There are two classes of observational studies: *retrospective* and *prospective*. Retrospective studies examine data for a sample of individuals. Sample surveys are retrospective studies. Prospective studies select a sample of individuals and follow their behavior as time goes on. Sometimes over many years.

> **Exam Tip:** The distinction between an experiment and an observational study is an important one. There is a reasonable chance that you will be asked to show you understand this distinction on the exam. Be sure this section makes sense to you.

example: In order to develop models for predicting risk for cardiovascular disease (CVD) more than 423,000 participants without CVD will be followed over many years to monitor their cardiovascular health. A machine learning algorithm based on 473 variables will be used to identify risk factors for CVD. This is a prospective observational study because participants were selected and will be followed into the future.

example: A group of 60 volunteers who do not exercise are randomly assigned to one of the two fitness programs. One group of 30 is enrolled in a daily walking program, and the other group is put into a running program. After a period of time, the two groups are compared based on their scores on a wellness index. This is an *experiment* because the researcher has imposed the treatment (walking or running) and then measured the effects of the treatment on a defined response.

It may be, even in a controlled experiment, that the measured response is a function of variables present in addition to the treatment variable. A **confounding variable** is one that has an effect on the outcomes of the study but whose effects cannot be separated from those of the treatment variable.

> **example:** A study is conducted to see if Yummy Kibble dog food results in shinier coats on golden retrievers. It's possible that the dogs with shinier coats have them because they have owners who are more conscientious in terms of grooming their pets. Both the dog food and the conscientious owners could contribute to the shinier coats. The variables are **confounded** because their effects cannot be separated.

A well-designed study attempts to anticipate confounding variables in advance and **control** for them. **Statistical control** refers to a researcher holding constant variables not under study that might have an influence on the outcomes.

> **example:** You are going to study the effectiveness of SAT preparation courses on SAT score. You know that better students tend to do well on SAT tests. You could control for the possible confounding effect of academic quality by running your study with groups of "A" students, "B" students, etc.

Control is often considered to be one of the three basic principles of experimental design. The other two basic principles are **randomization** and **replication**.

One purpose of *randomization* is to equalize groups so that the effects of extraneous variables are equalized among groups. *Randomization* involves the use of chance (like putting names in a bin, mixing them, and drawing them) to assign subjects to treatment and control groups. The hope is that the groups being studied will differ systematically *only* in the effects of the treatment variable. Although individuals within the groups may vary, the idea is to make the groups as alike as possible except for the treatment variable. Note that it isn't possible to produce, with certainty, groups free of any confounding variables. It is possible, through the use of randomization, to increase the probability of producing groups that are alike. The idea is to control for the effects of variables you aren't aware of but that might affect the response.

Replication involves applying the treatment to enough subjects (or units) to reduce the effects of chance variation on the outcomes. If each treatment is applied to only one subject, there is no indication of how much the response varies for the treatment (within-treatment variation). To see if there is a difference between treatments, we must be able to compare that to the difference within each treatment.

Completely Randomized Design

A completely randomized design for a study involves three essential elements: random allocation of subjects to treatment groups; administration of different treatments to each randomized group (if there is a control group, that is one of the treatment groups); and some sort of comparison of the outcomes from the various groups. A standard diagram of this situation is the following:

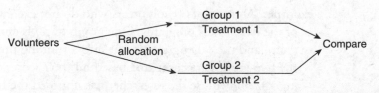

There may be several different treatment groups (different levels of a new drug, for example), in which case the diagram could be modified. The control group can either be an older treatment (like a medication currently on the market) or a **placebo**, a dummy treatment. (Note: as a control, placebos are used with human subjects.) A diagram for an experiment with more than two treatment groups might look something like this:

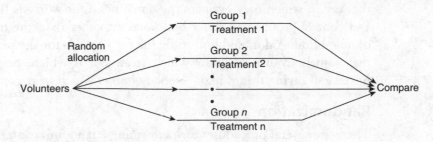

Remember that each group must have enough subjects so that the replication condition is met. The purpose of the *placebo* is to separate genuine treatment effects from possible subject responses due to simply being part of an experiment. Placebos are *not* necessary if a new treatment is being compared to a treatment whose effects have been previously experimentally established. In that case, the new treatment can be compared to the old treatment. A new cream to reduce acne (the treatment), for example, might be compared to an already-on-the-market cream whose effectiveness has long been established.

example: Three hundred graduate students in psychology volunteer to be subjects in an experiment whose purpose is to determine what dosage level of a new drug has the most positive effect on a performance test. There are three levels of the drug to be tested: 200 mg, 500 mg, and 750 mg. Include an additional treatment group to see whether the drug has any effect at all. Design a completely randomized study to test the effectiveness of the various drug levels.

solution: There are three levels of the drug to be tested: 200 mg, 500 mg, and 750 mg. A placebo control can be included although, strictly speaking, it isn't necessary as our purpose is to compare the three dosage levels. We need to randomly allocate the 300 students to each of four groups of 75 each: one group will receive the 200 mg dosage; one will receive the 500 mg dosage; one will receive the 750 mg dosage; and one will receive a placebo (if included). No group will know which treatment its members are receiving (all the pills look the same), nor will the test personnel who come in contact with them know which subjects received which pill (see the definition of "double-blind" given below). Each group will complete the performance test and the results of the various groups will be compared. This design can be diagrammed as follows:

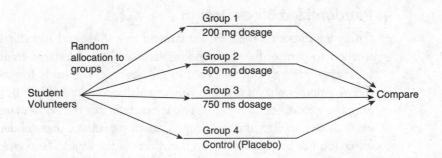

Double-Blind Experiments

In the example above, it was explained that neither the subjects nor the researchers knew who was receiving which dosage, or the placebo. A study is said to be **double-blind** when neither the subjects (or experimental units) nor the evaluators know which group(s) is/are receiving each treatment or control. The reason for this is that, on the part of subjects, simply knowing that they are part of a study may affect the way they respond, and, on the part of the evaluators, knowing which group is receiving which treatment can influence the way in which they evaluate the outcomes. Our worry is that the individual treatment and control groups will differ by something other than the treatment unless the study is double-blind. A double-blind study further controls for the placebo effect. Note, however, that somebody has to know which treatment is which. Sometimes students make the mistake of saying things like, "Nobody knows which treatment is which."

Randomization

There are several procedures for performing a randomization. They are:

- Tables of random digits. Most textbooks contain tables of random digits. These are usually tables where the digits 0, 1, 2, 3, 4, 5, 6, 7, 8, and 9 appear in random order. That means that, as you move through the table, each digit should appear with probability 1/10, and each entry is independent of the others (knowing what came before doesn't help you make predictions about what comes next).
- Electronic random number generators. Graphing calculators have several random functions: a basic random function (which generates a random real number between 0 and 1), a random integer function (which will generate a list of random integers in a specified range), a random normal function (which will generate random values from a normal distribution with a specified mean and standard deviation), and a random binomial function (which will generate random values from a binomial distribution with a specified number of trials and probability—see Chapter 9). Statistics software and some calculators also have the ability to generate random samples, both with and without replacement, from a fixed population.
- Physical probability generators. Dice, spinners, drawing slips of paper from a bag, and so on are also ways of representing random processes.

Exam questions could ask for descriptions of any of these approaches, and often the approach is left open, so familiarize yourself with these different methods. In general, if you are asked to describe a procedure for selecting a sample or assigning treatments to subjects, be sure to specify the method of randomization, and address whether you are sampling with replacement or without (equivalently, are repeated digits allowed or ignored). And if you are using a random digit table and are using, say, pairs of digits, be clear that 01 is how you represent the number 1, for example.

We will use tables of random digits and/or the calculator in Chapter 8 when we discuss simulation.

Randomized Block Design

Earlier we discussed the need for **control** in a study and identified **randomization** as the main method to control for confounding variables—variables that might influence the outcomes in some way but are not considered in the design of the study (usually because we aren't aware of them). Another type of control involves variables we think might influence the outcome of a study. Suppose we suspect, as in our previous example, that the performance test varies by gender as well as by dosage level of the test drug. That is, we suspect that gender is a *confounding variable* (its effects cannot be separated from the effects of the drug). To control for the effects of gender on the performance test, we utilize what is known as a **block design**. A block design involves doing a completely randomized experiment *within* each block. In this case, that means that each level of the drug would be tested within the group of females and within the group of males. To simplify the example, suppose that we were only testing one level (say 500 mg) of the drug versus a placebo. The experimental design, blocked by gender, could then be diagrammed as follows.

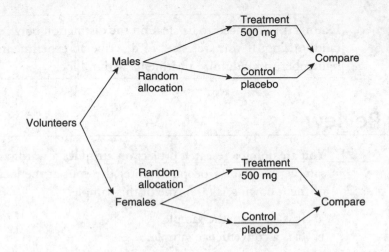

Randomization and block designs each serve a purpose. It's been said that you *block* to control for the variables you know about and *randomize* to control for the ones you don't. Note that your interest here is in studying the effect of the treatment within the population of males and within the population of females, *not* to compare the effects on men and women.

Matched-Pairs Design

A particular block design of interest is the **matched-pairs** design. One possible matched-pairs design involves before and after measurements on the same subjects. In this case, each subject becomes a block on which the experiment is conducted. Another type of matched pairs involves pairing the subjects in some way (matching on, say, height, race, age, etc.).

example: A study is instituted to determine the effectiveness of training teachers to teach AP Statistics. A pretest is administered to each of 23 prospective teachers who subsequently undergo a training program. When the program is finished, the teachers are given a posttest. A score for each teacher is arrived at by subtracting their pretest score from their posttest score. This is a matched-pairs design because two scores are paired for each teacher.

example: One of the questions on the 1997 AP Exam in Statistics asked students to design a study to compare the effects of differing formulations of fish food on fish growth. Students were given a room with eight fish tanks. The room had a heater at the back of the room, a door at the front center of the room, and windows at the front sides of the room. The most correct design involved blocking so that the two tanks nearest the heater in the back of the room were in a single block, the two away from the heater in a second block, the two in the front near the door in a third, and the two in the front near the windows in a fourth. This matching had the effect of controlling for known environmental variations in the room caused by the heater, the door, and the windows. Within each block, one tank was randomly assigned to receive one type of fish food and the other tank received the other. The blocking controlled for the known effects of the environment in which the experiment was conducted. The randomization controlled for unknown influences that might be present in the various tank locations.

You will need to recognize paired data, as distinct from two independent sets of data, later on when we study inference. Even though two sets of data are generated in a matched-pairs study, it is the differences between the matched values that form the one-sample data used for statistical analysis.

❭ Rapid Review

1. You are doing a research project on attitudes toward fast food and decide to use as your sample the first 25 people to enter the door at the local FatBurgers restaurant. Which of the following is (are) true of this sample?

   a. It is a systematic sample.
   b. It is a convenience sample.
   c. It is a random sample.
   d. It is a simple random sample.
   e. It is a self-selected sample.

   Answer: Only (b) is correct. (a), (c), and (d) are all probability samples, which rely on some random process to select the sample, and there is nothing random about the selection process in this situation. (e) is incorrect because, although the sample members voluntarily entered Fat Burgers, they haven't volunteered to respond to a survey.

2. How does an *experiment* differ from an *observational study*?

   Answer: In an experiment, the researcher imposes some treatment on the subjects (or experimental units) in order to observe a response. In an observational study, the researcher simply observes, compares, and measures, but does not impose a treatment.

3. What are the three key components of an experiment? Explain each.

   Answer: The three components are randomization, control, and replication. You randomize to be sure that the response does not systematically favor one outcome over another. The idea is to equalize groups as much as possible so that differences in response are attributable to the treatment variable alone. Control is designed to hold confounding variables constant (such as the placebo effect). Replication ensures that the experiment is conducted on sufficient numbers of subjects to minimize the effects of chance variation.

4. Your local pro football team has just suffered a humiliating defeat at the hands of its archrival. A local radio sports talk show conducts a call-in poll on whether or not the coach should be fired. What is the poll likely to find?

   Answer: The poll is likely to find that, overwhelmingly, respondents think the coach should be fired. This is a *voluntary* response poll, and we know that such a poll is most likely to draw a response from those who feel most strongly about the issue being polled. Fans who bother to vote in a call-in poll such as this are most likely upset at their team's loss and are looking for someone to blame—this gives them the opportunity. There is, of course, a chance that the coach may be very popular and draw support, but the point to remember is that this is a self-selecting nonrandom sample, and will probably exhibit response bias.

5. It is known that exercise and diet both influence weight loss. Your task is to conduct a study of the effects of diet on weight loss. Explain the concept of *blocking* as it relates to this study.

   Answer: If you did a completely randomized design for this study using diet as the treatment variable, it's very possible that your results would be confounded by the effects of exercise. Because you are aware of this, you would like to control for the effects of exercise. Hence, you *block* by exercise level. You might define, say, three blocks by level of exercise (very active, active, not very active) and do a completely randomized study within each of the blocks. Because exercise level is held constant, you can be confident that differences between treatment and control groups within each block are attributable to diet, not exercise.

6. Explain the concept of a *double-blind* study and why it is important.

   Answer: A study is *double-blind* if neither the subject of the study nor the researchers are aware of who is in the treatment group and who is in the control group. This is to control for the well-known effect of people to (subconsciously) attempt to respond in the way they think they should.

7. You are interested in studying the effects of preparation programs on SAT performance. Briefly describe a matched-pairs design and a completely randomized design for this study.

   Answer: Matched pairs: Choose, say, 100 students who have not participated in an SAT prep course. Have them take the SAT. Then have these students take a preparation course and retake the SAT. Do a statistical analysis of the difference between the pre- and postpreparation scores for each student. (*Note* that this design doesn't deal with the influence of retaking the SAT independent of any preparation course, which could be a confounding variable.)

   Completely randomized design: Select 100 students and randomly assign them to two groups, one of which takes the SAT with no preparation course and one of which has a preparation course before taking the SAT. Statistically, compare the average performance of each group.

Practice Problems

Multiple-Choice

1. Data were collected in 20 cities on the percentage of women in the workforce. Data were collected in 1990 and again in 1994. Gains, or losses, in this percentage were the measurement upon which the study's, conclusions were to be based. What kind of design was this?
   I.   A matched-pairs design
   II. An observational study
   III. An experiment using a block design

   (a) I only
   (b) II only
   (c) III only
   (d) I and III only
   (e) I and II only

2. You want to do a survey of members of the senior class at your school and want to select a *simple random sample*. You intend to include 40 students in your sample. Which of the following approaches will generate a simple random sample?

   (a) Write the name of each student in the senior class on a slip of paper and put the papers in a container. Then randomly select 40 slips of paper from the container.
   (b) Assuming that students are randomly assigned to classes, select two classes at random and include those students in your sample.

(c) From a list of all seniors, select one of the first 10 names at random. The select every *n*th name on the list until you have 40 people selected.

(d) Select the first 40 seniors to pass through the cafeteria door at lunch.

(e) Randomly select 10 students from each of the four senior calculus classes.

3. Which of the following is (are) important in designing an experiment?
   I. Control of all variables that might have an influence on the response variable
   II. Randomization of subjects to treatment groups
   III. Use of a large number of subjects to control for small-sample variability

   (a) I only
   (b) I and II only
   (c) II and III only
   (d) I, II, and III
   (e) II only

4. Your company has developed a new treatment for acne. You think men and women might react differently to the medication, so you separate them into two groups. Then the men are randomly assigned to two groups and the women are randomly assigned to two groups. One of the two groups is given the medication and the other is given a placebo. The basic design of this study is

   (a) completely randomized
   (b) blocked by gender
   (c) completely randomized, blocked by gender
   (d) randomized, blocked by gender and type of medication
   (e) a matched-pairs design

5. A *double-blind* design is important in an experiment because

   (a) There is a natural tendency for subjects in an experiment to want to please the researcher.
   (b) It helps control for the placebo effect.
   (c) Evaluators of the responses in a study can influence the outcomes if they know which subjects are in the treatment group and which are in the control group.
   (d) Subjects in a study might react differently if they knew they were receiving an active treatment or a placebo.
   (e) All of the above are reasons why an experiment should be *double-blind*.

6. Which of the following is not an example of a *probability sample*?

   (a) You are going to sample 10% of a group of students. You randomly select one of the first 10 students on an alphabetical list and then select every 10th student after than on the list.
   (b) You are a sports-talk radio host interested in opinions about whether or not Pete Rose should be elected to the Baseball Hall of Fame, even though he has admitted to betting on his own teams. You ask listeners to call in and vote.
   (c) A random sample of drivers is selected to receive a questionnaire about the manners of Department of Motor Vehicle employees.
   (d) In order to determine attitudes about the Medicare Drug Plan, a random sample is drawn so that each age group (65–70, 70–75, 75–80, 80–85) is represented in proportion to its percentage in the population.
   (e) In choosing respondents for a survey about a proposed recycling program in a large city, interviewers choose homes to survey based on rolling a die. If the die shows a 1, the house is selected. If the dies shows a 2–6, the interviewer moves to the next house.

7. Which of the following is true of an experiment but not of an observational study?

    (a) A cause-and-effect relationship can be more easily inferred.
    (b) The cost of conducting it is excessive.
    (c) More advanced statistics are needed for analysis after the data are gathered.
    (d) By law, the subjects need to be informed that they are part of a study.
    (e) Possible confounding variables are more difficult to control.

8. A study showed that persons who ate two carrots a day had significantly better eyesight than those who ate less than one carrot a week. Which of the following statements is (are) correct?
    I.   This study provides evidence that eating carrots contributes to better eyesight.
    II. The general health consciousness of people who eat carrots could be a confounding variable.
    III. This is an observational study and not an experiment.

    (a) I only
    (b) III only
    (c) I and II only
    (d) II and III only
    (e) I, II, and III

9. Which of the following situations is a cluster sample?

    (a) Survey five friends concerning their opinions of the local hockey team.
    (b) Take a random sample of five voting precincts in a large metropolitan area and do an exit poll at each voting site.
    (c) Measure the length of time each fifth person entering a restaurant has to wait to be seated.
    (d) From a list of all students in your school, randomly select 20 to answer a survey about Internet use.
    (e) Identify four different ethnic groups at your school. From each group, choose enough respondents so that the final sample contains roughly the same proportion of each group as the school population.

Free-Response

1. You are interested in the extent to which ingesting vitamin C inhibits getting a cold. You identify 300 volunteers, 150 of whom have been taking more than 1000 mg of vitamin C a day for the past month, and 150 of whom have not taken vitamin C at all during the past month. You record the number of colds during the following month for each group and find that the vitamin C group had significantly fewer colds. Is this an experiment or an observational study? Explain. What do we mean in this case when we say that the finding was *significant*?

2. Design an experiment that employs a *completely randomized design* to study the question of whether of not taking large doses of vitamin C is effective in reducing the number of colds.

3. A survey of physicians found that some doctors gave a placebo rather than an actual medication to patients who experienced pain symptoms for which no physical reason could be found. If the pain symptoms were reduced, the doctors concluded that there was no real physical basis for the complaints. Do the doctors understand *the placebo effect*? Explain.

4. Explain how you would use a table of random digits to help obtain a systematic sample of 10% of the names on a alphabetical list of voters in a community. Is this a random sample? Is it a simple random sample?

5. The *Literary Digest Magazine*, in 1936, predicted that Alf Landon would defeat Franklin Roosevelt in the presidential election that year. The prediction was based on questionnaires mailed to 10 million of its subscribers and to names drawn from other public lists. Those receiving the questionnaires were encouraged to mail back their ballot preference. The prediction was off by 19 percentage points. The magazine received back some 2.3 million ballots from the 10 million sent out. What are some of the things that might have caused the magazine to be so wrong (the same techniques had produced accurate predictions for several previous elections)? (Hint: Think about what was going on in the world in 1936.)

6. Interviewers, after the 9/11 attacks, asked a group of Arab Americans if they trust the administration to make efforts to counter anti-Arab activities. If the interviewer was of Arab descent, 42% responded "yes," and if the interviewer was of non-Arab descent, 55% responded "yes." What seems to be going on here?

7. There are three classes of statistics at your school, each with 30 students. You want to select a simple random sample of 15 students from the 90 students as part of an opinion-gathering project for your social studies class. Describe a procedure for doing this.

8. Question #1 stated, in part, "You are interested in the extent to which ingesting vitamin C inhibits getting a cold. You identify 300 volunteers, 150 of whom have been taking more than 1000 mg of vitamin C a day for the past month, and 150 of whom have not taken vitamin C at all during the past month. You record the number of colds during the following month for each group and find that the vitamin C group had significantly fewer colds." Explain the concept of *confounding* in the context of this problem and give an example of how it might have affected the finding that the vitamin C group had fewer colds.

9. A shopping mall wants to know about the attitudes of all shoppers who visit the mall. On a Wednesday morning, the mall places 10 interviewers at a variety of places in the mall and asks questions of shoppers as they pass by. Comment on any bias that might be inherent in this approach.

10. Question #2 asked you to design a *completely randomized experiment* for the situation presented in question #1. That is, to design an experiment that uses treatment and control groups to see if the groups differed in terms of the number of colds suffered by users of 1000 mg a day of vitamin C and those that didn't use vitamin C. Question #8 asked you about possible *confounding variables* in this study. Given that you believe that both general health habits and use of vitamin C might explain a reduced number of colds, design an experiment to determine the effectiveness of vitamin C taking into account general health habits. You may assume your volunteers vary in their history of vitamin C use.

11. You have developed a weight-loss treatment that involves a combination of exercise and diet pills. The treatment has been effective with subjects who have used a regular dose of the pill of 200 mg, when exercise level is held constant. There is some indication that higher doses of the pill will promote even better results, but you are worried about side effects if the dosage becomes too great. Assume you have 400 overweight volunteers for your study, who have all been on the same exercise program, but who have not been taking any kind of diet pill. Design a study to evaluate the relative effects of a 200 mg, 400 mg, 600 mg, and 800 mg daily dosage of the pill.

12. You are going to study the effectiveness of three different SAT preparation courses. You obtain 60 high school juniors as volunteers to participate in your study. You want to assign each of the 60 students, at random, to one of the three programs. Describe a procedure for assigning students to the programs if

    (a) you want there to be an equal number of students taking each course.
    (b) you want each student to be assigned independently to a group. That is, each student should have the same probability of being in any of the three groups.

13. A researcher wants to obtain a sample of 100 teachers who teach in high schools at various economic levels. The researcher has access to a list of teachers in several schools for each of the levels. She has identified four such economic levels (A, B, C, and D) that comprise 10%, 15%, 45%, and 30% of the schools in which the teachers work. Describe what is meant by a *stratified random sample* in this situation, and discuss how she might obtain it.

14. You are testing for sweetness in five varieties of strawberry. You have 10 plots available for testing. The 10 plots are arranged in two side-by-side groups of five. A river runs along the edge of one of the groups of five plots something like the diagram shown below (the available plots are numbered 1–10).

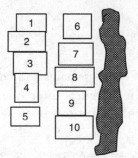 You decide to control for the possible confounding effect of the river by planting one of each type of strawberry in plots 1–5 and one of each type in plots 6–10 (that is, you block to control for the river). Then, within each block, you randomly assign one type of strawberry to each of the five plots within the block. What is the purpose of randomization in this situation?

15. Look at problem #14 again. It is the following year, and you now have only two types of strawberries to test. Faced with the same physical conditions you had in problem 14, and given that you are concerned that differing soil conditions (as well as proximity to the river) might affect sweetness, how might you block the experiment to produce the most reliable results?

16. A group of volunteers, who had never been in any kind of therapy, were randomly separated into two groups, one of which received an experimental therapy to improve self-concept. The other group, the control group, received traditional therapy. The subjects were not informed of which therapy they were receiving. Psychologists who specialize in self-concept issues evaluated both groups after training for self-concept, and the self-concept scores for the two groups were compared. Could this experiment have been *double-blind*? Explain. If it wasn't *double-blind*, what might have been the impact on the results?

17. You want to determine how students in your school feel about a new dress code for school dances. One group in the student council, call them group A, wants to word the question as follows: "As one way to help improve student behavior at school sponsored events, do you feel that there should be a dress code for school dances?" Another group, group B, prefers, "Should the school administration be allowed to restrict student rights by imposing a dress code for school dances?" Which group do you think favors a dress code, and which opposes it? Explain.

18. A study of reactions to different types of billboard advertising is to be carried out. Two different types of ads (call them Type I and Type II) for each product will be featured on numerous billboards. The organizer of the campaign is concerned that communities

representing different economic strata will react differently to the ads. The three communities where billboards will be placed have been identified as Upper Middle, Middle, and Lower Middle. Four billboards are available in each of the three communities. Design a study to compare the effectiveness of the two types of advertising taking into account the communities involved.

19. In 1976, Shere Hite published a book entitled *The Hite Report on Female Sexuality*. The conclusions reported in the book were based on 3000 returned surveys from some 100,000 sent out to, and distributed by, various women's groups. The results were that women were highly critical of men. In what way might the author's findings have been biased?

20. You have 26 women available for a study: Annie, Betty, Clara, Darlene, Edie, Fay, Grace, Helen, Ina, Jane, Koko, Laura, Mary, Nancy, Ophelia, Patty, Quincy, Robin, Suzy, Tina, Ulla, Vivien, Wanda, Xena, Yolanda, and Zoe. The women need to be divided into four groups for the purpose of the study. Explain how you could use a table of random digits to make the needed assignments.

Cumulative Review Problems

1. The five-number summary for a set of data is [52, 55, 60, 63, 85]. Is the mean most likely to be *less than* or *greater than* the median?

2. Pamela selects a random sample of 15 of her classmates and computes the mean and standard deviation of their pulse rates. She then uses these values to predict the mean and standard deviation of the pulse rates for the entire school. Which of these measures are *parameters* and which are *statistics*?

3. Consider the following set of values for a dataset: 15, 18, 23, 25, 25, 27, 28, 29, 35, 46, 55. Does this dataset have any *outliers* if we use an outlier rule that

   (a)   is based on the median?
   (b)   is based on the mean?

4. For the dataset of problem #3 above, what is z_{55}?

5. A study examining factors that contribute to a strong college GPA finds that 62% of the variation in college GPA can be explained by SAT score. What name is given to this statistic, and what is the correlation (r) between SAT score and college GPA?

Solutions to Practice Problems

Multiple-Choice

1. The correct answer is (e). The data are paired because there are two measurements for each city, so the data are not independent. There is no treatment being applied, so this is an observational study. Matched pairs is one type of block design, but this is NOT an experiment, so III is false.

2. The answer is (a). In order for this to be an SRS, all samples of size 40 must be equally likely. None of the other choices does this [and choice (d) isn't even random]. Note that (a), (b), and (c) are probability samples.

3. The correct answer is (d). These three items represent the three essential parts of an experiment: control, randomization, and replication.

4.   The correct answer is (b). You block men and women into different groups because you are concerned that differential reactions to the medication may confound the results. It is not completely randomized because it is blocked.

5.   The correct answer is (e).

6.   The correct answer is (b). This is an example of a voluntary response and is likely to be biased. Those who feel strongly about the issue are most likely to respond. The other choices all rely on some probability technique to draw a sample. In addition, responses (c) and (e) meet the criteria for a simple random sample (SRS).

7.   The correct answer is (a). If done properly, an experiment permits you to control the variable that might influence the results. Accordingly, you can argue that the only variable that influences the results is the treatment variable.

8.   The correct answer is (d). I isn't true because this is an observational study and, thus, shows a relationship but not necessarily a cause-and-effect one.

9.   The correct answer is (b). (a) is a convenience sample. (c) is a systematic sample. (d) is a simple random sample. (e) is a stratified random sample.

Free-Response

1.   It's an **observational study** because the researcher didn't provide a treatment, but simply observed different outcomes from two groups with at least one different characteristic. Participants self-selected themselves into either the vitamin C group or the nonvitamin C group. To say that the finding was significant in this case means that the difference between the number of colds in the vitamin C group and in the nonvitamin C group was too great to attribute to chance—it appears that something besides random variation may have accounted for the difference.

2.   Identify 300 volunteers for the study, preferably none of whom have been taking vitamin C. Randomly split the group into two groups of 150 participants each. One group can be randomly selected to receive a set dosage of vitamin C each day for a month and the other group to receive a placebo. Neither the subjects nor those who administer the medication will know which subjects received the vitamin C and which received the placebo (that is, the study should be *double-blind*). During the month following the giving of pills, you can count the number of colds within each group. Your measurement of interest is the difference in the number of colds between the two groups. Also, placebo effects often diminish over time.

3.   The doctors probably did not understand the placebo effect. We know that, sometimes, a real effect can occur even from a placebo. If people believe they are receiving a real treatment, they will often show a change. But without a control group, we have no way of knowing if the improvement would not have been even more significant with a real treatment. The *difference* between the placebo score and the treatment score is what is important, not one or the other.

4.   If you want 10% of the names on the list, you need every 10th name for your sample. Number the first ten names on the list 0, 1, 2, . . . , 9. Pick a random place to enter the table of random digits and note the first number. The first person in your sample is the person among the first 10 on the list that corresponds to the number chosen. Then pick every 10th name on the list after that name. This is a random sample to the extent that, before the first name was selected, every member of the population had an equal

chance to be chosen. It is not a simple random sample because not all possible samples of 10% of the population are equally likely. Adjacent names on the list, for example, could not both be part of the sample.

5. This is an instance of *voluntary response bias*. This poll was taken during the depths of the Depression, and people felt strongly about national leadership. Those who wanted a change were more likely to respond than those who were more or less satisfied with the current administration. Also, at the height of the Depression, people who subscribed to magazines and were on public lists were more likely to be wealthy and, hence, Republican (Landon was a Republican and Roosevelt was a Democrat).

6. Almost certainly, respondents are responding in a way they feel will please the interviewer. This is a form of response bias—in this circumstance, people may not give a truthful answer.

7. Many different solutions are possible. One way would be to put the names of all 90 students on slips of paper and put the slips of paper into a box. Then draw out 15 slips of paper at random. The names on the paper are your sample. Another way would be to identify each student by a two-digit number 01, 02, . . . , 90 and use a table of random digits to select 15 numbers. Or you could use the randInt function on your calculator to select 15 numbers between 1 and 90 inclusive. What you *cannot* do, if you want it to be an SRS, is to employ a procedure that selects five students randomly from each of the three classes.

8. Because the two groups were not selected randomly, it is possible that the fewer number of colds in the vitamin C group could be the result of some variable whose effects cannot be separated from the effects of the vitamin C. That would make this other variable a *confounding variable*. A possible confounding variable in this case might be that the group who take vitamin C might be, as a group, more health conscious than those who do not take vitamin C. This could account for the difference in the number of colds but could not be separated from the effects of taking vitamin C.

9. The study suffers from *undercoverage* of the population of interest, which was declared to be all shoppers at the mall. By restricting their interview time to a Wednesday morning, they effectively exclude most people who work on Wednesday. They essentially have a sample of the opinions of nonworking shoppers. There may be other problems with randomness, but without more specific information about how they gathered their sample, talking about it would only be speculation.

10. We could first administer a questionnaire to all 300 volunteers to determine differing levels of health consciousness. For simplicity, let's just say that the two groups identified are "health conscious" and "not health conscious." Then you would block by "health conscious" and "not health conscious" and run the experiment within each block. A diagram of this experiment might look like this:

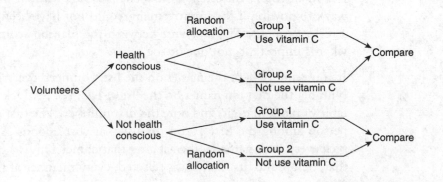

11. Because exercise level seems to be more or less constant among the volunteers, there is no need to block for its effect. Furthermore, because the effects of a 200 mg dosage are known, there is no need to have a placebo (although you could)—the 200 mg dosage will serve as the control. Randomly divide your 400 volunteers into four groups of 100 each. Randomly assign each group to one of the four treatment levels: 200 mg, 400 mg, 600 mg, or 800 mg. The study can be and should be double-blind. After a period of time, compare the weight loss results for the four groups.

12. (a) Many answers are possible. One solution involves putting the names of all 60 students on slips of paper, then randomly selecting the papers. The first student goes into program 1, the next into program 2, etc. until all 60 students have been assigned.

    (b) Use a random number generator to select integers from 1 to 3 (like `randInt(1,3)`) on the TI-83/84 or use a table of random numbers assigning each of the programs a range of values (such as 1–3, 4–6, 7–9, and ignore 0). Pick any student and generate a random number from 1 to 3. The student enters the program that corresponds to the number. In this way, the probability of a student ending up in any one group is 1/3, and the selections are independent. It would be unlikely to have the three groups come out completely even in terms of the numbers in each, but we would expect it to be close.

13. In this situation, a *stratified random sample* would be a sample in which the proportion of teachers from each of the four levels is the same as that of the population from which the sample was drawn. That is, in the sample of 100 teachers, 10 should be from level A, 15 from level B, 45 from level C, and 30 from level D. For level A, she could accomplish this by taking an SRS of 10 teachers from a list of all teachers who teach at that level. SRSs of 15, 45, and 30 would then be obtained from each of the other lists.

14. Remember that you block to control for the variables that might affect the outcome that you know about, and you randomize to control for the effect of those you don't know about. In this case, then, you randomize to control for any unknown systematic differences between the plots that might influence sweetness. An example might be that the plots on the northern end of the rows (plots 1 and 6) have naturally richer soil than those plots on the south side.

15. The idea is to get plots that are most similar in order to run the experiment. One possibility would be to match the plots the following way: close to the river north (6 and 7); close to the river south (9 and 10); away from the river north (1 and 2); and away from the river south (4 and 5). This pairing controls for both the effects of the river and possible north–south differences that might affect sweetness. Within each pair, you would randomly select one plot to plant one variety of strawberry, planting the other variety in the other plot.

    This arrangement leaves plots 3 and 8 unassigned. One possibility is simply to leave them empty. Another possibility is to randomly assign each of them to one of the pairs they adjoin. That is, plot 3 could be randomly assigned to join either plot 2 or plot 4. Similarly, plot 8 would join either plot 7 or plot 9.

16. The study could have been double-blind. The question indicates that the subjects did not know which treatment they were receiving. If the psychologists did not know which therapy the subjects had received before being evaluated, then the basic requirement of a double-blind study was met: neither the subjects nor the evaluators who come in contact with them are aware of who is in the treatment and who is in the control group.

If the study wasn't double-blind, it would be because the psychologists were aware of which subjects had which therapy. In this case, the attitudes of the psychologists toward the different therapies might influence their evaluations—probably because they might read more improvement into a therapy of which they approve.

17. Group A favors a dress code; group B does not. Both groups are hoping to bias the response in favor of their position by the way they have worded the question.

18. You probably want to block by community since it is felt that economic status influences attitudes toward advertising. That is, you will have three blocks: Upper Middle, Middle, and Lower Middle. Within each, you have four billboards. Randomly select two of the billboards within each block to receive the Type I ads, and put the Type II ads on the other two. After a few weeks, compare the differences in reaction to each type of advertising within each block.

19. With only 3000 of 100,000 surveys returned, *voluntary response bias* is most likely operating. That is, the 3000 women represented those who felt strongly enough (negatively) about men and were the most likely to respond. We have no way of knowing if the 3% who returned the survey were representative of the 100,000 who received it, but they most likely were not.

20. Assign each of the 26 women a two-digit number, say 01, 02, ..., 26. Then enter the table at a random location and note two-digit numbers. Ignore numbers outside of the 01–26 range. The first number chosen assigns the corresponding woman to the first group, the second to the second group, etc. until all 26 have been assigned. This method roughly equalizes the numbers in the group (not quite because 4 doesn't go evenly into 26), but does not assign them independently.

   If you wanted to assign the women independently, you would consider only the digits 1, 2, 3, or 4, which correspond to the four groups. As one of the women steps forward, one of the random digits is identified, and that woman goes into the group that corresponds to the chosen number. Proceed in this fashion until all 26 women are assigned a group. This procedure yields independent assignments to groups, but the groups most likely will be somewhat unequal in size. In fact, with only 26 women, group sizes might be quite unequal (a TI-83/84 simulation of this produced 4 1s, 11 2s, 4 3s, and 7 4s).

Solutions to Cumulative Review Problems

1. The dataset has an outlier at 85. Because the mean is not resistant to extreme values, it tends to be pulled in the direction of an outlier. Hence, we would expect the mean to be larger than the median.

2. *Parameters* are values that describe populations, and *statistics* are values that describe samples. Hence, the mean and standard deviation of the pulse rates of Pamela's sample are *statistics*, and the predicted mean and standard deviation for the entire school are *parameters*.

3. Putting the numbers in the calculator and doing 1-Var Stats, we find that $\bar{x} = 29.64$, $s = 11.78$, Q1 = 23, *Med* = 27, and Q3 = 35.

   (a) The interquartile range (IQR) = 35 − 23 = 12, 1.5(IQR) = 1.5(12) = 18. So the boundaries beyond which we find outliers are Q1 − 1.5(IQR) = 23 − 18 = 5 and Q3 + 1.5(IQR) = 35 + 18 = 53. Because 55 is beyond the boundary value of 53, it is an outlier, and it is the only outlier.

(b)   The usual rule for outliers based on the mean is $\bar{x} \pm 3s$. $\bar{x} \pm 3s = 29.64 \pm 3(11.78) = (-5.7, 64.98)$. Using this rule there are no outliers since there are no values less than -5.7 or greater than 64.98. Sometimes $\bar{x} \pm 2s$ is used to determine outliers. In this case, $\bar{x} \pm 2s = 29.64 \pm 2 (11.78) = (6.08, 53.2)$ Using this rule, 55 would be an outlier.

4.   For the given data, $\bar{x} = 29.64$ and $s = 11.78$. Hence,

$$z_{55} = \frac{55 - 29.64}{11.78} = 2.15.$$

Note that in doing problem #3, we could have computed this z-score and observed that because it is larger than 2, it represents an outlier by the $\bar{x} \pm 2s$ rule that is sometimes used.

5.   The problem is referring to the *coefficient of determination*—the proportion of variation in one variable that can be explained by the regression of that variable on another.

$r = \sqrt{\text{coefficient of determination}} = \sqrt{0.62} = 0.79$.

CHAPTER 8

Probability and Random Variables

IN THIS CHAPTER

Summary: We've completed the basics of data analysis, and we now begin the transition to inference. In order to do inference, we need to use the language of probability. In order to use the language of probability, we need an understanding of random variables and probabilities. The next two chapters lay the probability foundation for inference. In this chapter, we'll learn about the basic rules of probability, what it means for events to be independent, and about discrete and continuous random variables, simulation, and rules for combining random variables.

KEY IDEA

Key Ideas
✪ Probability
✪ Random Variables
✪ Discrete Random Variables
✪ Continuous Random Variables
✪ Probability Distributions
✪ Normal Probability
✪ Simulation
✪ Transforming and Combining Random Variables

Probability

The second major part of a course in statistics involves making *inferences* about populations based on sample data (the first was *exploratory data analysis*). The ability to do this is based on being able to make statements such as, "If we assume the null hypothesis to be true, then the probability of getting results such as ours (or more extreme) by chance alone is 0.06."

To make sense of this statement, you need to have an understanding of what is meant by the term "probability," as well as an understanding of some of the basics of probability theory.

A **chance event (random phenomenon)** is an activity whose outcome we can observe or measure but for which we cannot predict the outcome for any single trial. When examining the probability of a particular event, we refer to each occurrence of that event as a "success."

The **probability of an event** is the predicted long-run relative frequency of occurrences of that event. In other words, if we repeat a random process many times, the predicted proportion of outcomes that are "successes" is the probability of success. We can estimate probabilities in a couple of ways: experimentally and theoretically. Because it is a relative frequency, the probability of an event must be between 0 and 1. A probability of 0 means the event is impossible. A probability of 1 means the event is certain.

> **example:** If we roll a six-sided die, we know that we will get a 1, 2, 3, 4, 5, or 6, but we don't know *which* one of these we will get on the next trial. What are two ways we could estimate the probability of rolling a 6?

> **solution:** We could estimate the probability experimentally by rolling a die 300 times. The results of one such set of rolls are shown below.

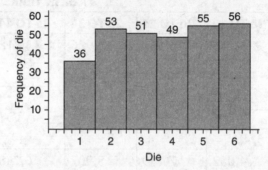

Of the 300 trials, 56 were "successes" so our estimate of the probability based on our experiment is $\frac{56}{300} = 0.187$. We could also estimate the probability theoretically by assuming each possible result is equally likely. If that is the case, then the probability of each number appearing is $\frac{1}{6} = 0.167$. Notice the estimates from the two approaches agree fairly well.

Outcome: One of the possible results of a chance process. Generally we refer to the most basic kind of *occurrences* as outcomes.

Event: A collection of outcomes or simple events.

> **example:** The possible outcomes for the roll of a single die are 1, 2, 3, 4, 5, and 6. Rolling an even number would be an *event* that consists of the *outcomes* 2, 4, and 6.

Sample Spaces and Events

A **sample space** is a **complete** list of **disjoint** outcomes or events. "Complete" means that you have all the possibilities listed. "Disjoint" (also called **mutually exclusive**) means that the events have no outcomes in common. Only one event can happen in a particular trial.

> **example:** Consider the experiment of flipping two coins and noting whether each coin lands heads or tails. A sample space is HH, HT, TH, TT. The list is complete because there is no other possibility for tossing two coins. The list is disjoint because a pair of coin tosses can result in only one of those outcomes.

If Event B = "at least one coin shows a head," then B is composed of events HH, HT, and TH. Event B is a subset of the sample space S.

example: We could also define a sample space for flipping two coins based on the number of heads. In that case, the sample space is 0, 1, and 2. In this case, however, the outcomes are not equally likely. As the previous example shows, the equally likely outcomes HT and TH both result in one head.

If we let E = the event of interest, and we have defined the sample space with equally likely outcomes, then the probability of E, written $P(E)$, is given by

$$P(E) = \frac{\text{Number of outcomes in E}}{\text{Number of outcomes in the sample space}}$$

The sum of the probabilities of all possible outcomes in a sample space is 1.

The table below shows summaries of employment data by type of occupation and age for 2016. This table of data will be referred to several times over the next couple pages. The numbers are in thousands of employees in the U.S.

	AGE IN YEARS							
	16 TO 19	**20 TO 24**	**25 TO 34**	**35 TO 44**	**45 TO 54**	**55 TO 64**	**65+**	**TOTAL**
Management, professional, and related occupations	3,118	13,561	13,926	13,818	10,689	3,931	44	**59,087**
Service occupations	4,007	6,077	4,889	4,821	3,655	1,292	38	**24,779**
Sales and office occupations	4,052	7,159	5,907	6,748	5,754	2,237	42	**31,899**
Natural resources, construction, and maintenance occupations	1,162	3,245	3,294	3,194	2,177	515	42	**13,629**
Production, transportation, and material-moving occupations	1,687	3,681	3,547	4,138	3,249	942	43	**17,287**
Total employed	14,026	33,723	31,563	32,719	25,524	8,917	208	**146,680**

example: If we were to randomly select one person from this population of employees (our repeatable random process), how would you verify that the outcomes in the table define a sample space?

solution: You could check that each employee falls into exactly one category in the table. One check for this is to see that the row and column totals are actually the sum of the numbers in the cells. (They are.) If they were not, it would mean that either some people fall into more than one category (the list would not be disjoint) or there would be employees not included in the table (the list would not be complete).

example: Using the employment data above, if you randomly select a person from the population of employees, what is the probability the person is 45 years old or older?

solution: P(45 or older) $\frac{Number\ 45\ or\ older}{Number\ of\ employees} = \frac{25,524+8,917+208}{146,680} \approx 0.236$. There is a 23.6% probability that a randomly selected person in the labor force is 45 years old or older.

The numbers in the total row and the total column are called **marginal frequencies**. The cells in the middle of the table give the **joint frequencies**. Dividing a frequency in a cell by the total gives the proportion of cases in that cell. This is also called the **relative frequency**. If you do this to the marginal frequencies and joint frequencies, you get the **marginal relative frequencies** and the **joint relative frequencies**. If you look at randomly selecting an individual from the population, those relative frequencies give the probability of selecting a person from that cell.

Probabilities of Combined Events

P(**A or B**): The probability that **either** event A **or** event B occurs (or both). Using set notation, P(A or B) can be written $P(A \cup B)$. $A \cup B$ is spoken as, "A union B." You won't see set notation much, but it is the way the formula appears on your formula sheet, so it's worth mentioning.

P(**A and B**): The probability that **both** event A **and** event B occur. Using set notation, P(A and B) can be written $P(A \cap B)$. $A \cap B$ is spoken as, "A intersect B."

example: Using the employment data above, if you randomly select a person from the population of employees, what is the probability the person is 45 to 54 years old **or** is in a service occupation?

solution: P(*45 to 54* or *Service occupations*) is the sum of all joint relative frequencies that are either in the column for *45 to 54* or in the row for *Service occupations*. The cell that is in both must be counted only once.

	16 TO 19	20 TO 24	25 TO 34	35 TO 44	45 TO 54	55 TO 64	65+	TOTAL
Management, professional, and related occupations	3,118	13,561	13,926	13,818	10,689	3,931	44	**59,087**
Service occupations	**4,007**	**6,077**	**4,889**	**4,821**	**3,655**	**1,292**	**38**	**24,779**
Sales and office occupations	4,052	7,159	5,907	6,748	5,754	2,237	42	**31,899**
Natural resources, construction, and maintenance occupations	1,162	3,245	3,294	3,194	2,177	515	42	**13,629**
Production, transportation, and material-moving occupations	1,687	3,681	3,547	4,138	3,249	942	43	**17,287**
Total employed	**14,026**	**33,723**	**31,563**	**32,719**	**25,524**	**8,917**	**208**	**146,680**

The sum of these cells is 46,648. P(*45 to 54* or *Service occupations*) $= \frac{46,648}{146,680} \approx 0.318$.

A shortcut for the previous example is to recognize that you could add the total for *45 to 54* and the total for *Service occupations*. But this would count the cell with 3,655 employees twice, so you must subtract that amount from the sum. This is called the **addition rule**.

Addition rule: $P(\text{A or B}) = P(\text{A}) + P(\text{B}) - P(\text{A and B})$

example: If you randomly select one employee from the population above, use the addition rule to find $P(45 \text{ to } 54 \text{ or } \textit{Service occupations})$.

solution: $P(45 \text{ to } 54 \text{ or } \textit{Service occupations}) = P(45 \text{ to } 54) + P(\textit{Service occupations}) - P(45 \text{ to } 54 \text{ and } \textit{Service occupations}) = \frac{25,524}{146,680} + \frac{24,779}{146,680} - \frac{3,655}{146,680} = \frac{46,648}{146,680} \approx 0.318$.

example: In the employment data example above, if you are to randomly select one employee, what is $P(\textit{Service occupations} \text{ or } \textit{Sales and office occupations})$?

solution: Because these are disjoint events, $P(\textit{Service occupations} \text{ and } \textit{Sales and office occupations}) = 0$. So the addition rule simplifies to the **addition rule for disjoint events:** $P(\text{A or B}) = P(\text{A}) + P(\text{B})$. In this case, $P(\textit{Service occupations} \text{ or } \textit{Sales and office occupations}) = P(\textit{Service occupations}) + P(\textit{Sales and office occupations}) = \frac{24,779}{146,680} + \frac{31,899}{146,680} = \frac{56,678}{146,680} \approx 0.386$.

Complement of an event A: Events in the sample space that are not in event A. The complement of an event A is symbolized by $\bar{\text{A}}$, or A^c. Furthermore, $P(\bar{\text{A}}) = 1 - P(\text{A})$. This can be a useful way to calculate certain probabilities.

example: If you randomly select an employee, what is the probability that the employee is at least 20 years old?

solution: You could add the marginal probabilities for all the age categories starting with 20 to 24, or you could simply subtract the marginal probability for the one remaining category, 16 to 19, and subtract that from 1. $P(20 \text{ or more years old}) = 1 - P(16 \text{ to } 19) = 1 - \frac{14,026}{146,680} \approx 0.904$.

Conditional Probability

Conditional probability: "The probability of A given B" assumes we have knowledge of an event B having occurred before we compute the probability of event A. This is symbolized by $P(\text{A} \mid \text{B})$.

example: If you randomly select an employee that is 16 to 19 years old, what is the probability that the employee is in a service occupation?

solution: You know the employee is 16 to 19 years old. That means we do not need to consider the entire population of employees. We need look only at the relevant column.

	16 to 19
Management, professional, and related occupations	3,118
Service occupations	4,007
Sales and office occupations	4,052
Natural resources, construction, and maintenance occupations	1,162
Production, transportation, and material-moving occupations	1,687
Total employed	14,026

Within the *16 to 19* column, 4,007 of 14,026 16- to 19-year-olds are employed in service occupations. So $P(\textit{Service occupations} \mid \textit{16 to 19}) = \frac{4,007}{14,026} \approx 0.286$.

Some conditional probability problems can be solved by using a **tree diagram**. A tree diagram is a schematic way of looking at all possible outcomes.

> **example:** Suppose a computer company has manufacturing plants in three states. Fifty percent of its computers are manufactured in California, and 85% of these are desktops; 30% of computers are manufactured in Washington, and 40% of these are laptops; and 20% of computers are manufactured in Oregon, and 40% of these are desktops. All computers are first shipped to a distribution site in Nebraska before being sent out to stores. If you picked a computer at random from the Nebraska distribution center, what is the probability that it is a laptop?

solution:

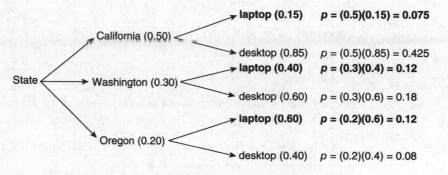

Note that the final probabilities add to 1 so we know we have considered all possible outcomes. Now, $P(\textit{laptop}) = 0.075 + 0.12 + 0.12 = 0.315$.

Independent Events

Independent events: Events A and B are said to be *independent* if and only if the knowledge of one event having occurred does not change the probability that the other event occurs. That is, A and B are independent if the probability of A, given that B occurred, is the same as the probability that A occurred, and vice versa. In symbols, A and B are independent if and only if $P(A|B) = P(A)$ or $P(B|A) = P(B)$.

> **example:** Consider drawing one card from a standard deck of 52 playing cards.
>
> Let A = "the card drawn is an ace." $P(A) = 4/52 = 1/13$.
>
> Let B = "the card drawn is a 10, J, Q, K, or A." $P(B) = 20/52 = 5/13$.
>
> Let C = "the card drawn is a diamond." $P(C) = 13/52 = 1/4$.
>
> (i) Are A and B independent?
>
> **solution:** $P(A|B) = P$(the card drawn is an ace | the card is a 10, J, Q, K, or A) = 4/20 = 1/5 (there are 20 cards to consider, 4 of which are aces). Since $P(A)$ = 1/13, knowledge of B has changed what we know about A. That is, in this case, $P(A) \neq P(A|B)$, so events A and B are *not* independent.
>
> (ii) Are A and C independent?
>
> **solution:** $P(A|C) = P$(the card drawn is an ace | the card drawn is a diamond) = 1/13 (there are 13 diamonds, one of which is an ace). So, in this case, $P(A) = P(A|C)$, so that the events "the card drawn is an ace" and "the card drawn is a diamond" are independent.

example: If we randomly select an employee from the population in the earlier examples, are the events *20 to 24 years old* and *Management, professional, and related occupations* independent?

solution: $P(20\ to\ 24) = \frac{33{,}723}{146{,}680} \approx 0.230$. $P(20\ to\ 24\ |\ Management,\ professional,$ $and\ related\ occupations) = \frac{6{,}077}{24{,}779} \approx 0.245$. Because these probabilities are not the same, the events *20 to 24 years old* and *Management, professional, and related occupations* are not independent.

Probability of A and B or A or B

The Multiplication Rule: $P(A\ and\ B) = P(A) \cdot P(B|A)$.

Special case of *the multiplication rule:* If A and B are *independent,*

$P(B|A) = P(B)$, so $P(A\ and\ B) = P(A) \cdot P(B)$.

example: A and B are two mutually exclusive events for which $P(A) = 0.3$ and $P(B) = 0.25$. Find $P(A\ or\ B)$.

solution: $P(A\ or\ B) = 0.3 + 0.25 = 0.55$.

example: A basketball player has a 0.6 probability of making a free throw. What is his probability of making two consecutive free throws if

(a) he gets very nervous after making the first shot and his probability of making the second shot drops to 0.4?

solution: P(making the first shot) = 0.6, P(making the second shot | he made the first) = 0.4. So, P(making both shots) = (0.6)(0.4) = 0.24.

(b) the events "he makes his first shot" and "he makes the succeeding shot" are independent?

solution: Since the events are independent, his probability of making each shot is the same. Thus, P(he makes both shots) = (0.6)(0.6) = 0.36.

Random Variables

Recall our earlier definition of a **probability experiment (random phenomenon)**: An activity whose outcome we can observe and measure, but for which we can't predict the result of any single trial. A **random variable**, X, is a numerical value assigned to an outcome of a random phenomenon. Particular values of the random variable X are often given lowercase names, such as x to represent a general value, or k to represent a specific value. It is common to see expressions of the form $P(X = x)$ or $P(X = k)$, which refer to the probability that the random variable X takes on the particular value x.

example: If we roll a fair die, the random variable X could be the face-up value of the die. The possible values of X are {1, 2, 3, 4, 5, 6}. $P(X = 2) = 1/6$.

example: The score a college-hopeful student gets on her SAT test can take on values from 200 to 800. These are the possible values of the random variable X, the score a randomly selected student gets on his or her test.

There are two types of random variables: **discrete random variables** and **continuous random variables**.

Discrete Random Variables

A **discrete random variable (DRV)** is a random variable with a countable number of outcomes. You can think of a discrete random variable as one whose values are separated by gaps. The number of left-handed people in a sample of 100 people, for example, can only take on whole number values from 0 to 100. This is an example of a discrete random variable.

> **example:** the number of votes earned by different candidates in an election.

> **example:** the number of successes in 25 trials of an event whose probability of success on any one trial is known to be 0.3.

Continuous Random Variables

A **continuous random variable (CRV)** is a random variable that assumes values associated with one or more intervals on the number line. The continuous random variable X can assume infinitely many outcomes within an interval. Heights of people, for example, are continuous, although the reported heights may be discrete if they are reported to the nearest inch.

> **example:** Consider the *uniform* distribution $y = 3$ defined on the interval $1 \leq x \leq 5$. The area under $y = 3$ and above the x axis for any interval corresponds to a continuous random variable. For example, if $2 \leq x \leq 3$, then $X = 3$. If $2 \leq x \leq 4.5$, then $X = (4.5 - 2)(3) = 7.5$. Note that there are an infinite number of possible outcomes for X.

Probability Distribution of a Random Variable

A **probability distribution for a random variable** is the possible values of the random variable X together with the probabilities corresponding to those values.

A **probability distribution for a discrete random variable (DRV)** is a list of the possible values of the DRV together with their respective probabilities.

> **example:** Let X be the number of boys in a three-child family. Assuming that the probability of a boy on any one birth is 0.5, the probability distribution for X is

X	0	1	2	3
$P(X)$	1/8	3/8	3/8	1/8

The probabilities P_i of a DRV satisfy two conditions:

(1) $0 \leq P_i \leq 1$ (that is, every probability is between 0 and 1).

(2) $\Sigma P_i = 1$ (that is, the sum of all probabilities is 1).

(Are these conditions satisfied in the above example?)

The **mean** of a discrete random variable, also called the **expected value**, is given by

$$\mu_X = \sum x \cdot P(x).$$

The **variance of a discrete random variable** is given by

$$\sigma_X^2 = \sum (x - \mu_X)^2 \cdot P(x).$$

The **standard deviation of a discrete random variable** is given by

$$\sigma_X = \sqrt{\sum (x - \mu_X)^2 \cdot P(x)}.$$

example: Given that the following is the probability distribution for a DRV, find $P(X = 3)$.

X	2	3	4	5	6
P(X)	0.15		0.2	0.2	0.35

solution: Since $\sum P_i = 1$, $P(3) = 1 - (0.15 + 0.2 + 0.2 + 0.35) = 0.1$.

example: For the probability distribution given above, find μ_x and σ_x.

solution:

$$\mu_X = 2(0.15) + 3(0.1) + 4(0.2) + 5(0.2) + 6(0.35) = 4.5.$$

$$\sigma_X = \sqrt{(2 - 4.5)^2(0.15) + (3 - 4.5)^2(0.1) + \cdots + (6 - 4.5)^2(0.35)} = 1.432.$$

 Calculator Tip: While it's helpful to know the formulas given above, they are given on your formula sheet. In practice it's easier to use your calculator to do the computations. Most calculators have a shortcut for calculation summary statistics of values in a list. For probability distributions you can usually enter the values in one list and the probabilities in the other and set the probability list as your frequency. This feature may make it easier to do homework problems, but on the AP exam you are expected to show your work as above.

example: Redo the previous example using the TI-83/84, or equivalent, calculator.

solution: Enter the x values in a list (say, L1) and the probabilities in another list (say, L2). Then enter "1-Var Stats L1,L2" and press ENTER. The calculator will read the probabilities in L2 as relative frequencies and return 4.5 for the mean and 1.432 for the standard deviation.

Probability Histogram

A **probability histogram** of a DRV is a way to picture the probability distribution. The following is a TI-83/84 histogram of the probability distribution we used in a couple of the examples above.

x	2	3	4	5	6
$p(x)$	0.15	0.1	0.2	0.2	0.35

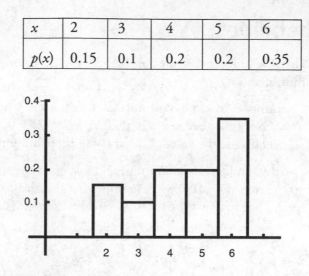

Probability Distribution for a Continuous Random Variable. When we randomly select a value from a population modeled with a normal distribution, we are using a continuous random variable. Here are a few things to note:

- The probability of any individual value is 0. That is, if a is a point on the horizontal axis, $P(X = a) = 0$.
- To find the probability of an event, you must find the probability that x falls in some given interval. To do this, use the `normalcdf` function on your calculator.
- We will learn about a couple more continuous distributions during the course. The normal distribution is not the only one!

In a normal distribution, the tails of the curve extend to infinity, although there is very little area under the curve when we get more than, say, three standard deviations away from the mean. (The 68-95-99.7 rule stated that about 99.7% of the terms in a normal distribution are within three standard deviations of the mean. Thus, only about 0.3% lie beyond three standard deviations of the mean.)

Areas between two values on the number line and under the normal probability distribution correspond to probabilities. In Chapter 4, we found the proportion of terms falling within certain intervals. Because the total area under the curve is 1, in this chapter we will consider those proportions to be probabilities.

Remember that we *standardized* the normal distribution by converting the data to z-scores $\left(z = \dfrac{x - \bar{x}}{s_x} \right)$.

We learned in Chapter 4 that a standardized normal distribution has a mean of 0 and a standard deviation of 1.

It is generally easiest to use your calculator for calculating normal probabilities, but you can also use the table of Standard Normal Probabilities. It would be best to memorize the values corresponding to the 68-95-99.7 rule.

Normal Probabilities

When we know a distribution is approximately normal, we can solve many types of problems.

example: In a standard normal distribution, what is the probability that $z < 1.5$? (Note that because z is a CRV, $P(X = a) = 0$, so this problem could have been equivalently stated, "What is the probability that $z \leq 1.5$?")

solution: The standard normal table gives areas to the left of a specified z-score. From the table, we determine that the area to the left of $z = 1.5$ is 0.9332. That is, $P(z < 1.5) = 0.9332$. This can be visualized as follows:

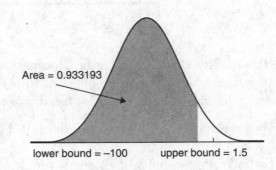

Area = 0.933193

lower bound = –100   upper bound = 1.5

example: It is known that the heights (X) of students at Downtown College are approximately normally distributed with a mean of 68 inches and a standard deviation of 3 inches. That is, X has $N(68,3)$. Determine

(a) $P(X < 65)$.

solution: $P(X < 65) = P\left(z < \frac{65-68}{3} = -1\right) = 0.1587$ (the area to the left of $z = -1$ from Table A). On the TI-83/84, the corresponding calculation is `normalcdf` `(-100,-1)` = `normalcdf(-1000,65,68,30)=0.1586552596`.

(b) $P(X > 65)$.

solution: From part (a) of the example, we have $P(X < 65) = 0.1587$. Hence, $P(X > 65) = 1 - P(X < 65) = 1 - 0.1587 = 0.8413$. On the TI-83/84, the corresponding calculation is `normalcdf(-1,100)` = `normalcdf(65,1000,68,3)=0.8413447404`.

(c) $P(65 < X < 70)$.

solution: $P(65 < X < 70) = P\left(\frac{65-68}{3} < z < \frac{70-68}{3}\right) = P(-1 < z < 0.667) =$

0.7486 − 0.1587 = 0.5899 (from Table A, the geometry of the situation dictates that the area to the left of $z = -1$ must be subtracted from the area to the left of $z = 0.667$). Using the TI-83/84, the calculation is `normal-cdf(-1,0.67) = normalcdf(65,70,68,3) = 0.5889`. This situation is pictured below.

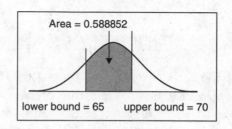

Note that there is some rounding error when using Table A (see Appendix). In part (c), $z = 0.66667$, but we must use 0.67 to use the table.

(d) $P(70 < X < 75)$

solution: Now we need the area between 70 and 75. The geometry of the situation dictates that we subtract the area to the left of 70 from the area to the left of 75. This is pictured below.

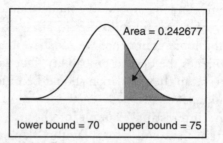

We saw from part (c) that the area to the left of 70 is 0.7486. In a similar fashion, we find that the area to the left of 75 is 0.9901 (based on $z = 2.33$). Thus $P(70 < X < 75) = 0.9901 - 0.7486 = 0.2415$. The calculation on the TI-83/84 is: `normalcdf (70,75,68,3) = 0.2427`. The difference in the answers is due to rounding.

example: SAT scores are approximately normally distributed with a mean of about 500 and a standard deviation of 100. Laurie needs to be in the top 15% on the SAT in order to ensure her acceptance by Giant U. What is the minimum score she must earn to be able to start packing her bags for college?

solution: This is somewhat different from the previous examples. Up until now, we have been given, or have figured out, a z-score, and have needed to determine an area. Now, we are given an area and are asked to determine a particular score. If we are using the table of normal probabilities, it is a situation in which we must read from inside the table out to the z-scores rather than from the outside in. If the particular value of X we are looking for is the lower bound for the top 15% of scores, then there are 85% of the scores to the left of x. We look through the table and find the closest entry to 0.8500 and determine it to be 0.8508. This corresponds to a *z*-score of 1.04. Another way to write the *z*-score of the desired value of *X* is

$$z = \frac{x - 500}{100}.$$

Thus, $z = \frac{x - 500}{100} = 1.04$.

Solving for *x*, we get *x* = 500 + 1.04(100) = 604. So, Laurie must achieve an SAT score of at least 604. While this problem can be done on the calculator as follows: `invNorm(0.85,500,100)`, recent rubrics have required that you list the z-score even if you use this command on your calculator.

Most problems of this type can be solved in the same way: express the *z*-score of the desired value in two different ways (from the definition, finding the actual value from Table A, or by using the `invNorm` function on the calculator), then equate the expressions and solve for *x*.

Simulation and Random Number Generation

Sometimes probability situations do not lend themselves easily to analytical solutions. In some situations, an acceptable approach might be to run a **simulation**. A simulation utilizes some random process to conduct numerous trials of the situation and then counts the number of successful outcomes to arrive at an estimated probability. In general, the more trials, the more confidence we can have that the relative frequency of successes accurately approximates the desired probability. The **law of large numbers** states that the proportion of successes in the simulation should become, over time, close to the true proportion in the population.

One interesting example of the use of simulation has been in the development of certain "systems" for playing Blackjack. The number of possible situations in Blackjack is large but finite. A computer was used to conduct thousands of simulations of each possible playing decision for each of the possible hands. In this way, certain situations favorable to the player were identified and formed the basis for the published systems.

> **example:** Suppose there is a small Pacific Island society that places a high value on families having a baby girl. Suppose further that *every* family in the society decides to keep having children until they have a girl and then they stop. If the first child is a girl, they are a one-child family, but it may take several tries before they succeed. Assume that when this policy was decided on that the proportion of girls in the population was 0.5 and the probability of having a girl is 0.5 for each birth. Would this behavior change the proportion of girls in the population? Design a simulation to answer this question.

> **solution:** Use a random number generator, say a fair coin, to simulate a birth. Let heads = "have a girl" and tails = "have a boy." Flip the coin and note whether it falls heads or tails. If it falls heads, the trial ends. If it falls tails, flip again because this represents having a boy. The outcome of interest is the number of trials (births) necessary until a girl is born (if the third flip gives the first head,

then $x = 3$). Repeat this many times and determine how many girls and how many boys have been born.

If flipping a coin many times seems a bit tedious, you can also use your calculator to simulate flipping a coin. Let 1 be a head and let 2 be a tail. Then enter MATH PRB rand-Int(1,2) and press ENTER to generate a random 1 or 2. Continue to press ENTER to generate additional random integers 1 or 2. Enter randInt(1,2,n) to generate n random integers, each of which is a 1 or a 2. Enter randInt(a,b,n) to generate n random integers X such that $a \le X \le b$.

The following represents a few trials of this simulation (actually done using the random number generator on the TI-83/84 calculator):

TRIAL #	TRIAL RESULTS (H = "GIRL")	# FLIPS UNTIL FIRST GIRL	TOTAL # OF GIRLS AFTER TRIAL IS FINISHED	TOTAL # OF BOYS AFTER TRIAL IS FINISHED
1	TH	2	1	1
2	H	1	2	1
3	TTTH	4	3	4
4	H	1	4	4
5	TH	2	5	5
6	H	1	6	5
7	H	1	7	5
8	H	1	8	5
9	TH	2	9	6
10	H	1	10	6
11	TTTH	4	11	9
12	H	1	12	9
13	H	1	13	9
14	TTTTH	5	14	13
15	TTH	3	15	15

In this limited simulation the number of boys and girls in the population of 15 families are equal. Based on our simulation, if every familiy kept having children until they have a girl, it would not change the proportion of girls in the population. If the simulation were to be run again, it is unlikely that it would turn out exactly the same. For more reliable estimates, do more trials.

Exam Tip: If you are asked to do a simulation on the AP Statistics exam (and there have been such questions), you may be asked to use a table of random numbers rather than the random number generator on your calculator. This allows the grader to know exactly which numbers you used in your simulation and, therefore, whether you are using them correctly. It is also possible that the question will describe a simulation and you will be asked to interpret the results of the simulation based on a plot of the simulated results.

The following gives 200 outcomes of a typical random number generator separated into groups of 5 digits:

79692	51707	73274	12548	91497	11135	81218	79572	06484	87440
41957	21607	51248	54772	19481	90392	35268	36234	90244	02146
07094	31750	69426	62510	90127	43365	61167	53938	03694	76923
59365	43671	12704	87941	51620	45102	22785	07729	40985	92589

example: A coin is known to be biased in such a way that the probability of getting a head is 0.4. If the coin is flipped 50 times, how many heads would you expect to get?

solution: Let 0, 1, 2, 3 be a head and 4, 5, 6, 7, 8, 9 be a tail. If we look at 50 digits beginning with the first row, we see that there are 18 heads (bold-faced below), so the proportion of heads is 18/50 = 0.36. This is close to the expected value of 0.4.

796**92** 51**7**07 73274 **12**548 91497 **111**3**5** 81218 79**572** 06484 87440

Sometimes the simulation will be a **wait-time simulation**. In the example above, we could have asked how long it would take, on average, until we get five heads. In this case, using the same definitions for the various digits, we would proceed through the table until we noted five numbers with digits 0–3. We would then write down how many digits we had to look at. Three trials of that simulation might look like this (individual trials are separated by \\):

796**92** 51**7**07 **732**\\74 **12**548 91497 **11****135** **8121**\\.

So, it took 13, 14, and 7 trials to get our five heads, or an average of 11.3 trials (the theoretical expected number of trials is 12.5).

Transforming and Combining Random Variables

If X is a random variable, we can transform the data by adding a constant to each value of X, multiplying each value by a constant, or some linear combination of the two. We may do this to make numbers more manageable. For example, if values in our dataset ranged from 8500 to 9000, we could subtract, say, 8500 from each value to get a dataset that ranged from 0 to 500. We would then be interested in the mean and standard deviation of the new dataset as compared to the old dataset.

Some facts from algebra can help us out here. Let μ_x and σ_x be th e mean and standard deviation of the random variable X. Each of the following statements can be algebraically verified if we add or subtract the same constant, a, to or from each term in a dataset

$(X \pm a)$, or multiply each term by the same constant b (bX), or some combination of these ($a \pm bX$):

- $\mu_{a \pm bX} = a \pm b\mu_x.$
- $\sigma_{a \pm bX} = b\sigma_X$ ($\sigma^2_{a \pm bx} = b^2 \sigma^2_X$).

> **example:** Consider a distribution with $\mu_X = 14$, $\sigma_X = 2$. Multiply each value of X by 4 and then add 3 to each. Then $\mu_{3+4X} = 3 + 4(14) = 59$, $\sigma_{3+4X} = 4(2) = 8$.

Rules for the Mean and Standard Deviation of Combined Random Variables

Sometimes we need to combine two random variables. For example, suppose one contractor can finish a particular job, on average, in 40 hours ($\mu_x = 40$). Another contractor can finish a similar job in 35 hours ($\mu_y = 35$). If they work on two separate jobs, how many hours, on average, will they bill for completing both jobs? It should be clear that the average of $X + Y$ is just the average for X plus the average for Y. That is,

$$\mu_{X \pm Y} = \mu_X \pm \mu_Y.$$

The situation is somewhat less clear when we combine variances. In the contractor example above, suppose that

$$\sigma^2_X = 5 \text{ and } \sigma^2_Y = 4.$$

Does the variance of the sum equal the sum of the variances? Well, yes and no. Yes, if the random variables X and Y are independent. No, if the random variables are not independent, but are dependent in some way. Furthermore, it doesn't matter if the random variables are added or subtracted, we are still combining the variances. That is,

- $\sigma^2_{X \pm Y} = \sigma^2_X + \sigma^2_Y$, if and only if X and Y are independent.

- $\sigma_{X \pm Y} = \sqrt{\sigma^2_X + \sigma^2_Y}$, if and only if X and Y are independent.

> **Exam Tip:** The rules for means and variances when you combine random variables may seem a bit obscure, but there have been questions on more than one occasion that depend on your knowledge of how this is done.

The rules for means and variances generalize. That is, no matter how many random variables you have: $\mu_{X1 \pm X2 \pm \ldots \pm Xn} = \mu_{X1} \pm \mu_{X2 \pm \ldots} + \mu_{Xn}$ and, if $X_1, X_2, \ldots, X_n$ are all independent, $\sigma^2_{X_1 \pm X_2 \pm \ldots \pm X_n} = \sigma^2_{X_1} + \sigma^2_{X_2 +} \ldots + \sigma^2_{X_n}$.

example: A prestigious private school offers an admission test on the first Saturday of November and the first Saturday of December each year. In 2008, the mean score for hopeful students taking the test in November (X) was 156 with a standard deviation of 12. For those taking the test in December (Y), the mean score was 165 with a standard deviation of 11. What are the mean and standard deviation of the total score $X + Y$ of all students who took the test in 2008?

solution: We have no reason to think that scores of students who take the test in December are influenced by the scores of those students who took the test in November. Hence, it is reasonable to assume that X and Y are independent. Accordingly,

$$\mu_{X+Y} = \mu_X + \mu_Y = 156 + 165 = 321,$$

$$\sigma_{X+Y} = \sqrt{\sigma^2_X + \sigma^2_Y} = \sqrt{12^2 + 11^2} = \sqrt{265} = 16.28.$$

› Rapid Review

1. A bag has 8 green marbles and 12 red marbles. If you draw one marble from the bag, what is P(draw a green marble)?

   Answer: Let s = number of ways to draw a green marble.

   Let f = number of ways to draw a red marble.

   $$P(E) = \frac{s}{s+f} = \frac{8}{8+12} = \frac{8}{20} = \frac{2}{5}.$$

2. A married couple has three children. At least one of their children is a boy. What is the probability that the couple has exactly two boys?

   Answer: The sample space for having three children is {BBB, BBG, BGB, GBB, BGG, GBG, GGB, GGG}. Of these, there are seven outcomes that have at least one boy. Of these, three have two boys and one girl. Thus, P(the couple has exactly two boys | they have at least one boy) = 3/7.

3. Does the following table represent the probability distribution for a discrete random variable?

X	1	2	3	4
$P(X)$	0.2	0.3	0.3	0.4

Answer: No, because

$$\sum P_i = 1.2.$$

4. In a standard normal distribution, what is $P(z > 0.5)$?

   Answer: From the table, we see that $P(z < 0.5) = 0.6915$. Hence, $P(z > 0.5) = 1 - 0.6915 = 0.3085$. By calculator, `normalcdf (0.5,100) = 0.3085375322`.

5. A random variable X has $N(13, 0.45)$. Describe the distribution of $2 - 4X$ (that is, each datapoint in the distribution is multiplied by 4, and that value is subtracted from 2).

   Answer: We are given that the distribution of X is normal with $\mu_X = 13$ and $\sigma_X = 0.45$. Because $\mu_{a \pm bX} = a \pm b\mu_X$, $\mu_{2-4X} = 2 - 4\mu_X = 2 - 4(13) = -50$. Also, because $\sigma_{a \pm bX} = b\sigma_X$, $\sigma_{2-4X} = 4\sigma_X = 4(0.45) = 1.8$.

Practice Problems

Multiple-Choice

1.

	D	E	Total
A	15	12	27
B	15	23	38
C	32	28	60
Total	62	63	125

In the table above, what are $P(A \text{ and } E)$ and $P(C|E)$?

(a) 12/125, 28/125
(b) 12/63, 28/60
(c) 12/125, 28/63
(d) 12/125, 28/60
(e) 12/63, 28/63

2.

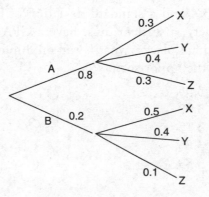

For the tree diagram pictured above, what is $P(B|X)$?

(a) 1/4
(b) 5/17
(c) 2/5
(d) 1/3
(e) 4/5

3. It turns out that 25 seniors at Fashionable High School took both the AP Statistics exam and the AP Spanish Language exam. The mean score on the Statistics exam for the 25 seniors was 2.4 with a standard deviation of 0.6, and the mean score on the Spanish Language exam was 2.65 with a standard deviation of 0.55. We want to combine the scores into a single score. What are the correct mean and standard deviation of the combined scores?

   (a) 5.05; 1.15
   (b) 5.05; 1.07
   (c) 5.05; 0.66
   (d) 5.05; 0.81
   (e) 5.05; you cannot determine the standard deviation from this information.

4. The GPAs (grade point averages) of students who take the AP Statistics exam are approximately normally distributed with a mean of 3.4 and a standard deviation of 0.3. Using Table A, what is the probability that a student selected at random from this group has a GPA lower than 3.0?

   (a) 0.0918
   (b) 0.4082
   (c) 0.9082
   (d) −0.0918
   (e) 0

5. The 2000 Census identified the ethnic breakdown of the state of California to be approximately as follows: White: 46%, Latino: 32%, Asian: 11%, Black: 7%, and Other: 4%. Assuming that these are mutually exclusive categories (this is not a realistic assumption), what is the probability that a randomly selected person from the state of California is of Asian or Latino descent?

   (a) 46%
   (b) 32%
   (c) 11%
   (d) 43%
   (e) 3.5%

6. The students in problem #4 above were normally distributed with a mean GPA of 3.4 and a standard deviation of 0.3. In order to qualify for the school honor society, a student must have a GPA in the top 5% of all GPAs. Accurate to two decimal places, what is the minimum GPA Norma must have in order to qualify for the honor society?

   (a) 3.95
   (b) 3.92
   (c) 3.75
   (d) 3.85
   (e) 3.89

7. The following are the probability distributions for two random variables, X and Y:

X	$P(X = x)$
3	$\frac{1}{3}$
5	$\frac{1}{2}$
7	$\frac{1}{6}$

Y	$P(Y = y)$
1	$\frac{1}{8}$
3	$\frac{3}{8}$
4	?
5	$\frac{3}{16}$

If X and Y are independent, what is $P(X = 5$ and $Y = 4)$?

(a) $\frac{5}{16}$

(b) $\frac{13}{16}$

(c) $\frac{5}{32}$

(d) $\frac{3}{32}$

(e) $\frac{3}{16}$

8. The following table gives the probabilities of various outcomes for a gambling game.

Outcome	Lose $1	Win $1	Win $2
Probability	0.6	0.25	0.15

What is the player's expected return on a bet of $1?

(a) $0.05

(b) −$0.60

(c) −$0.05

(d) −$0.10

(e) You can't answer this question since this is not a complete probability distribution.

9. You own an unusual die. Three faces are marked with the letter "X," two faces with the letter "Y," and one face with the letter "Z." What is the probability that at least one of the first two rolls is a "Y"?

(a) $\dfrac{1}{6}$

(b) $\dfrac{2}{3}$

(c) $\dfrac{1}{3}$

(d) $\dfrac{5}{9}$

(e) $\dfrac{2}{9}$

10. You roll two six-sided dice. What is the probability that the sum is 6 given that one die shows a 4?

(a) $\dfrac{2}{12}$

(b) $\dfrac{2}{11}$

(c) $\dfrac{11}{36}$

(d) $\dfrac{2}{36}$

(e) $\dfrac{12}{36}$

Free-Response

1. Find μ_X and σ_X for the following discrete probability distribution:

X	2	3	4
$P(X)$	1/3	5/12	1/4

2. Given that $P(A) = 0.6$, $P(B) = 0.3$, and $P(B|A) = 0.5$.
   (a) $P(A \text{ and } B) = ?$
   (b) $P(A \text{ or } B) = ?$
   (c) Are events A and B independent?

3. Consider a set of 9000 scores on a national test that is known to be approximately normally distributed with a mean of 500 and a standard deviation of 90.

    (a) What is the probability that a randomly selected student has a score greater than 600?
    (b) How many scores are there between 450 and 600?
    (c) Rachel needs to be in the top 1% of the scores on this test to qualify for a scholarship. What is the minimum score Rachel needs?

4. Consider a random variable X with $\mu_X = 3$, $\sigma^2_X = 0.25$. Find

    (a) μ_{3+6X}
    (b) σ_{3+6X}

5. Harvey, Laura, and Gina take turns throwing spitwads at a target. Harvey hits the target 1/2 the time, Laura hits it 1/3 of the time, and Gina hits the target 1/4 of the time. Given that somebody hit the target, what is the probability that it was Laura?

6. Consider two discrete, independent, random variables X and Y with $\mu_X = 3$, $\sigma^2_X = 1$, $\mu_Y = 5$, and $\sigma^2_Y = 1.3$. Find μ_{Y-X} and σ_{X-Y}.

7. Which of the following statements is (are) true of a normal distribution?

    I. Exactly 95% of the data are within two standard deviations of the mean.
    II. The mean = the median = the mode.
    III. The area under the normal curve between $z = 1$ and $z = 2$ is greater than the area between $z = 2$ and $z = 3$.

8. Consider the experiment of drawing two cards from a standard deck of 52 cards. Let event A = "draw a face card on the first draw," B = "draw a face card on the second draw," and C = "the first card drawn is a diamond."

    (a) Are the events A and B *independent?*
    (b) Are the events A and C *independent?*

9. A normal distribution has mean 700 and standard deviation 50. The probability is 0.6 that a randomly selected term from this distribution is above x. What is x?

10. Suppose 80% of the homes in Lakeville have a desktop computer and 30% have both a desktop computer and a laptop computer. What is the probability that a randomly selected home will have a laptop computer given that it has a desktop computer?

11. Consider a probability density curve defined by the line $y = 2x$ on the interval [0,1] (the area under $y = 2x$ on [0,1] is 1). Find $P(0.2 \le X \le 0.7)$.

12. Half Moon Bay, California, has an annual pumpkin festival at Halloween. A prime attraction to this festival is a "largest pumpkin" contest. Suppose that the weights of these giant pumpkins are approximately normally distributed with a mean of 125 pounds and a standard deviation of 18 pounds. Farmer Harv brings a pumpkin that is at the 90th percentile of all the pumpkins in the contest. What is the approximate weight of Harv's pumpkin?

13. Consider the following two probability distributions for independent discrete random variable X and Y:

X	2	3	4
$P(X)$	0.3	0.5	?

Y	3	4	5	6
$P(Y)$	?	0.1	?	0.4

If $P(X = 4$ and $Y = 3) = 0.03$, what is $P(Y = 5)$?

14. A contest is held to give away a free pizza. Contestants pick an integer at random from the integers 1 through 100. If the number chosen is divisible by 24 or by 36, the contestant wins the pizza. What is the probability that a contestant wins a pizza?

Use the following excerpt from a random number table for questions 15 and 16:

79692	51707	73274	12548	91497	11135	81218	79572	06484	87440
41957	21607	51248	54772	19481	90392	35268	36234	90244	02146
07094	31750	69426	62510	90127	43365	61167	53938	03694	76923
59365	43671	12704	87941	51620	45102	22785	07729	40985	92589
91547	03927	92309	10589	22107	04390	86297	32990	16963	09131

15. Men and women are about equally likely to earn degrees at City U. However, there is some question whether or not women have equal access to the prestigious School of Law. This year, only 4 of the 12 new students are female. Describe and conduct five trials of a simulation to help determine if this is evidence that women are underrepresented in the School of Law.

16. Suppose that, on a planet far away, the probability of a girl being born is 0.6, and it is socially advantageous to have three girls. How many children would a couple have to have, on average, until they had three girls? Describe and conduct five trials of a simulation to help answer this question.

17. Consider a random variable X with the following probability distribution:

X	20	21	22	23	24
$P(X)$	0.2	0.3	0.2	0.1	0.2

(a) Find $P(X \leq 22)$.
(b) Find $P(X > 21)$.
(c) Find $P(21 \leq X < 24)$.
(d) Find $P(X \leq 21$ or $X > 23)$.

18. In the casino game of roulette, a ball is rolled around the rim of a circular bowl while a wheel containing 38 slots into which the ball can drop is spun in the opposite direction from the rolling ball; 18 of the slots are red, 18 are black, and 2 are green. A player bets a set amount, say $1, and wins $1 (and keeps her $1 bet) if the ball falls into the color slot the player has wagered on. Assume a player decides to bet that the ball will fall into one of the red slots.

(a) What is the probability that the player will win?
(b) What is the expected return on a single bet of $1 on red?

19. A random variable X is normally distributed with mean μ, and standard deviation σ (that is, X has $N(\mu,\sigma)$). What is the probability that a term selected at random from this population will be more than 2.5 standard deviations from the mean?

20. The normal random variable X has a standard deviation of 12. We also know that $P(x > 50) = 0.90$. Find the mean μ of the distribution.

Cumulative Review Problems

1. Consider the following histogram:

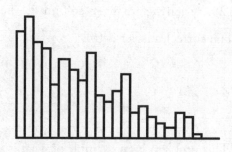

Which of the following statements is true and why?
I. The mean and median are approximately the same value.
II. The mean is probably greater than the median.
III. The median is probably greater than the mean.

2. You are going to do an opinion survey in your school. You can sample 100 students and desire that the sample accurately reflects the ethnic composition of your school. The school data clerk tells you that the student body is 25% Asian, 8% African American, 12% Latino, and 55% Caucasian. How could you sample the student body so that your sample of 100 would reflect this composition, and what is such a sample called?

3. The following data represent the scores on a 50-point AP Statistics quiz:
   46, 36, 50, 42, 46, 30, 46, 32, 50, 32, 40, 42, 20, 47, 39, 32, 22, 43, 42, 46, 48, 34, 47, 46, 27, 50, 46, 42, 20, 23, 42

   Determine the five-number summary for the quiz and draw a boxplot of the data.

4. The following represents some computer output that can be used to predict the number of manatee deaths from the number of powerboats registered in Florida.

Predictor	Coef	St Dev	t ratio	P
Constant	−41.430	7.412	−5.59	.000
Boats	0.12486	0.01290	9.68	.000

   (a) Write the least-square regression line for predicting the number of manatee deaths from the number of powerboat registrations.
   (b) Interpret the slope of the line in the context of the problem.

5. Use the *68-95-99.7 rule* to state whether it seems reasonable that the following sample data could have been drawn from a normal distribution: 12.3, 6.6, 10.6, 9.4, 9.1, 13.7, 12.2, 9, 9.4, 9.2, 8.8, 10.1, 7.0, 10.9, 7.8, 6.5, 10.3, 8.6, 10.6, 13, 11.5, 8.1, 13.0, 10.7, 8.8.

Solutions to Practice Problems

Multiple-Choice

1. The correct answer is (c). There are 12 values in the A *and* E cell out of the total of 125. When we are given column E, the total is 63. Of those, 28 are C.

2. The correct answer is (b).

$$P(X) = (0.8)(0.3) + (0.2)(0.5) = 0.34.$$

$$P(B|X) = \frac{(0.2)(0.5)}{(0.8)(0.3)+(0.2)(0.5)} = \frac{0.10}{0.34} = \frac{5}{17}.$$

(This problem is an example of what is known as Bayes's rule. It's still conditional probability, but sort of backwards. That is, rather than being given a path and finding the probability of going along that path—$P(X|B)$ refers to the probability of first traveling along B and then along X—we are given the outcome and asked for the probability of having gone along a certain path to get there—$P(B|X)$ refers to the probability of having gotten to X by first having traveled along B. You don't need to know Bayes's rule by name for the AP exam, but you may have to solve a problem like this one.)

3. The correct answer is (e). If you knew that the variables "Score on Statistics Exam" and "Score on Spanish Language Exam" were independent, then the standard deviation would be given by

$$\sqrt{\sigma_1^2 + \sigma_2^2} = \sqrt{(0.6)^2 + (0.55)^2} \approx 0.82.$$

However, you cannot assume that they are independent in this situation. In fact, they aren't because we have two scores on the same people. Hence, there is not enough information.

4. The correct answer is (a).

$$P(X < 3.0) = P\left(z < \frac{3-3.4}{0.3} = -1.33 \right) = 0.0918.$$

The calculator answer is `normalcdf(-100,3,3.4,0.3) = 0.0912`. Note that answer (d) makes no sense since probability values must be nonnegative (and, of course, less than or equal to 1).

5. The correct answer is (d). Because ethnic group categories are assumed to be mutually exclusive, P(Asian or Latino) = P(Asian) + P(Latino) = 32% + 11% = 43%.

6. The correct answer is (e). The situation is as pictured below:

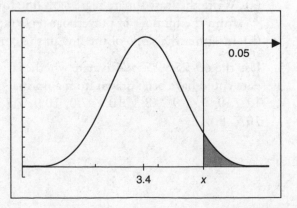

From Table A, $z_x = 1.645$ (also, invNorm(0.95) = 1.645).

Hence, $z_x = 1.645 = \dfrac{x - 3.4}{0.3} \Rightarrow x = 3.4 + (0.3)(1.645) = 3.89$. Norma would need a minimum GPA of 3.89 in order to qualify for the honor society.

7. The correct answer is (c). $P(Y = 4) = 1 - \left(\dfrac{1}{8} + \dfrac{3}{8} + \dfrac{3}{16}\right) = \dfrac{5}{16}$. Since they are independent,

$$P(X = 5 \text{ and } Y = 4) = P(X = 5) \cdot P(Y = 4) = \dfrac{1}{2} \cdot \dfrac{5}{16} = \dfrac{5}{32}.$$

8. The correct answer is (c). The expected value is $(-1)(0.6) + (1)(0.25) + (2)(0.15) = -0.05$.

9. The correct answer is (d). P(at least one of the first two rolls is "Y") = P(the first roll is "Y") + P(the second roll is "Y") − P(both rolls are "Y") = $\dfrac{1}{3} + \dfrac{1}{3} - \left(\dfrac{1}{3}\right)\left(\dfrac{1}{3}\right) = \dfrac{5}{9}$.

    Alternatively, P(at least one of the first two rolls is "Y") = 1 − P(neither roll is "Y") = $1 - \left(\dfrac{2}{3}\right)^2 = \dfrac{5}{9}$.

10. The correct answer is (b). The possible outcomes where one die shows a 4 are highlighted in the table of all possible sums:

	1	2	3	4	5	6
1	2	3	4	**5**	6	7
2	3	4	5	**6**	7	8
3	4	5	6	**7**	8	9
4	**5**	**6**	**7**	**8**	**9**	**10**
5	6	7	8	**9**	10	11
6	7	8	9	**10**	11	12

There are 11 cells for which one die is a 4 (be careful not to count the **8** twice), 2 of which are 6s.

Free-Response

1. $\mu_x = 2\left(\dfrac{1}{3}\right) + 3\left(\dfrac{5}{12}\right) + 4\left(\dfrac{1}{4}\right) = \dfrac{35}{12} \approx 2.92$

$$\sigma_X = \sqrt{\left(2 - \dfrac{35}{12}\right)^2 \left(\dfrac{1}{3}\right) + \left(3 - \dfrac{35}{12}\right)^2 \left(\dfrac{5}{12}\right) + \left(4 - \dfrac{35}{12}\right)^2 \left(\dfrac{1}{4}\right)} = 0.759.$$

This can also be done on the TI-83/84 by putting the X values in L1 and the probabilities in L2. Then 1-Var Stats L1,L2 will give the above values for the mean and standard deviation.

2. (a) $P(A \text{ and } B) = P(A) \cdot P(B|A) = (0.6)(0.5) = 0.30$.
    (b) $P(A \text{ or } B) = P(A) + P(B) - P(A \text{ and } B) = 0.6 + 0.3 - 0.3 = 0.6$
        (Note that the 0.3 that is subtracted came from part (a).)
    (c) $P(B) = 0.3$, $P(B|A) = 0.5$. Since $P(B) \neq P(B|A)$, events A and B are not independent.

3. (a) Let X represent the score a student earns. We know that X has approximately $N(500, 90)$. What we are looking for is shown in the following graph.

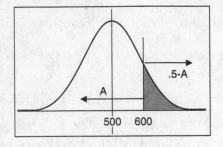

$$P(X > 600) = 1 - 0.8667 = 0.1333.$$

The calculator answer is `normalcdf(600,10,000,500.900)=0.1333` (remember that the upper bound must be "big"—in this exercise, 10,000 was used to get a sufficient number of standard deviations above 600).

(b) We already know, from part (a), that the area to the left of 600 is 0.8667. Similarly we determine the area to the left of 450 as follows:

$$Z_{450} = \frac{450 - 500}{90} = -0.56 \Rightarrow A = 0.2877.$$

Then

$$P(450 < X < 600) = 0.8667 - 0.2877 = 0.5790.$$

There are $0.5790(9000) \approx 5211$ scores.
[This could be done on the calculator as follows: `normalcdf(450,600, 500,90) = 0.5775`.]

(c) This situation could be pictured as follows.

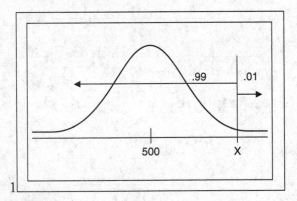

The z-score corresponding to an area of 0.99 is 2.33 (`invNorm(0.99)` on the calculator). So, $z_x = 2.33$. But, also,

$$z_x = \frac{x - 500}{90}.$$

Thus,

$$\frac{x - 500}{90} = 2.33.$$

Solving algebraically for x, we get $x = 709.7$. Rachel needs a score of 710 or higher.

Remember that this type of problem is usually solved by expressing z in two ways (using the definition and finding the area) and solving the equation formed by equating them. On the TI-83/84, the answer could be found as follows: `invNorm(0.99,500,90)=` `709.37`.

4. (a) $\mu_{3+6X} = 3 + 6\mu_X = 3 + 6(3) = 21$.
   (b) Because $\sigma^2_{a+bx} = b^2\sigma^2, \sigma^2_{3+6x} = 6^2\sigma^2_x = 36(0.25) = 9$.
   Thus,

$$\sigma_{3+6x} = \sqrt{\sigma^2_{3+6X}} = \sqrt{9} = 3.$$

5.

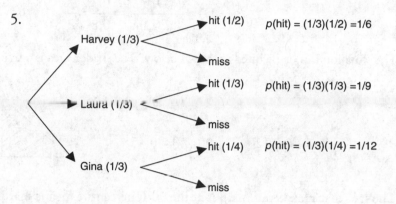

hit (1/2) $p(\text{hit}) = (1/3)(1/2) = 1/6$
Harvey (1/3)
miss

hit (1/3) $p(\text{hit}) = (1/3)(1/3) = 1/9$
Laura (1/3)
miss

hit (1/4) $p(\text{hit}) = (1/3)(1/4) = 1/12$
Gina (1/3)
miss

Because we are given that the target was hit, we only need to look at those outcomes. P(the person who hit the target was Laura|the target was hit)

$$= \frac{\frac{1}{9}}{\frac{1}{6} + \frac{1}{9} + \frac{1}{12}} = \frac{4}{13}.$$

6. $\mu_{X-Y} = \mu_X - \mu_Y = 3 - 5 = -2$.
   Since X and Y are independent, we have $\sigma_{X-Y} = \sqrt{\sigma^2_X + \sigma^2_Y} = \sqrt{1+1.3} = 1.52$. Note that the variances add even though we are subtracting one random variable from another.

7. I is not true. This is an approximation based on the *68-95-99.7 rule*. The actual value proportion within two standard deviations of the mean, to 4 decimal places, is 0.9545.
   II is true. This is a property of the normal curve.
   III is true. This is because the bell shape of the normal curve means that there is more area under the curve for a given interval length for intervals closer to the center.

8. (a) No, the events are not independent. The probability of B changes depending on what happens with A. Because there are 12 face cards, if the first card drawn is a face card, then $P(B) = 11/51$. If the first card is not a face card, then $P(B) = 12/51$. Because the probability of B is affected by the outcome of A, A and B are not independent.
   (b) $P(A) = 12/52 = 3/13$. $P(A|C) = 3/13$ (3 of the 13 diamonds are face cards). Because these are the same, the events "draw a face card on the first draw" and "the first card drawn is a diamond" are independent.

9. The area to the right of x is 0.6, so the area to the left is 0.4. From the table of Standard Normal Probabilities, $A = 0.4 \Rightarrow z_x = -0.25$. Also

$$z_x = \frac{x-700}{50}.$$

So,

$$z_x = \frac{x-700}{50} = -0.25 \Rightarrow x = 687.5.$$

60% of the area is to the right of 687.5. The calculator answer is given by `invNorm(0.4,700,50)=687.33`.

10. Let D = "a home has a desktop computer"; L = "a home has a laptop computer." We are given that $P(D) = 0.8$ and $P(D \text{ and } L) = 0.3$. Thus,

$$P(L \mid D) = \frac{P(D \cap L)}{P(D)} = \frac{0.3}{0.8} = \frac{3}{8}.$$

11. The situation can be pictured as shown below. The shaded area is a trapezoid whose area is

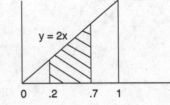

$$\frac{1}{2}(0.7-0.2)[2(0.2)+2(0.7)] = 0.45.$$

12. The fact that Harv's pumpkin is at the 90th percentile means that it is larger than 90% of the pumpkins in the contest. From the table of Standard Normal Probabilities, the area to the left of a term with a z-score of 1.28 is about 0.90. Thus,

$$z_x = 1.28 = \frac{x-125}{18} \Rightarrow x = 148.04.$$

So, Harv's pumpkin weighed about 148 pounds (for your information, the winning pumpkin at the Half Moon Bay Pumpkin Festival in 2008 weighed over 1528 pounds!). Seven of the 62 pumpkins weighed more than 1000 pounds!

13. Since $\sum P(X) = 1$, we have $P(X = 4) = 1 - P(X = 2) - P(X = 3) = 1 - 0.3 - 0.5 = 0.2$. Thus, filling in the table for X, we have

X	2	3	4
$P(X)$	0.3	0.5	**0.2**

Since X and Y are independent, $P(X = 4 \text{ and } Y = 3) = P(X = 4) \cdot P(X = 3)$. We are given that $P(X = 4 \text{ and } Y = 3) = 0.03$. Thus, $P(X = 4) \cdot P(Y = 3) = 0.03$. Since we now know that $P(X = 4) = 0.2$, we have $(0.2) \cdot P(Y = 3) = 0.03$, which gives us $P(Y = 3) = \frac{0.03}{0.2} = 0.15$.

Now, since $\sum P(Y) = 1$, we have $P(Y = 5) = 1 - P(Y = 3) - P(Y = 4) - P(Y = 6) = 1 - 0.15 - 0.1 - 0.4 = 0.35$.

14. Let A = "the number is divisible by 24" = {24, 48, 72, 96}. Let B = "the number is divisible by 36" = {36, 72}.

Note that $P(A \text{ and } B) = \dfrac{1}{100}$ (72 is the only number divisible by *both 24 and 36*).

$$P(\text{win a pizza}) = P(A \text{ or } B) = P(A) + P(B) - P(A \text{ and } B) = \frac{4}{100} + \frac{2}{100} - \frac{1}{100} = \frac{5}{100} = 0.05.$$

15. Because the numbers of men and women in the school are about equal (that is, $P(\text{women}) = 0.5$), let an even number represent a female and an odd number

represent a male. Begin on the first line of the table and consider groups of 12 digits. Count the even numbers among the 12. This will be the number of females among the group. Repeat five times. The relevant part of the table is shown below, with even numbers underlined and groups of 12 separated by two slanted bars (\\):

796<u>92</u> 517<u>0</u>7 73\\<u>274</u> 12<u>548</u> 91<u>4</u>9\\7 11135 <u>81218</u>

7\\9572 <u>06484</u> <u>874</u>\\<u>40</u> 41957 <u>21607</u>\\

In the five groups of 12 people, there were 3, 6, 3, 8, and 6 women. (*Note:* The result will, of course, vary if a different assignment of digits is made. For example, if you let the digits 0 – 4 represent a female and 5 – 9 represent a male, there would be 4, 7, 7, 5, and 7 women in the five groups.) So, in 40% of the trials there were 4 or fewer women in the class even though we would expect the average to be 6 (the average of these 5 trials is 5.2). Hence, it seems that getting only 4 women in a class when we expect 6 really isn't too unusual because it occurs 40% of the time. (It is shown in the next chapter that the theoretical probability of getting 4 or fewer women in a group of 12 people, assuming that men and women are equally likely, is about 0.19.)

16. Because $P(\text{girl}) = 0.6$, let the random digits 1, 2, 3, 4, 5, 6 represent the birth of a girl and 0, 7, 8, 9 represent the birth of a boy. Start on the second row of the random digit table and move across the line until you find the third digit that represents a girl. Note the number of digits needed to get three successes. Repeat 5 times and compute the average. The simulation is shown below (each success, i.e., girl, is underlined and separate trials are delineated by \\).

Start: <u>4195</u>\\7   21607   5\\<u>124</u>\\8   <u>54772</u>   \\<u>19481</u>\\   90392

It took 4, 7, 3, 6, and 5 children before they got their three girls. The average wait was 5. (The theoretical average is exactly 5—we got lucky this time!) As with Exercise 15, the result will vary with different assignment of random digits.

17. (a) $P(x \le 22) = P(x = 20) + P(x = 21) + P(x = 22) = 0.2 + 0.3 + 0.2 = 0.7$.
(b) $P(x > 21) = P(x = 22) + P(x = 23) + P(x = 24) = 0.2 + 0.1 + 0.2 = 0.5$.
(c) $P(21 \le x < 24) = P(x = 21) + P(x = 22) + P(x = 23) = 0.3 + 0.2 + 0.1 = 0.6$.
(d) $P(x \le 21 \text{ or } x > 23) = P(x = 20) + P(x = 21) + P(x = 24) = 0.2 + 0.3 + 0.2 = 0.7$.

18. (a) 18 of the 38 slots are winners, so $P(\text{win if bet on red}) = \dfrac{18}{38} = 0.474$.

(b) The probability distribution for this game is

Outcome	Win	Lose
X	1	-1
$P(X)$	$18/38$	$20/38$

$$E(X) = \mu_X = \left(\frac{18}{38}\right) + (-1)\left(\frac{20}{38}\right) = -0.052 \text{ or } -5.2¢.$$

The player will lose 5.2¢, on average, for each dollar bet.

19. From the tables, we see that $P(z < -2.5) = P(z > 2.5) = 0.0062$. So the probability that we are more than 2.5 standard deviations from the mean is $2(0.0062) = 0.0124$. (This can be found on the calculator as follows: 2 `normalcdf (2.5,1000)`.)

20. The situation can be pictured as follows:

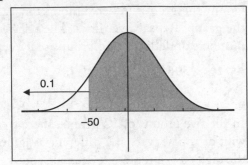

If 90% of the area is to the *right* of 50, then 10% of the area is to the left. So,

$$z_{50} = -1.28 = \frac{50 - \mu}{12} \rightarrow \mu = 65.36$$

Solutions to Cumulative Review Problems

1. II is true: the mean is most likely greater than the median. This is because the mean, being nonresistant, is pulled in the direction of outliers or skewness. Because the given histogram is clearly skewed to the right, the mean is likely to be to the right of (that is, greater than) the median.

2. The kind of sample you want is a *stratified random sample*. The sample should have 25 Asian students, 8 African American students, 12 Latino students, and 55 Caucasian students. You could get a list of all Asian students from the data clerk and randomly select 25 students from the list. Repeat this process for percentages of African American, Latino, and Caucasian students. Now the proportion of each ethnic group in your sample is the same as its proportion in the population.

3. The five-number summary is [20, 32, 42, 46, 50]. The box plot looks like this:

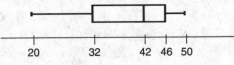

4. (a) The LSRL line is: $\overline{\text{\# Manatee deaths}} = -41.430 + 0.12486(\text{\#boats})$.
    (b) For each additional registered powerboat, the number of manatee deaths is predicted to increase by 0.12. You could also say that the number increases, on average, by 0.12.

5. For this set of data, $\bar{x} = 9.9$ and $s = 2.0$. Examination of the 25 points in the dataset yields the following.

	Percentage expected in each interval by the 68-95-99.7 rule	Number of terms from dataset in each interval
$9.9 \pm 1(2) = (7.9, 11.9)$	68% (17/25)	16/25 = 64%
$9.9 \pm 2(2) = (5.9, 13.9)$	95% (23.8/25)	25/25 = 100%
$9.9 \pm 3(2) = (3.9, 15.9)$	99.7% (24.9/25)	25/25 = 100%

The actual values in the dataset (16, 25, 25) are quite close to the expected values (17, 23.8, 24.9) if this truly were data from a normal population. Hence, it seems reasonable that the data could have been a random sample drawn from a population that is approximately normally distributed.

CHAPTER 9

Binomial Distributions, Geometric Distributions, and Sampling Distributions

IN THIS CHAPTER

Summary: In this chapter we finish laying the mathematical (probability) basis for inference by considering the binomial and geometric situations that occur often enough to warrant our study. In the last part of this chapter, we begin our study of inference by introducing the idea of a sampling distribution, one of the most important concepts in statistics. Once we've mastered this material, we will be ready to plunge into a study of formal inference (Chapters 10–13).

Key Ideas

✪ Binomial Distributions
✪ Normal Approximation to the Binomial
✪ Geometric Distributions
✪ Sampling Distributions
✪ Central Limit Theorem

Binomial Distributions

A **binomial experiment** has the following properties:

- The experiment consists of a fixed number, n, of identical trials.
- There are only two possible outcomes (that's the "bi" in "binomial"): success (S) or failure (F).

- The probability of success, p, is the same for each trial.
- The trials are independent (that is, knowledge of the outcomes of earlier trials does not affect the probability of success of the next trial).
- Our interest is in a **binomial random variable** X, which is the count of successes in n trials. The probability distribution of X is the **binomial distribution**.

(Taken together, the second, third, and fourth bullets above are called *Bernoulli trials*. One way to think of a binomial setting is as a fixed number n of Bernoulli trials in which our random variable of interest is the count of successes X in the n trials. You do not need to know the term Bernoulli trials for the AP exam.)

The short version of this is to say that a *binomial experiment* consists of n independent trials of an experiment that has two possible outcomes (success or failure), each trial having the same probability of success (p). The *binomial random variable X* is the count of successes.

In practice, we may consider a situation to be binomial when, in fact, the independence condition is not quite satisfied. This occurs when the probability of occurrence of a given trial is affected only slightly by prior trials. For example, suppose that the probability of a defect in a manufacturing process is 0.0005. That is, there is, on average, only 1 defect in 2000 items. Suppose we check a sample of 10,000 items for defects. When we check the first item, the proportion of defects remaining changes slightly for the remaining 9,999 items in the sample. We would expect 5 out of 10,000 (0.0005) to be defective. But if the first one we look at is *not* defective, the probability of the next one being defective has changed to 5/9999 or 0.0005005. It's a small change, but it means that the trials are not, strictly speaking, independent. A common rule of thumb is that we will consider a situation to be binomial if the population size is at least 10 times the sample size.

Symbolically, for the *binomial random variable X*, we say X has $B(n, p)$.

> **example:** Suppose Maria is a 65% free throw shooter. If we assume that repeated shots are independent, we could ask, "What is the probability that Maria makes exactly 7 of her next 10 free throws?" If X is the binomial random variable that gives us the count of successes for this experiment, then we say that X has $B(10, 0.65)$. Our question is then: $P(X = 7) = ?$

> We can think of $B(n, p, x)$ as a particular binomial probability. In this example, then, $B(10, 0.65, 7)$ is the probability that there are exactly 7 successes in 10 repetitions of a binomial experiment where $p = 0.65$. This is handy because it is the same syntax used by the TI-83/84 calculator (`binompdf(n,p,x)`) when doing binomial problems.

If X has $B(n, p)$, then X can take on the values 0, 1, 2, ..., n. Then,

$$B(n, p, x) = P(X = x) = \binom{n}{x} p^x (1-p)^{n-x}$$

gives the *binomial probability* of exactly x successes for a binomial random variable X that has $B(n, p)$.

Now,

$$\binom{n}{x} = \frac{n!}{x!(n-x)!}.$$

Some other notations include:

$$\binom{n}{x} = {}_nC_r = C(n,r),$$

Graphing and calculators will have a command for this.

example: Find $B(15, .3, 5)$. That is, find $P(X = 5)$ for a 15 trials of a binomial random variable X that succeeds with probability 0.3.

solution:

$$P(X = 5) = \binom{15}{5}(0.3)^5(1 - 0.3)^{15-5}$$

$$= \frac{15!}{5!10!}(0.3)^5(0.7)^{10} = .206.$$

Calculator Tip: The solution to the previous example can be found using the *binomial pdf* command on your calculator.

example: Consider once again our free-throw shooter (Maria) from the earlier example. Maria is a 65% free-throw shooter and each shot is independent. If X is the count of free throws made by Maria, then X has $B(10, 0.65)$ if she shoots 10 free throws. What is $P(X = 7)$?

solution:

$$P(X = 7) = \binom{10}{7}(0.65)^7(0.35)^3 = \frac{10!}{7!3!}(0.65)^7(0.65)^3$$

$$= 0.252.$$

Exam Tip: On binomial probability questions, the expectation is that you identify the distribution (by name or by formula) and parameters. That means, for example, it is enough to write the formula with numbers substituted and the answer. The formula identifies the procedure and parameters. You can also just write "Binomial, $n = 10$, $p = 0.65$. $P(x = 7) = 0.252$."

example: What is the probability that Maria makes *no more than 5* free throws? That is, what is $P(X \le 5)$?

solution:

$$P(X \le 5) = P(X = 0) + P(X = 1) + P(X = 2) + P(X = 3)$$

$$+ P(X = 4) + P(X = 5) = \binom{10}{0}(0.65)^0(0.35)^{10} + \binom{10}{1}(0.65)^1(0.35)^9$$

$$+ \cdots + \binom{10}{5}(0.65)^5(0.35)^5 = 0.249.$$

There is about a 25% chance that she will make 5 or fewer free throws. Graphing calculators have a command to answer such cumulative binomial questions.

example: What is the probability that Maria makes at least 6 free throws?

solution: $P(X \ge 6) = P(X = 6) + P(X = 7) + \cdots + P(X = 10)$
$= 1 - P(X \le 5) = 0.751.$

(Note that, in a multi-part question on the AP Exam, if you are using the same distribution and parameters as a previous part, you will not need to identify them again. But if the parameters are different, you will need to identify the distribution and parameters again.)

The **mean and standard deviation of a binomial random variable** X are given by $\mu_X = np; \sigma_X = \sqrt{np(1-p)}$. A binomial distribution for a given n and p (meaning you have all possible values of x along with their corresponding probabilities) is an example of a *probability distribution* as defined in Chapter 6. The mean and standard deviation of a binomial random variable X could be found by using the formulas from Chapter 8:

$$\mu_x = \sum x_i p_i \text{ and } \sigma_x = \sqrt{\sum (x - \mu_x)^2 p_i},$$

but clearly the formulas for the binomial are easier to use. Be careful that you don't try to use the formulas for the mean and standard deviation of a binomial random variable for a discrete random variable that is *not* binomial.

> **example:** Find the mean and standard deviation of a binomial random variable X that has $B(85, 0.6)$.
>
> **solution:** $\mu_X = (85)(0.6) = 51$; $\sigma_X = \sqrt{85(0.6)(0.4)} = 4.52$.

Normal Approximation to the Binomial

It should be noted that the normal approximation to the binomial is not, strictly speaking, an AP Exam topic. However, the normal approximation to the sampling distribution of the sample proportion is necessary and is based on this idea. Realize, though, that binomial probability problems do not need to be done using a normal approximation.

Under the proper conditions, the shape of a binomial distribution is approximately normal, and binomial probabilities can be estimated using normal probabilities. Generally, this is true when $np \geq 10$ and $n(1 - p) \geq 10$ (some books use $np \geq 5$ and $n(1 - p) \geq 5$; that's OK). These conditions are not satisfied in Graph A (X has $B(20, 0.1)$) below, but they are satisfied in Graph B (X has $B(20, 0.5)$).

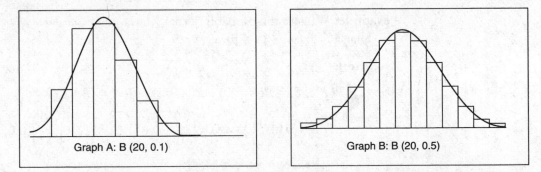

Graph A: B (20, 0.1)   Graph B: B (20, 0.5)

It should be clear that Graph A is noticeably skewed to the right, and Graph B is approximately normal in shape, so it is reasonable that a normal curve would approximate Graph B better than Graph A. The approximating normal curve clearly fits the binomial histogram better in Graph B than in Graph A.

When np and $n(1 - p)$ are sufficiently large (that is, they are both greater than or equal to 10 or 5), the binomial random variable X has approximately a normal distribution with

$$\mu = np \text{ and } \sigma = \sqrt{np(1-p)}.$$

Another way to say this is: If X has $B(n, p)$, then X has approximately $N(np, \sqrt{np(1-p)})$, provided that $np \geq 10$ and $n(1 - p) \geq 10$ (or $np \geq 5$ and $n(1 - p) \geq 5$).

example: Nationally, 15% of community college students live more than 6 miles from campus. Data from a simple random sample of 400 students at one community college are analyzed.

(a) What are the mean and standard deviation for the number of students in the sample who live more than 6 miles from campus?

(b) Use a normal approximation to calculate the probability that at least 65 of the students in the sample live more than 6 miles from campus.

solution: If X is the number of students who live more than 6 miles from campus, then X has $B(400, 0.15)$.

(a) $\mu = 400(0.15) = 60$; $\sigma = \sqrt{400(0.15)(0.85)} = 7.14$.

(b) Because $400(0.15) = 60$ and $400(0.85) = 340$, we can use the normal approximation to the binomial with mean 60 and standard deviation 7.14. The situation is pictured below:

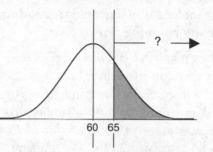

Using Table A, we have $P(X \geq 65) = P\left(z \geq \dfrac{65-60}{7.14} = 0.70\right) = 1 - 0.7580 = 0.242$.

By calculator, this can be found as `normalcdf(65,1000,60,7.14) = 0.242`.

The exact binomial solution to this problem is given by

$$\text{1-binomcdf}(400,0.15,64) = 0.261 \text{ (you use } x = 64 \text{ since } P(X \geq 65)$$
$$= 1 - P(X \leq 64)).$$

In reality, you will need to use a normal approximation to the binomial only in limited circumstances. In the example above, the answer can be arrived at quite easily using the exact binomial capabilities of your calculator. The only time you might want to use a normal approximation is if the size of the binomial exceeds the capacity of your calculator (for example, enter `binomcdf(50000000,0.7,3250000)`. You'll most likely see `ERR:DOMAIN`, which means you have exceeded the capacity of your calculator, and you didn't have access to a computer. The real concept that you need to understand the normal approximation to a binomial is that another way of looking at binomial data is in terms of the *proportion* of successes rather than the count of successes. We will approximate a distribution of sample proportions with a normal distribution and the concepts and conditions for it are the same.

Geometric Distributions

In the Binomial Distributions section of this chapter, we defined a binomial setting as an experiment in which the following conditions are present:

- The experiment consists of a fixed number, n, of identical trials.
- There are only two possible outcomes: success (S) or failure (F).
- The probability of success, p, is the same for each trial.
- The trials are independent (that is, knowledge of the outcomes of earlier trials does not affect the probability of success of the next trial).
- Our interest is in a **binomial random variable** X, which is the count of successes in n trials. The probability distribution of X is the **binomial distribution**.

There are times we are interested not in the count of successes out of n fixed trials, but in the probability that the first success occurs on a given trial, or in the average number of trials until the first success. A **geometric setting** is defined as follows.

- There are only two possible outcomes: success (S) or failure (F).
- The probability of success, p, is the same for each trial.
- The trials are independent (that is, knowledge of the outcomes of earlier trials does not affect the probability of success of the next trial).
- Our interest is in a **geometric random variable** X, which is the number of trials necessary to obtain the *first* success.

Note that if X is a *binomial*, then X can take on the values 0, 1, 2, … , n. If X is *geometric*, then it takes on the values 1, 2, 3, …. There can be zero successes in a binomial, but the earliest a first success can come in a geometric setting is on the first trial.

If X is geometric, the probability that the first success occurs on the *nth trial* is given by $P(X = n) = p(1 - p)^{n-1}$. The value of $P(X = n)$ in a geometric setting can be found on the TI-83/84 calculator, in the `DISTR` menu, as `geometpdf(p,n)` (note that the order of p and n are, for reasons known only to the good folks at TI, reversed from the binomial). Given the relative simplicity of the formula for $P(X = n)$ for a geometric setting, it's probably just as easy to calculate the expression directly. There is also a `geometcdf` function that behaves analogously to the `binomcdf` function, but is not much needed in this course.

> **example:** Remember Maria, the basketball player whose free-throw shooting percentage was 0.65? What is the probability that the first free throw she manages to hit is on her fourth attempt?

> **solution:** $P(X = 4) = (0.65)(1 - 0.65)^{4-1} = (0.65)(0.35)^3 = 0.028$. This can be done on the TI-83/84 as follows: `geometpdf(p,n) = geometpdf(0.65,4) = 0.028`.

> **example:** In a standard deck of 52 cards, there are 12 face cards. So the probability of drawing a face card from a full deck is 12/52 = 0.231.

> (a) If you draw cards with replacement (that is, you replace the card in the deck before drawing the next card), what is the probability that the first face card you draw is the 10th card?

> (b) If you draw cards without replacement, what is the probability that the first face card you draw is the 10th card?

> **solution:**

> (a) $P(X = 10) = (0.231)(1 - 0.231)^9 = 0.022$. On the TI-83/84: `geometpdf(0.231,10)=0.0217`.

(b) If you don't replace the card each time, the probability of drawing a face card on each trial is different because the proportion of face cards in the deck changes each time a card is removed. Hence, this is not a geometric setting and cannot be answered by the techniques of this section. It *can* be answered, but not easily, by the techniques of the previous chapter.

Rather than the probability that the first success occurs on a specified trial, we may be interested in the average wait until the first success. The average wait until the first success of a geometric random variable is $1/p$. (This can be derived by summing $(1) \cdot P(X = 1) + (2) \cdot P(X = 2) + (3) \cdot P(X = 3) + \ldots = 1p + 2p(1 - p) + 3p(1 - p)^2 + \ldots$, which can be done using algebraic techniques for summing an infinite series with a common ratio less than 1.)

> **example:** On average, how many free throws will Maria have to take before she makes one (remember, $p = 0.65$)?

> **solution:** $1/p = \dfrac{1}{0.65} = 1.54$.

Since, in a geometric distribution $P(X = n) = p(1 - p)^{n-1}$, the probabilities become less likely as n increases since we are multiplying by $1 - p$, a number less than one. The geometric distribution has a step-ladder graph that looks like this:

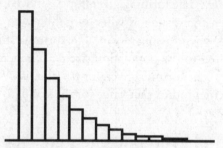

Sampling Distributions

Suppose we drew a sample of size 10 from an approximately normal population with unknown mean and standard deviation and got $\bar{x} = 18.87$. Two questions arise: (1) what does this sample tell us about the population from which the sample was drawn, and (2) what would happen if we drew more samples?

Suppose we drew 5 more samples of size 10 from this population and got $\bar{x} = 20.35$, $\bar{x} = 20.04$, $\bar{x} = 19.20$, $\bar{x} = 19.02$, and $\bar{x} = 20.35$. In answer to question (1), we might believe that the population from which these samples was drawn had a mean around 20 because these averages tend to group there (in fact, the six samples were drawn from a normal population whose mean is 20 and whose standard deviation is 4). The mean of the 6 samples is 19.64, which supports our feeling that the mean of the original population might have been 20.

The standard deviation of the 6 samples is 0.68, and you might not have any intuitive sense about how that relates to the population standard deviation, although you might suspect that the standard deviation of the samples should be less than the standard deviation of the population because the chance of an extreme value for an average should be less than that for an individual term (it just doesn't seem very likely that we would draw a *lot* of extreme values in a single sample).

Suppose we continued to draw samples of size 10 from this population until we were exhausted or until we had drawn *all possible samples of size 10*. If we did succeed in drawing all possible samples of size 10, and computed the mean of each sample, the distribution of these sample means would be the **sampling distribution of $\bar{x}$**.

Remembering that a "statistic" is a value that describes a sample, the **sampling distribution of a statistic** is the distribution of that statistic for all possible samples of a given size. It's important to understand that a dotplot of a few samples drawn from a population is not a distribution (it's a *simulation* of a distribution)—it becomes a distribution only when all possible samples of a given size are drawn.

Sampling Distribution of a Sample Mean

Suppose we have the sampling distribution of $\bar{x}$. That is, we have formed a distribution of the means of all possible samples of size n from an unknown population (thus, we know little about its shape, center, or spread). Let $\mu_{\bar{x}}$ and $\sigma_{\bar{x}}$ represent the mean and standard deviation of the sampling distribution of $\bar{x}$, respectively.

Then

$$\mu_{\bar{x}} = \mu \text{ and } \sigma_{\bar{x}} = \frac{\sigma}{\sqrt{n}}$$

for any population with mean μ and standard deviation σ.

(*Note:* The formula given for $\sigma_{\bar{x}}$ above is correct only if the sample has true independence between trials, such as when sampling with replacement, or when the population is infinite, such as when tossing a coin. The formula still works well enough if the sample size n is small compared to the population size N. A general rule is that n should be no more than 10% of N to use the formula given for $\sigma_{\bar{x}}$ (that is, $N > 10n$). This is because, when sampling without replacement, the exact value for the standard deviation of the sampling distribution is actually

$$\sigma_{\bar{x}} = \frac{\sigma}{\sqrt{n}} \sqrt{\frac{N-n}{N-1}}.$$

In practice this usually isn't a major issue because this **Finite Population Correction Factor**

$$\sqrt{\frac{N-n}{N-1}}$$

is close to 1 whenever N is large in comparison to n. (You don't have to know this formula for the AP exam. However, you do need to understand that when our sample size is larger than 10% of the population, an adjustment must be made to the standard deviation formula.)

> **example:** A large population is known to have a mean of 23 and a standard deviation of 2.5. What are the mean and standard deviation of the sampling distribution of means of samples of size 20 drawn from this population?

solution:

$$\mu_{\bar{x}} = \mu = 23, \ \sigma_{\bar{x}} = \frac{\sigma}{\sqrt{n}} = \frac{2.5}{\sqrt{20}} = 0.559.$$

Central Limit Theorem

The discussion above gives us measures of center and spread for the sampling distribution of $\bar{x}$ but tells us nothing about the *shape* of the sampling distribution. It turns out that the shape of the sampling distribution is determined by (a) the shape of the original population and (b) n, the sample size. If the original population is approximately normal, then it's easy: the shape of the sampling distribution will be approximately normal if the population is approximately normal.

If the shape of the original population is not normal, or unknown, and the sample size is small, then the shape of the sampling distribution will be similar to that of the original population. For example, if a population is skewed to the right, we would expect the sampling distribution of the mean for small samples also to be somewhat skewed to the right, although not as much as the original population.

When the sample size is large, we have the following result, known as the **central limit theorem**: For large n, the sampling distribution of $\bar{x}$ will be approximately normal. The larger is n, the more normal will be the shape of the sampling distribution.

A rough rule of thumb for using the central limit theorem is that n should be at least 30, although the sampling distribution may be approximately normal for much smaller values of n if the population doesn't depart much from normal. The central limit theorem allows us to use normal calculations to do problems involving sampling distributions without having to have knowledge of the original population. Note that calculations involving z-procedures require that you know the value of σ, the population standard deviation. Since you will rarely know σ, the large sample size essentially says that the sampling distribution is *approximately*, but not exactly, normal. That is, technically you should not be using z-procedures unless you know σ but, as a practical matter, z-procedures are numerically close to correct for large n. Given that the population size (N) is large in relation to the sample size (n), the information presented in this section can be summarized in the following table:

	POPULATION	SAMPLING DISTRIBUTION
Mean	μ	$\mu_{\bar{x}} = \mu$
Standard Deviation	σ	$\sigma_{\bar{x}} = \dfrac{\sigma}{\sqrt{n}}$
Shape	Normal	Normal
	Undetermined (skewed, etc.)	If n is "small" shape is similar to shape of the population *OR* As n gets "larger," shape becomes more approximately normal (central limit theorem). If $n \geq 30$, the shape of the sampling distribution will generally be approximately normal.

example: Describe the sampling distribution of $\bar{x}$ for samples of size 15 drawn from a normal population with mean 65 and standard deviation 9.

solution: Because the original population is normal, $\bar{x}$ is normal with mean 65 and standard deviation $\dfrac{9}{\sqrt{15}} = 2.32$. That is, $\bar{x}$ has $N\left(65, \dfrac{9}{\sqrt{15}}\right)$.

example: Describe the sampling distribution of $\bar{x}$ for samples of size 15 drawn from a population that is strongly skewed to the left (like the scores on a very easy test) with mean 65 and standard deviation 9.

solution: $\mu_{\bar{x}} = 65$ and $\sigma_{\bar{x}} = 2.32$ as in the above example. However this time the population is skewed to the left. The sample size is reasonably large, but not large enough to argue, based on our rule of thumb ($n \geq 30$), that the sampling distribution is normal. The best we can say is that the sampling distribution is probably more mound-shaped than the original but might still be somewhat skewed to the left.

example: The average adult has completed an average of 11.25 years of education with a standard deviation of 1.75 years. A random sample of 90 adults is obtained. What is the probability that the sample will have a mean

(a) greater than 11.5 years?

(b) between 11 and 11.5 years?

solution: The sampling distribution of $\bar{x}$ has $\mu_{\bar{x}} = 11.25$ and

$$\sigma_{\bar{x}} = \frac{1.75}{\sqrt{90}} = 0.184.$$

Because the sample size is large ($n = 90$), the central limit theorem tells us that large sample techniques are appropriate. Accordingly,

(a) The graph of the sampling distribution is shown below:

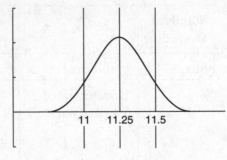

$$P(\bar{x} > 11.5) = P\left(z > \frac{11.5 - 11.25}{1.75/\sqrt{90}} = \frac{0.25}{0.184} = 1.36 \right) = 0.0869.$$

(b) From part (a), the area to the left of 11.5 is $1 - 0.0869 = 0.9131$. Since the sampling distribution is approximately normal, it is symmetric. Since 11 is the same distance to the left of the mean as 11.5 is to the right, we know that $P(\bar{x} < 11) = P(\bar{x} > 11.5) = 0.0869$. Hence, $P(11 < \bar{x} < 11.5) = 0.9131 - 0.0869 = 0.8262$. The calculator solution is `normalcdf(11,11.5,11, 0.184)=0.8258`.

example: Over the years, the scores on the final exam for AP Calculus have been normally distributed with a mean of 82 and a standard deviation of 6. The instructor thought that this year's class was quite dull and, in fact, they only averaged 79 on their final. Assuming that this class is a random sample of

32 students from AP Calculus, what is the probability that the average score on the final for this class is no more than 79? Do you think the instructor was right?

solution:

$$P(\bar{x} \leq 79) = P\left(z \leq \frac{79-82}{6 / \sqrt{32}} = \frac{-3}{1.06} = -2.83\right) = 0.0023.$$

If this group really were typical, there is less than a 1% probability of getting an average this low by chance alone. That seems unlikely, so we have good evidence that the instructor was correct.

(The calculator solution for this problem is `normalcdf(-1000,79, 82,1.06)`.)

Sampling Distributions of a Sample Proportion

If X is the count of successes in a sample of n trials of a binomial random variable, then the **proportion of success** is given by $\hat{p} = X/n$. $\hat{p}$ is what we use for the sample proportion (a statistic). The true population proportion would then be given by p.

> **Digression:** Before introducing $\hat{p}$, we have used $\bar{x}$ and s as statistics, and μ and σ as parameters. Often we represent statistics with English letters and parameters with Greek letters. However, we depart from that convention here by using $\hat{p}$ as a statistic and p as a parameter. There are texts that are true to the English/Greek convention by using p for the sample proportion and Π for the population proportion.

We learned in the first section of this chapter that, if X is a binomial random variable, the mean and standard deviation of the sampling distribution of X are given by

$$\mu_X = np, \ \sigma_X = \sqrt{np(1-p)}.$$

We know that if we divide each term in a dataset by the same value n, then the mean and standard deviation of the transformed dataset will be the mean and standard deviation of the original dataset divided by n. Doing the algebra, we find that the mean and standard deviation of the sampling distribution of $\hat{p}$ are given by

$$\mu_{\hat{p}} = p, \ \sigma_{\hat{p}} = \sqrt{\frac{p(1-p)}{n}}.$$

Like the binomial, the sampling distribution of $\hat{p}$ will be approximately normally distributed if n and p are large enough. The test is exactly the same as it was for the binomial: If X has $B(n, p)$, and $\hat{p} = X/n$, then $\hat{p}$ has approximately

$$N\left(p, \sqrt{\frac{p(1-p)}{n}}\right),$$

provided that $np \geq 10$ and $n(1-p) \geq 10$ (or $np \geq 5$ and $n(1-p) \geq 5$).

example: Harold fails to study for his statistics final. The final has 100 multiple-choice questions, each with 5 choices. Harold has no choice but to guess randomly at all 100 questions. What is the probability that Harold will get at least 30% on the test?

solution: Since 100(0.2) and 100(0.8) are both greater than 10, we can use the normal approximation to the sampling distribution of $\hat{p}$. Since

$$p = 0.2, \; \mu_{\hat{p}} = 0.2 \text{ and } \sigma_{\hat{p}} = \sqrt{\frac{0.2(1-0.2)}{100}} = 0.04.$$

Therefore,

$$P(\hat{p} \geq 0.3) = P\left(z \geq \frac{0.3 - 0.2}{0.04} = 2.5\right) = 0.0062. \text{ The TI-83/84 solution is given by}$$

`normalcdf(0.3,100,0.2,0.040)=0.0062`.

Harold should have studied.

› Rapid Review

1. A coin is known to be unbalanced in such a way that heads only comes up 0.4 of the time.

   (a) What is the probability the first head appears on the 4th toss?
   (b) How many tosses would it take, on average, to flip two heads?

   Answer:

   (a) P(first head appears on fourth toss) $= 0.4 \, (1 - 0.4)^{4-1} = 0.4(0.6)^3 = 0.0864$.

   (b) Average wait to flip two heads $= 2$(average wait to flip one head) $= 2\left(\dfrac{1}{0.4}\right) = 5$.

2. The coin of problem #1 is flipped 50 times. Let X be the number of heads. What is

   (a) the probability of *exactly* 20 heads?
   (b) the probability of *at least* 20 heads?

   Answer:

   (a) $P(X=20) = \dbinom{50}{20}(0.4)^{20}(0.6)^{30} = 0.115$ [on the TI-83/84: `binompdf(50, 0.4,20)`].

   (b) $P(X \geq 20) = \dbinom{50}{20}(0.4)^{20}(0.6)^{30} + \dbinom{50}{21}(0.4)^{21}(0.6)^{29} + \ldots + \dbinom{50}{50}(0.4)^{50}(0.6)^0$

   $= $ `1-binomcdf(50,0.4,19)=0.554`.

3. A binomial random variable X has $B(300, 0.2)$. Describe the sampling distribution of $\hat{p}$.

   Answer: Since $300(0.2) = 60 \geq 10$ and $300(0.8) = 240 \geq 10$, $\hat{p}$ has approximately a normal distribution with $\mu_{\hat{p}} = 0.2$ and $\sigma_{\hat{p}} = \sqrt{\dfrac{0.2(1-0.2)}{300}} = 0.023$.

4. A distribution is known to be highly skewed to the left with mean 25 and standard deviation 4. Samples of size 10 are drawn from this population, and the mean of each sample is calculated. Describe the sampling distribution of $\bar{x}$.

Answer: $\mu_{\bar{x}} = 25$, $\sigma_{\bar{x}} = \dfrac{4}{\sqrt{10}} = 1.26$.

Since the samples are small, the shape of the sampling distribution would probably show some left-skewness but would be more mound-shaped than the original population.

5. What is the probability that a sample of size 35 drawn from a population with mean 65 and standard deviation 6 will have a mean less than 64?

Answer: The sample size is large enough that we can use large-sample procedures. Hence,

$$P(\bar{x} < 64) = P\left(z < \frac{64 - 65}{6 / \sqrt{35}} = -0.99 \right) = 0.1611.$$

On the TI-83/84, the solution is given by `normalcdf(-100,64, 65,` $6/\sqrt{35}$ `)`.

Practice Problems

Multiple-Choice

1. A binomial event has $n = 60$ trials. The probability of success on each trial is 0.4. Let X be the count of successes of the event during the 60 trials. What are μ_x and σ_x?

   (a) 24, 3.79
   (b) 24, 14.4
   (c) 4.90, 3.79
   (d) 4.90, 14.4
   (e) 2.4, 3.79

2. Consider repeated trials of a binomial random variable. Suppose the probability of the first success occurring on the second trial is 0.25. What is the probability of success on the first trial?

   (a) $1/4$
   (b) 1
   (c) $1/2$
   (d) $1/8$
   (e) $3/16$

3. To use a normal approximation to the binomial, which of the following does *not* have to be true?

   (a) $np \geq 10$, $n(1 - p) \geq 10$ (or: $np \geq 5$, $n(1 - p) \geq 5$).
   (b) The individual trials must be independent.
   (c) The sample size in the problem must be too large to permit doing the problem on a calculator.
   (d) For the binomial, the population size must be at least 10 times as large as the sample size.
   (e) All of the above are true.

4. You form a distribution of the means of all samples of size 9 drawn from an infinite population that is skewed to the left (like the scores on an easy Stats quiz!). The population from which the samples are drawn has a mean of 50 and a standard deviation of 12. Which one of the following statements is true of this distribution?

   (a) $\mu_{\bar{x}} = 50$, $\sigma_{\bar{x}} = 12$, the sampling distribution is skewed somewhat to the left.
   (b) $\mu_{\bar{x}} = 50$, $\sigma_{\bar{x}} = 4$, the sampling distribution is skewed somewhat to the left.
   (c) $\mu_{\bar{x}} = 50$, $\sigma_{\bar{x}} = 12$, the sampling distribution is approximately normal.
   (d) $\mu_{\bar{x}} = 50$, $\sigma_{\bar{x}} = 4$, the sampling distribution is approximately normal.
   (e) $\mu_{\bar{x}} = 50$, $\sigma_{\bar{x}} = 4$, the sample size is too small to make any statements about the shape of the sampling distribution.

5. A 12-sided die has faces numbered from 1–12. Assuming the die is fair (that is, each face is equally likely to appear each time), which of the following would give the exact probability of getting at least 10 3s out of 50 rolls?

   (a) $\binom{50}{0}(0.083)^0(0.917)^{50} + \binom{50}{1}(0.083)^1(0.917)^{49} + \ldots + \binom{50}{9}(0.083)^9(0.917)^{41}$.

   (b) $\binom{50}{11}(0.083)^{11}(0.917)^{39} + \binom{50}{12}(0.083)^{12}(0.917)^{38} + \ldots + \binom{50}{50}(0.083)^{50}(0.917)^0$.

   (c) $1 - \left[\binom{50}{0}(0.083)^0(0.917)^{50} + \binom{50}{1}(0.083)^1(0.917)^{49} + \ldots + \binom{50}{10}(0.083)^{10}(0.917)^{40}\right]$.

   (d) $1 - \left[\binom{50}{0}(0.083)^0(0.917)^{50} + \binom{50}{1}(0.083)^1(0.917)^{49} + \ldots + \binom{50}{9}(0.083)^9(0.917)^{41}\right]$.

   (e) $\binom{50}{0}(0.083)^0(0.917)^{50} + \binom{50}{1}(0.083)^1(0.917)^{49} + \ldots + \binom{50}{10}(0.083)^{10}(0.917)^{40}$.

6. In a large population, 55% of the people get a physical examination at least once every two years. An SRS of 100 people are interviewed and the sample proportion is computed. The mean and standard deviation of the sampling distribution of the sample proportion are

   (a) 55, 4.97
   (b) 0.55, 0.002
   (c) 55, 2
   (d) 0.55, 0.0497
   (e) The standard deviation cannot be determined from the given information.

7. Which of the following best describes the sampling distribution of a sample mean?

   (a) It is the distribution of all possible sample means of a given size.
   (b) It is the particular distribution in which $\mu_{\bar{x}} = \mu$ and $\sigma_{\bar{x}} = \sigma$.
   (c) It is a graphical representation of the means of all possible samples.
   (d) It is the distribution of all possible sample means from a given population.
   (e) It is the probability distribution for each possible sample size.

8. Which of the following is not a common characteristic of binomial and geometric experiments?

    (a) There are exactly two possible outcomes: success or failure.
    (b) There is a random variable X that counts the number of successes.
    (c) Each trial is independent (knowledge about what has happened on previous trials gives you no information about the current trial).
    (d) The probability of success stays the same from trial to trial.
    (e) $P(\text{success}) + P(\text{failure}) = 1$.

9. A school survey of students concerning which band to hire for the next school dance shows 70% of students in favor of hiring The Greasy Slugs. What is the approximate probability that, in a random sample of 200 students, at least 150 will favor hiring The Greasy Slugs?

    (a) $\binom{200}{150}(0.7)^{150}(0.3)^{50}$.

    (b) $\binom{200}{150}(0.3)^{150}(0.7)^{50}$.

    (c) $P\left(z > \dfrac{150-140}{\sqrt{200(0.7)(0.3)}}\right)$.

    (d) $P\left(z > \dfrac{150-140}{\sqrt{150(0.7)(0.3)}}\right)$.

    (e) $P\left(z > \dfrac{140-150}{\sqrt{200(0.7)(0.3)}}\right)$.

Free-Response

1. A factory manufacturing tennis balls determines that the probability that a single can of three balls will contain at least one defective ball is 0.025. What is the probability that a case of 48 cans will contain at least two cans with a defective ball?

2. A population is highly skewed to the left. Describe the shape of the sampling distribution of $\bar{x}$ drawn from this population if the sample size is (a) 3 or (b) 30.

3. Suppose you had lots of time on your hands and decided to flip a fair coin 1,000,000 times and note whether each flip was a head or a tail. Let X be the count of heads. What is the probability that there are at least 1000 more heads than tails? (*Note*: This is a binomial, but your calculator may not be able to do the binomial computation because the numbers are too large for it.)

4. In Chapter 8, we had an example in which we asked if it would change the proportion of girls in the population (assumed to be 0.5) if families continued to have children until they had a girl and then they stopped. That problem was to be done by simulation. How could you use what you know about the geometric distribution to answer this same question?

5. At a school better known for football than academics (a school its football team can be proud of), it is known that only 20% of the scholarship athletes graduate within 5 years. The school is able to give 55 scholarships for football. What are the expected mean and standard deviation of the number of graduates for a group of 55 scholarship athletes?

6. Consider a population consisting of the numbers 2, 4, 5, and 7. List all possible samples of size two from this population and compute the mean and standard deviation of the sampling distribution of $\bar{x}$. Compare this with the values obtained by relevant formulas for the sampling distribution of $\bar{x}$. Note that the sample size is large relative to the population—this may affect how you compute $\sigma_{\bar{x}}$ by formula.

7. Approximately 10% of the population of the United States is known to have blood type B. What is the probability that between 11% and 15%, inclusive, of a random sample of 500 adults will have type B blood?

8. Which of the following is (are) true of the central limit theorem? (More than one answer might be true.)

   I. $\mu_{\bar{x}} = \mu$.

   II. $\sigma_{\bar{x}} = \dfrac{\sigma}{\sqrt{n}}$ (if $N \geq 10n$).

   III. The sampling distribution of a sample mean will be approximately normally distributed for sufficiently large samples, regardless of the shape of the original population.

   IV. The sampling distribution of a sample mean will be normally distributed if the population from which the samples are drawn is brakes.

9. A brake inspection station reports that 15% of all cars tested have brakes in need of replacement pads. For a sample of 20 cars that come to the inspection station,

   (a) what is the probability that exactly 3 have defective brakes?
   (b) what is the mean and standard deviation of the number of cars that need replacement pads?

10. A tire manufacturer claims that his tires will last 40,000 miles with a standard deviation of 5000 miles.

    (a) Assuming that the claim is true, describe the sampling distribution of the mean lifetime of a random sample of 160 tires. Remember that "describe" means discuss center, spread, and shape.
    (b) What is the probability that the mean life time of the sample of 160 tires will be less than 39,000 miles? Interpret the probability in terms of the truth of the manufacturer's claim.

11. The probability of winning a bet on red in roulette is 0.474. The binomial probability of winning money if you play 10 games is 0.31, and drops to 0.27 if you play 100 games. Use a normal approximation to the binomial to estimate your probability of coming out ahead (that is, winning more than $^1/_2$ of your bets) if you play 1000 times. Justify being able to use a normal approximation for this situation.

12. Crabs off the coast of Northern California have a mean weight of 2 lb with a standard deviation of 5 oz. A large trap captures 35 crabs.

    (a) Describe the sampling distribution for the average weight of a random sample of 35 crabs taken from this population.
    (b) What would the mean weight of a sample of 35 crabs have to be in order to be in the top 10% of all such samples?

13. The probability that a person recovers from a particular type of cancer operation is 0.7. Suppose 8 people have the operation. What is the probability that

    (a) exactly 5 recover?
    (b) they all recover?
    (c) at least one of them recovers?

14. A certain type of lightbulb is advertised to have an average life of 1200 hours. If, in fact, lightbulbs of this type only average 1185 hours with a standard deviation of 80 hours, what is the probability that a sample of 100 bulbs will have an average life of at least 1200 hours?

15. Your task is to explain to your friend Gretchen, who knows virtually nothing (and cares even less) about statistics, just what the sampling distribution of the mean is. Explain the idea of a sampling distribution in such a way that even Gretchen, if she pays attention, will understand.

16. Consider the distribution shown at the right. Describe the shape of the sampling distribution of $\bar{x}$ for samples of size n if

    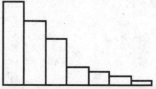

    (a) $n = 3$.
    (b) $n = 40$.

17. After the *Challenger* disaster of 1986, it was discovered that the explosion was caused by defective O-rings. The probability that a single O-ring was defective and would fail (with catastrophic consequences) was 0.003, and there were 12 of them (6 outer and 6 inner). What was the probability that at least one of the O-rings would fail (as it actually did)?

18. Your favorite cereal has a little prize in each box. There are 5 such prizes. Each box is equally likely to contain any one of the prizes. So far, you have been able to collect 2 of the prizes. What is:

    (a) the probability that you will get the third different prize on the next box you buy?
    (b) the probability that it will take three more boxes to get the next prize?
    (c) the average number of boxes you will have to buy before getting the third prize?

19. We wish to approximate the binomial distribution $B(40, 0.8)$ with a normal curve $N(\mu, \sigma)$. Is this an appropriate approximation and, if so, what are μ and σ for approximating the normal curve?

20. Opinion polls in 2002 showed that about 70% of the population had a favorable opinion of President Bush. That same year, a simple random sample of 600 adults living in the San Francisco Bay Area showed only 65% had a favorable opinion of President Bush. What is the probability of getting a rating of 65% or less in a random sample of this size if the true proportion in the population was 0.70?

Cumulative Review Problems

1. An unbalanced coin has $p = 0.6$ of turning up heads. Toss the coin three times and let X be the count of heads among the three coins. Construct the probability distribution for this experiment.

2. You are doing a survey for your school newspaper and want to select a sample of 25 seniors. You decide to do this by randomly selecting 5 students from each of the 5 senior-level classes, each of which contains 28 students. The school data clerk assures you that students have been randomly assigned, by computer, to each of the 5 classes. Is this sample

    (a) a random sample?
    (b) a simple random sample?

3. Data are collected in an experiment to measure a person's reaction time (in seconds) as a function of the number of milligrams of a new drug. The least squares regression line (LSRL) for the data is $\overline{Reaction\ Time} = 0.2 + 0.8(mg)$. Interpret the slope of the regression line in the context of the situation.

4. If $P(A) = 0.5$, $P(B) = 0.3$, and $P(A \text{ or } B) = 0.65$, are events A and B independent?

5. Which of the following is (are) examples of *quantitative data* and which is (are) examples of *categorical data*?

   (a) The height of an individual, measured in inches.
   (b) The color of the shirts in my closet.
   (c) The outcome of a flip of a coin described as "heads" or "tails."
   (d) The value of the change in your pocket.
   (e) Individuals, after they are weighed, are identified as thin, normal, or heavy.
   (f) Your pulse rate.
   (g) Your religion.

Solutions to Practice Problems

Multiple-Choice

1. The correct answer is (a).

$$\mu_X = (60)(0.4) = 24,\ \sigma_X = \sqrt{60(0.4)(0.6)} = \sqrt{14.4} = 3.79.$$

2. The correct answer is (c). If it is a binomial random variable, the probability of success, p, is the same on each trial. The probability of not succeeding on the first trial and then succeeding on the second trial is $(1 - p)(p)$. Thus, $(1 - p)p = 0.25$. Solving algebraically, $p = \frac{1}{2}$.

3. The correct answer is (c). Although you probably wouldn't need to use a normal approximation to the binomial for small sample sizes, there is no reason (except perhaps accuracy) that you couldn't.

4. The answer is (b).

$$\mu_{\bar{x}} = \mu,\ \sigma_{\bar{x}} = \frac{\sigma}{\sqrt{n}}.$$

For small samples, the shape of the sampling distribution of $\bar{x}$ will resemble the shape of the sampling distribution of the original population. The shape of the sampling distribution of $\bar{x}$ is approximately normal for n sufficiently large.

5. The correct answer is (d). Because the problem stated "at least 10," we must include the term where $x = 10$. If the problem had said "more than 10," the correct answer would have been (b) or (c) (they are equivalent). The answer could also have been given as

$$\binom{50}{10}(0.083)^{10}(0.917)^{40} + \binom{50}{11}(0.083)^{11}(0.917)^{39} + \ldots + \binom{50}{50}(0.083)^{50}(0.917)^{0}.$$

6. The correct answer is (d). $\mu_{\hat{p}} = p = 0.55,\ \sigma_{\hat{p}} = \sqrt{\frac{(0.55)(0.45)}{100}} = 0.0497.$

7. The correct answer is (a).

8. The correct answer is (b). This is a characteristic of a binomial experiment. The analogous characteristic for a geometric experiment is that there is a random variable X that is the number of trials needed to achieve the first success.

9. The correct answer is (c). This is actually a binomial situation. If X is the count of students "in favor," then X has $B(200, 0.70)$. Thus, $P(X \geq 150) = P(X = 150) + P(X = 151) + \ldots + P(X = 200)$. Using the TI-83/84, the exact binomial answer equals `1-binomcdf(200,0.7.0,149)=0.0695`. None of the listed choices shows a sum of several binomial expressions, so we assume this is to be done as a normal approximation.

   We note that $B(200, 0.7)$ can be approximated by $N(200(0.7), \sqrt{200(0.7)(0.3)}) = N(140, 6.4807)$. A normal approximation is OK since $200(0.7)$ and $200(0.3)$ are both much greater than 10. Since 75% of 200 is 150, we have $P(X \geq 150) =$

   $$P\left(z \geq \frac{150-140}{6.487} = 1.543\right) = 0.614.$$

Free-Response

1. If X is the count of cans with at least one defective ball, then X has $B(48, 0.025)$.
   $$P(X \geq 2) = 1 - P(X = 0) - P(X = 1) =$$
   $$1 - \binom{48}{0}(0.025)^0(0.975)^{48} - \binom{48}{1}(0.025)^1(0.975)^{47} = 0.338.$$

   On the TI-83/84, the solution is given by `1-binomcdf(48,0.025,1)`.

2. We know that the sampling distribution of $\bar{x}$ will be similar to the shape of the original population for small n and approximately normal for large n (that's the central limit theorem). Hence,

   (a) if $n = 3$, the sampling distribution would probably be somewhat skewed to the left.
   (b) if $n = 30$, the sampling distribution should be approximately normal.

   Remember that using $n \geq 30$ as a rule of thumb for deciding whether to assume normality is for a sampling distribution just that: a rule of thumb. This is probably a bit conservative. Unless the original population differs markedly from mound-shaped and symmetric, we would expect to see the sampling distribution of $\bar{x}$ be approximately normal for considerably smaller values of n.

3. Since the `binomcdf` function can't be used due to calculator overflow, we will use a normal approximation to the binomial. Let X = the count of heads. Then $\mu_X = (1{,}000{,}000)(0.5) = 500{,}000$ (assuming a fair coin) and $\sigma_X = \sqrt{(1{,}000{,}000)(0.5)(0.5)} = 500$. Certainly both np and $n(1 - p)$ are greater than 10, so the conditions needed to use a normal approximation are present. If we are to have at least 1000 more heads than tails, then there must be at least 500,500 heads (and, of course no more than 499,500 tails). Thus, P(there are at least 1000 more heads than tails) $= P(X) \geq 500500 = P(z \geq \dfrac{500{,}500 - 500{,}000}{500} = 1) = 0.1587.$

4. The average wait for the first success to occur in a geometric setting is $1/p$, where p is the probability of success on any one trial. In this case, the probability of a girl for any one birth is $p = 0.5$. Hence, the average wait for the first girl is $\dfrac{1}{0.5} = 2$. So, we have one boy and one girl, on average, for each two children. The proportion of girls in the population would not change.

5. If X is the count of scholarship athletes that graduate from any sample of 55 players, then X has $B(55, 0.20)$. $\mu_X = 55(0.20) = 11$ and $\sigma_X = \sqrt{55(0.20)(0.80)} = 2.97$.

6. Putting the numbers 2, 4, 5, and 7 into a list in a calculator and doing 1-Var Stats, we find $\mu = 4.5$ and $\sigma = 1.802775638$. The set of all samples of size 2 is $\{(2,4), (2,5), (2,7), (4,5), (4,7), (5,7)\}$ and the means of these samples are $\{3, 3.5, 4.5, 4.5, 5.5, 6\}$. Putting the means into a list and doing 1-Var Stats to find $\mu_{\bar{x}}$ and $\sigma_{\bar{x}}$, we get $\mu_{\bar{x}} = 4.5$ (which agrees with the formula) and $\sigma_{\bar{x}} = 1.040833$ (which does not agree with $\sigma_{\bar{x}} = \dfrac{\sigma}{\sqrt{n}} = \dfrac{1.802775638}{\sqrt{2}} = 1.27475878$). Since the sample is large compared with the population (that is, the population isn't at least 10 times as large as the sample), we use $\sigma_{\bar{x}} = \dfrac{\sigma}{\sqrt{n}}\sqrt{\dfrac{N-n}{N-1}} = \dfrac{1.802775638}{\sqrt{2}}\sqrt{\dfrac{4-2}{4-1}} = 1.040833$, which does agree with the computed value. (Note, students have not been asked to use this formula on the AP exam.)

7. There are three different ways to do this problem: exact binomial, using proportions, or using a normal approximation to the binomial. The last two are essentially the same.

   (i) **Exact binomial.** Let X be the count of persons in the sample that have blood type B. Then X has $B(500, 0.10)$. Also, 11% of 500 is 55 and 15% of 500 is 75. Hence, $P(55 \leq X \leq 75) = P(X \leq 75) - P(X \leq 54) = $ binomcdf(500,0.10,75)-binomcdf(500,0.10,54)=0.2475.

   (ii) **Proportions.** We note that $\mu_X = np = 500(0.1) = 50$ and $n(1 - p) = 500(0.9) = 90$, so we are OK to use a normal approximation. Also, $\mu_{\hat{p}} = p = 0.10$ and $\sigma_{\hat{p}} = \sqrt{\dfrac{(0.1)(0.9)}{500}} = 0.0134$. $P(0.11 \leq \hat{p} \leq 0.15) = P\left(\dfrac{0.11 - 0.10}{0.0134} \leq z \leq \dfrac{0.15 - 0.10}{0.0134}\right) = P(0.7463 \leq z \leq 3.731) = 0.2276$. On the TI 83/84: normalcdf(0.7463,3.731).

   (iii) **Normal approximation to the binomial.** The conditions for doing a normal approximation were established in part (ii). Also, $\mu_X = 500(0.1) = 50$ and $\sigma_X = \sqrt{500(0.1)(0.9)} = 6.7082$. $P(55 \leq X \leq 75) = P\left(\dfrac{55 - 50}{6.7082} \leq z \leq \dfrac{75 - 50}{6.7082}\right) = P(0.7454 \leq z \leq 3.7268) = 0.2279$.

8. All four of these statement are true. However, only III is a statement of the central limit theorem. The others are true of sampling distributions in general.

9. If X is the count of cars with defective pads, then X has $B(20, 0.15)$.

   (a) $P(X = 3)\dbinom{20}{3}(0.15)^3(0.85)^{17} = 0.243$. On the TI-83/84, the solution is given by binompdf(20,0.15,3).

   (b) $\mu_X = np = 20(0.15) = 3$, $\sigma_X = \sqrt{np(1-p)} = \sqrt{20(0.15)(1-0.15)} = 1.597$.

10. $\mu_{\bar{x}} = 40{,}000$ miles and $\sigma_{\bar{x}} = \dfrac{5000}{\sqrt{160}} = 395.28$ miles.

    (a) With $n = 160$, the sampling distribution of $\bar{x}$ will be approximately normally distributed with mean equal to 40,000 miles and standard deviation 395.28 miles.

    (b) $P(\bar{x} < 39{,}000) = P\left(z < \dfrac{39{,}000 - 40{,}000}{395.28} = -2.53\right) = 0.006$.

If the manufacturer is correct, there is only about a 0.6% chance of getting an average this low or lower. That makes it unlikely to be just a chance occurrence, and we should have some doubts about the manufacturer's claim.

11. If X is the number of times you win, then X has $B(1000, 0.474)$. To come out ahead, you must win more than half your bets. That is, you are being asked for $P(X > 500)$. Because $(1000)(0.474) = 474$ and $1000(1 - 0.474) = 526$ are both greater than 10, we are justified in using a normal approximation to the binomial. Furthermore, we find that

$$\mu_x = 1000(0.474) = 474 \text{ and } \sigma_x = \sqrt{1000(0.474)(0.526)} = 15.79.$$

Now,

$$P(X > 500) = P\left(z > \frac{500 - 474}{15.79} = 1.65\right) = 0.05.$$

That is, you have slightly less than a 5% chance of making money if you play 1000 games of roulette.

Using the TI-83/84, the normal approximation is given by `normalcdf(500, 10000,474,15.79) = 0.0498`. The exact binomial solution using the calculator is `1-binomcdf(1000,0.474,500)=0.0467`.

12. $\mu_{\bar{x}} = 2 \text{ lbs} = 32 \text{ oz and } \sigma_{\bar{x}} = \dfrac{5}{\sqrt{35}} = 0.845 \text{ oz.}$

(a) With samples of size 35, the central limit theorem tells us that the sampling distribution of $\bar{x}$ is approximately normal. The mean is 32 oz and standard deviation is 0.845 oz.

(b) In order for $\bar{x}$ to be in the top 10% of samples, it would have to be at the 90th percentile, which tells us that its z-score is 1.28 [that's `InvNorm(0.9)` on your calculator]. Hence,

$$z_{\bar{x}} = 1.28 = \frac{\bar{x} - 32}{0.845}.$$

Solving, we have $\bar{x} = 33.08$ oz. The mean weight of a sample of 35 crabs has to be at least 33.08 oz, or about 2 lb 1 oz, to be in the top 10% of samples of this size.

13. If X is the number that recover, then X has $B(8, 0.7)$.

(a) $P(X = 5) = \dbinom{8}{5}(0.7)^5(0.3)^3 = 0.254$. On the TI-83/84, the solution is given by

`binompdf(8,0.7,5)`.

(b) $P(X = 8) = \dbinom{8}{8}(0.7)^8(0.3)^0 = 0.058$. On the TI-83/84, the solution is given by

`binompdf(8,0.7,8)`.

(c) $P(X \geq 1) = 1 - P(X = 0) = 1 - \dbinom{8}{0}(0.7)^0(0.3)^8 = 0.999$. On the TI-83/84, the

solution is given by `1-binompdf(8,0.7,0)`.

14. $\mu_{\bar{x}} = 1185$ hours, and $\sigma_{\bar{x}} = \dfrac{80}{\sqrt{100}} = 8$ hours.

$$P(\bar{x} \geq 1200) = P\left(z \geq \frac{1200 - 1185}{8} = 1.875\right) = 0.03.$$

15. The first thing Gretchen needs to understand is that a distribution is just the set of all possible values of some variable. For example the distribution of SAT scores for the current senior class is just the values of all the SAT scores. We can draw samples from that population if, say, we want to estimate the average SAT score for the senior class but don't have the time or money to get all the data. Suppose we draw samples of size n and compute $\bar{x}$ for each sample. Imagine drawing ALL possible samples of size n from the original distribution (that was the set of SAT scores for everybody in the senior class). Now consider the distribution (all the values) of means for those samples. That is what we call the sampling distribution of $\bar{x}$ (the short version: the sampling distribution of $\bar{x}$ is the set of all possible values of $\bar{x}$ computed from samples of size n.)

16. The distribution is skewed to the right.

    (a) If $n = 3$, the sampling distribution of $\bar{x}$ will have some right skewness, but will be more mound-shaped than the parent population.
    (b) If $n = 40$, the central limit theorem tells us that the sampling distribution of $\bar{x}$ will be approximately normal.

17. If X is the count of O-rings that failed, then X has $B(12, 0.003)$.

    P (at least one fails) $= P(X = 1) + P(X = 2) + \cdots + P(X = 12)$

    $$= 1 - P(X = 0) = 1 - \binom{12}{0}(0.003)^0 (0.997)^{12} = 0.035.$$

    On the TI-83/84, the solution is given by `1-binompdf(12,0.003,0)`.
    The clear message here is that even though the probability of any one failure seems remote (0.003), the probability of at least one failure (3.5%) is large enough to be worrisome.

18. Because you already have 2 of the 5 prizes, the probability that the next box contains a prize you don't have is $3/5 = 0.6$. If n is the number of trials until the first success, then $P(X = n) = (0.6) \cdot (0.4)^{n-1}$.

    (a) $P(X = 1) = (0.6)(0.4)^{1-1} = (0.6)(1) = 0.6$. On the TI-83/84 calculator, the answer can be found by `geometpdf(0.6,1)`.
    (b) $P(X = 3) = (0.6)(0.4)^{3-2} = 0.096$. On the calculator: `geometpdf(0.6,3)`.
    (c) The average number of boxes you will have to buy before getting the third prize is
    $$\frac{1}{0.6} = 1.67.$$

19. $40(0.8) = 32$ and $40(0.2) = 8$. The rule we have given is that both np and $n(1 - p)$ must be greater than 10 to use a normal approximation. However, as noted in earlier in this chapter, many texts allow the approximation when $np \geq 5$ and $n(1 - p) \geq 5$. Whether the normal approximation is valid or not depends on the standard applied. Assuming that, in this case, the conditions necessary to do a normal approximation are present, we have

    $$\mu_X = 40(0.8) = 32, \text{ and } \sigma_X = \sqrt{40(0.8)(0.2)} = 2.53.$$

20. If $p = 0.70$, then $\mu_{\hat{p}} = 0.70$ and $\sigma_{\hat{p}} = \sqrt{\dfrac{0.70(1-0.70)}{600}} = 0.019$. Thus, $P(\hat{p} \leq 0.65) =$
$P\left(z \leq \dfrac{0.65 - 0.70}{0.019} = -2.63 \right) = 0.004$. Since there is a very small probability of getting

a sample proportion as small as 0.65 if the true proportion is really 0.70, it appears that the San Francisco Bay Area may not be representative of the United States as a whole (that is, it is unlikely that we would have obtained a value as small as 0.65 if the true value were 0.70).

Solutions to Cumulative Review Problems

1. The sample space for this event is {HHH, **HHT**, **HTH**, **THH**, HTT, HTH, THH, TTT}. One way to do this problem, using techniques developed in Chapter 8, is to compute the probability of each event. Let $X =$ the count of heads. Then, for example (bold faced in the list above), $P(X = 2) = (0.6)(0.6)(0.4) + (0.6)(0.4)(0.6) + (0.4)(0.6)(0.6) = 3(0.6)^2(0.4) = 0.432$. Another way is to take advantage of the techniques developed in this chapter (noting that the possible values of X are 0, 1, 2, and 3):

$$P(X=0)=(0.4)^3 =0.064; P(X=1)=\binom{3}{1}(0.6)^1(0.4)^2 = \texttt{binompdf(3,0.6,1)} =$$

$$0.288; P(X=2)=\binom{3}{2}(0.6)^2(0.4)^1 = \texttt{binompdf(3,0.6,2)} = 0.432; \text{ and } P(X = 3) =$$

$$\binom{3}{3}(0.6)^3(0.4)^0 = \texttt{binompdf(3,0.6,3)} = 0.216. \text{ Either way, the probability distribu-}$$

tion is then

X	0	1	2	3
$P(X)$	0.064	0.288	0.432	0.216

Be sure to check that the sum of the probabilities is 1 (it is!).

2. (a) Yes, it is a random sample because each student in any of the 5 classes is equally likely to be included in the sample.
   (b) No, it is not a simple random sample (SRS) because not all samples of size 25 are equally likely. For example, in an SRS, one possible sample is having all 25 come from the same class. Because we only take 5 from each class, this isn't possible.

3. The slope of the regression line is 0.8. For each additional milligram of the drug, reaction time is *predicted* to increase by 0.8 seconds. Or you could say for each additional milligram of the drug, reaction time will increase by 0.8 seconds, *on average*.

4. $P(A \text{ or } B) = P(A \cup B) = P(A) + P(B) - P(A \cap B) = 0.5 + 0.3 - P(A \cap B) = 0.65 \Rightarrow$ $P(A \cap B) = 0.15$. Now, A and B are independent if $P(A \cap B) = P(A) \cdot P(B)$. So, $P(A) \cdot P(B) = (0.3)(0.5) = 0.15 = P(A \cap B)$. Hence, A and B are independent.

5. (a) Quantitative
   (b) Categorical
   (c) Categorical
   (d) Quantitative
   (e) Categorical
   (f) Quantitative
   (g) Categorical

CHAPTER > 10

Inference: Estimating with Confidence Intervals

IN THIS CHAPTER

Summary: In this chapter we begin our formal study of inference by estimating a population parameter. We will learn about confidence intervals, a way of identifying a range of values that we think might contain our parameter of interest. We will develop intervals to estimate a single population proportion and the difference between two population proportions. Then we will introduce the t statistic (as an adjustment to z) and talk about estimates for means and the difference between two population means.

Key Ideas

- ✪ Estimation
- ✪ Confidence Intervals and Margin of Error
- ✪ Confidence Intervals for a Proportion
- ✪ Confidence Intervals for a Difference in Proportions
- ✪ t Procedures
- ✪ Confidence Intervals for a Mean
- ✪ Confidence Intervals for a Difference in Means
- ✪ Choosing a Sample Size for a Confidence Interval

Estimation and Confidence Intervals

We have previously described a *statistic* as a value that describes a sample and a *parameter* as a value that describes a population. Now we want use a *statistic* as an **estimate** of a *parameter*. In Chapter 9, we explored sampling distributions to see how a statistic behaves when we repeatedly select random samples from a known population. This helps us understand the tools we use when, in reality, we select *one* sample and we use it to make inferences about the population. What we will do now is develop a process by which we will use our sample to generate a range of plausible population values for the parameter. The statistic itself is called a **point estimate**, and the range of plausible population values from which we might have obtained our estimate is called a **confidence interval**.

> **example:** We survey a random sample of 50 students from a large high school, and find that 42% of the students in the sample say they plan to vote for Charlotte for student body treasurer. That is, $\hat{p} = 0.42$ is our *point estimate* for the proportion of *all* students at this school who would say they plan to vote for Charlotte for student body treasurer. But this estimate is almost certainly incorrect. It is extremely unlikely that *exactly* 42% of all students would say this.
>
> So, what is the point of taking a survey if we are almost certain to get an incorrect estimate? To answer that question, we need to go back to what we learned about the sampling distribution of $\hat{p}$. For example, we learned that, with a large enough sample size, about 95% of sample proportions would be within 1.96 standard deviations of p. Knowing this allows us to *quantify* our uncertainty in that estimate. In particular, would we be surprised to learn that p is actually, say, 0.39?
>
> If p were 0.39, we would expect about 95% of samples to have a $\hat{p}$ within 1.96 standard deviations of 0.39 as represented by the shaded area below. Our sample proportion 0.42 (shown by the vertical line) is clearly in this region, so we would not be surprised to learn that p is 0.39. (Note that here we are defining "surprise" based on whether 0.42 is in the middle 95% of sample proportions.)

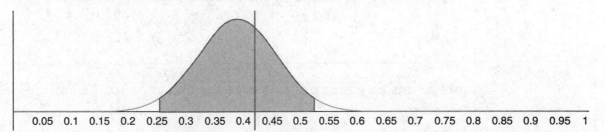

| 0.05 | 0.1 | 0.15 | 0.2 | 0.25 | 0.3 | 0.35 | 0.4 | 0.45 | 0.5 | 0.55 | 0.6 | 0.65 | 0.7 | 0.75 | 0.8 | 0.85 | 0.9 | 0.95 | 1 |

> What other values of p would contain $\hat{p} = 0.42$ in the middle 95%? As shown below, if p is anywhere between about 0.30 and about 0.55, then 0.42 is in the middle 95% of values.

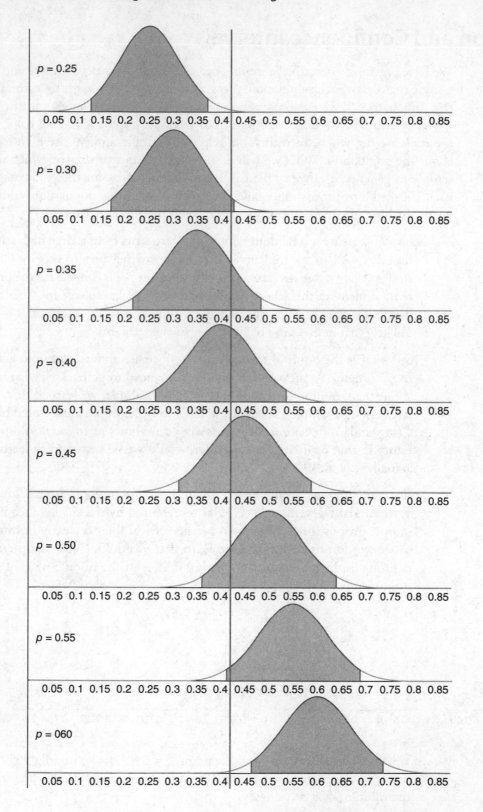

That range of *plausible values* of p is called the **95% confidence interval** for p. So we cannot be very certain at all that p is 0.42, but we can be 95% confident that p is between about 0.30 and 0.55. 95% here is referred to as the **confidence level**.

Formula for a Confidence Interval for a Proportion

In the diagram above, any value of p that contains 0.42 in the middle 95% of sample proportions is considered a plausible value for p. To be in the middle 95%, $\hat{p} = 0.42$ is within 1.96 standard deviations of that plausible value. That, of course, means that the plausible value is within 1.96 standard deviations of $\hat{p}$. (Those standard deviations are slightly different because they depend on the value we use for p in the formula, but not different enough to worry about, so we'll use $\hat{p}$.) We will say that the plausible values of p are in the confidence interval $\hat{p} \pm 1.96\sqrt{\dfrac{\hat{p}(1-\hat{p})}{n}}$. (The 1.96 is called the **critical value**. It is the z-score that coincides with the desired confidence level, and is denoted by z^*.)

This allows us to create a more precise interval estimate for p, the proportion of all students at this school who plan to vote for Charlotte for student body treasurer. The interval is

$$\hat{p} \pm z^* \cdot \sqrt{\frac{\hat{p}(1-\hat{p})}{n}} = 0.42 \pm 1.96\sqrt{\frac{0.42(1-0.42)}{50}} = 0.42 \pm 0.137 = (0.28, 0.55).$$

We interpret that interval as follows: *We are 95% confident that the proportion of all students at this school who plan to vote for Charlotte for student body treasurer is between 0.28 and 0.55.*

(Note: The part of the formula following the $\pm$ is called the *margin of error*. In this example, we have a margin of error of 0.137, or 13.7 percentage points.)

example: Find the critical value of z required to construct a 98% confidence interval for a population proportion.

solution: We are reading from Table A, the table of Standard Normal Probabilities. Remember that table entries are areas to the left of a given z-score. With $C = 0.98$, we want $\dfrac{1-0.98}{2} = 0.01$ in each tail, or 0.99 to the left if z^*. Finding $p = 0.9900$ in the table (the closest we can get is 0.9901), we have $z^* = 2.33$. With technology, you would use an Inverse Normal function with 0.99 (or 0.01) as the area.

Conditions for a Confidence Interval for a Proportion

Notice that in the figure on the previous page, all the sampling distributions were represented by normal models. Normal models are appropriate here because

$$np = 50 \cdot 0.25 = 12.5 \geq 10 \text{ and } n(1 - p) = 50 \cdot 0.75 =) 37.5 \geq 10$$

and that is the most extreme of the proportions used. And it is that normal model that allows us to use 1.96 standard deviations to mark the middle 95% of sample proportions for each possible value of p. Whenever we use a normal model as a basis for our procedure, we have to ensure that it is appropriate to do so. In addition, our model assumes simple random sampling from an infinitely large population (equivalent to sampling with replacement).

Now, we know the sampling distribution is not exactly normal. Ever. We also know that we are not sampling from an infinitely large population and are rarely sampling with replacement. This means the model we are using is wrong. But, as well-known statistician George Box said, "All models are wrong. But some of them are useful." To see if the model we will use is useful, we check some conditions to see if those assumptions are "close enough" to being met for the model to be useful.

ASSUMPTION	CONDITION
We have a simple random sample. This ensures that our model is centered in the correct place, because simple random samples are unbiased.	**Check that we have a simple random sample or its equivalent.** This requires us to look at how the data were collected.
We have a normally distributed sampling distribution. This assures the correct shape of our model.	$n\hat{p}$ **and** $n(1-\hat{p})$ **are both at least 10.** This guideline, along with random sampling, ensures that our sampling distribution is close enough to normal for the normal model to be useful.
We are sampling with replacement or from an infinitely large population. This ensures we have the correct standard deviation of our sampling distribution.	**We are essentially sampling without replacement or our sample size is less than 10% of the population size.** This, in conjunction with random sampling, allows us to use $\sqrt{\dfrac{\hat{p}(1-\hat{p})}{n}}$ as the standard error of the sampling distribution, ignoring the Finite Population Correction Factor described in the previous chapter.

Construct and Interpret a Confidence Interval for a Proportion

Let's revisit the example about Charlotte to see a model response to the following question:

example: We survey a random sample of 50 students from a large high school, and find that 42% of the students in the sample say they plan to vote for Charlotte for student body treasurer. Construct and interpret a 95% confidence interval for the proportion of all students at this school who plan to vote for Charlotte for student body treasurer.

solution:

Identify the procedure and check conditions:
We will construct a confidence interval for a proportion. (Note: This identifies the procedure by name.)

• We are told they selected a random sample of 50 students from the school.
• $n\hat{p} = 50 \cdot 0.42 = 21 \geq 10$ and $n(1 - \hat{p}) = 50 \cdot 0.58 = 29 \geq 10$
• We don't know the population of the high school, but we are told it is a large high school. So the population is almost certainly greater than $10n = 10 \cdot 50 = 500$.

Conditions are met.

Computations:
The 95% confidence interval is

$$\hat{p} \pm z^* \sqrt{\frac{\hat{p}(1-\hat{p})}{n}} = 0.42 \pm 1.96 \sqrt{\frac{0.42(1-0.42)}{50}} = (0.28, 0.55).$$

(Note: The expression above identifies the procedure by formula.)
z^* above is called the **critical value** for z. It is the z-score corresponding to the middle 95% in this case, or whatever confidence level you use.

Interpretation:
We are 95% confident that the proportion of all students at this school who plan to vote for Charlotte for student body treasurer is between 0.28 and 0.55.

Notice that the prompt did not specify to check conditions. But you *must* check conditions as part of the entire process. Learn to do that automatically for every inference procedure.

Confidence Level as a Capture Rate

We saw that we referred to a 95% confidence interval as the set of population proportions that would contain $\hat{p}$ in their middle 95% of sample proportions. There is another important way to think about what that 95% confidence level refers to.

Assume in the situation above that 37% of the student body plans to vote for Charlotte. Notice that the interval above, when $\hat{p}$ was 0.42, contains 0.37. We could say that this interval captured the population proportion. But a different sample would result in a different $\hat{p}$ and a different interval. The figure below shows 100 possible samples from this population in which 37% of students plan to vote for Charlotte. The dot in the center shows $\hat{p}$ and the vertical bars show the interval based on that sample proportion. The dark squares represent values of $\hat{p}$ whose intervals captured 0.37. The circles represent values whose intervals did not capture 0.37. We can see that six of the intervals did not capture 0.37, so 94 did. In the long run, we expect 95% of these intervals will capture 0.37.

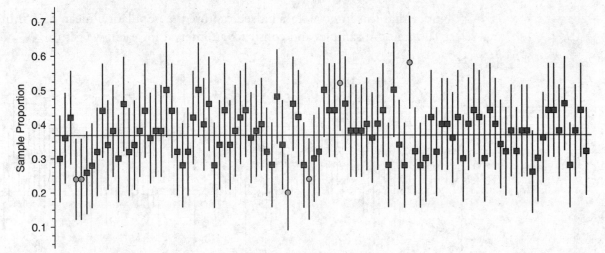

This gives us a way to interpret a 95% confidence level: "If we were to select many samples of 50 students, and construct a confidence interval for each sample, we expect about 95% of the intervals to contain the proportion of all students who plan to vote for Charlotte."

Important: When we say, "We are 95% confident that the true population value lies in an interval," we mean that the process used to generate the interval will capture the true population value 95% of the time. We must be careful not to make a probability statement about the interval that has been constructed. Our "confidence" is in the process that generated the interval, but if we are discussing the particular interval, there is no longer a repeatable random event. That's why we use the word "confidence" instead of "probability."

A Confidence Interval for a Mean (*t* Procedures)

General Form of a Confidence Interval

A confidence interval is composed of two parts: a point estimate of a population value and a margin of error. We specify a level of confidence to communicate how certain we are that the interval contains the true population parameter.

A **level *C* confidence interval** has the following form: (estimate) ± (margin of error). In turn, the *margin of error* for a confidence interval is composed of two parts: the critical value of *z* (or *t* as we shall see) and the standard error. Hence, all confidence intervals take the form

(estimate) ± (margin of error) = (estimate) ± (critical value)(standard error)

When we discussed the sampling distribution of $\bar{x}$ in Chapter 9, we knew the population mean and standard deviation. We know the shape of the sampling distribution of $\bar{x}$ is approximately normal if the population is approximately normal, or if n is large enough. The mean of the sampling distribution is μ and the standard error is $\frac{\sigma}{\sqrt{n}}$. In this chapter, we are estimating the population mean from a sample.

For example, pennies (one-cent coins) produced in the United States now have an average mass of 2.5 grams and a standard deviation of 0.05 grams. The masses are approximately normally distributed. Say we randomly select 5 pennies from this population and the masses are 2.47 g, 2.48 g, 2.51 g, 2.59 g, and 2.45 g. We could construct a confidence interval for a mean using a process similar to that for proportions. The mean and standard deviation of this sample are 2.499 g and 0.056 g, respectively. The 90% confidence interval is

$$\bar{x} \pm z^* \frac{\sigma}{\sqrt{n}} = 2.499 \pm 1.645 \frac{0.05}{\sqrt{5}} = (2.462, 2.536).$$

Notice that this interval does indeed capture the population mean of 2.5. In fact, if we select 100 random samples and construct an interval for each, we get this:

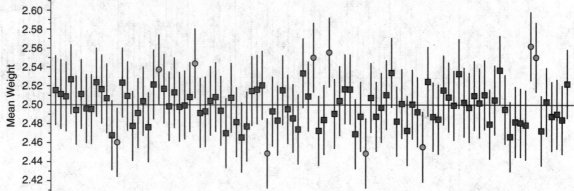

Here 90 of our intervals captured $\mu = 2.5$. In the long run we'd expect about 90% of our intervals to capture μ.

Unfortunately, we pretty much never have this situation in which we would know the value of σ. If we knew enough about the population to know σ, we'd probably also know μ. In fact, we *need* μ to calculate σ. But in reality we generally have to use the sample standard deviation s to estimate σ. That turns our formula for the confidence interval into $\bar{x} \pm z^* \frac{s}{\sqrt{n}}$. But let's see what happens to our capture rate when we do this:

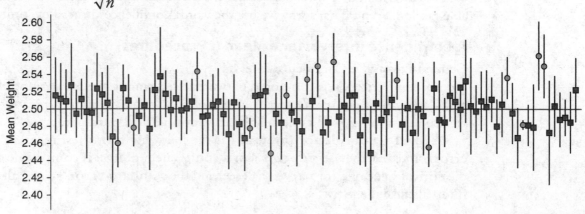

Now only 85 of our intervals captured $\mu = 2.5$. This is because s tend to *underestimate* σ and it means our process doesn't capture the mean as often as we say it will. The way to fix this (as discovered by William Sealy Gosset over 100 years ago!) is to adjust the critical value. The new critical value is called t, and it will always be larger than z. This makes the intervals longer and gets us to the correct capture rate. Here we need a critical value of 2.132.

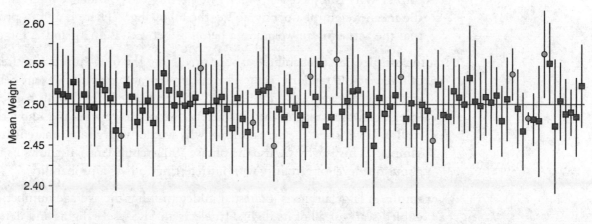

Now we have captured 89 of the intervals. The long-run capture rate is back to the 90% we claimed.

To find the appropriate value of t, we need to know two things: the confidence level and the sample size n, because the larger the sample size, the smaller the adjustment we need to make. (Actually, we need a number called the *degrees of freedom*, which for now is $n - 1$. More on degrees of freedom later!)

A confidence interval for a population mean would take on the form

$$\bar{x} \pm t^* \frac{s}{\sqrt{n}}$$

where t^* depends on the confidence level C and the degrees of freedom, s is the sample standard deviation, and n is the sample size.

Conditions for a Confidence Interval for a Mean

The conditions for a confidence interval for a mean are similar to those for a proportion. The difference is in the condition for normality. The population should be approximately normally distributed or we need a large enough sample size that the Central Limit Theorem can be counted on to give us an approximately normal sampling distribution of the $\bar{x}$. We still need a simple random sample and a sample that is less than 10% of the population size.

The confidence level is often expressed as a percent: a 95% confidence interval means that $C = 0.95$, or a 99% confidence interval means that $C = 0.99$. Although any value of C can be used as a confidence level, typical levels are 0.90, 0.95, and 0.99.

> **example:** Find the critical value of t required to construct a 99% confidence interval for a population mean based on a sample size of 15.

> **solution:** First, we need to know that $df = 15 - 1 = 14$. Using the t distribution table (Table B in the Appendices) we can look at the bottom of the table for the column for 99% confidence, and move up the table to the row for 14 df. We find $t^* = 2.977$.

Using technology, you would use an Inverse T command and input the area below the critical value you want. A confidence of 99% refers to the *middle* 99%, which goes from the 0.5 percentile to the 99.5 percentile. So the area would be either 0.005 (for the negative value) or 0.995 (for the positive value). The arguments you would enter would be *area* = 0.005 (or 0.995) and *df* = 14.

example: Why is it incorrect, when interpreting the confidence interval above for the mean weight of a penny, to say the following? "There is a 90% probability that the actual mean weight of a penny is between 2.462 g and 2.536 g."

solution: Because probability is about predicting the long-run relative frequency of a repeatable random event. There is no repeatable random event here. The interval is determined, and the actual mean weight is fixed. This particular interval always captures the parameter. (Although we wouldn't know that in reality. If we knew the actual mean, why would we try to estimate it from a sample?) This is why we use the phrase "90% confident." It allows us to talk about our level of certainty without referring to it as a probability.

example: A large airline is interested in determining the average number of unoccupied seats for all of its flights. It selects an SRS of 81 flights and determines that the average number of unoccupied seats for the sample is 12.5 seats with a sample standard deviation of 3.9 seats. Construct and interpret a 95% confidence interval for the true number of unoccupied seats for all flights.

solution:

Conditions:

The problem states that we have a simple random sample of 81 flights. The large sample size justifies the construction of a one-sample confidence interval for the population mean. There are definitely more than $10 \times 81 = 810$ flights for a large airline. Conditions are met for a confidence interval for a mean.

Calculations:

The 95% confidence interval is $\bar{x} \pm t^* \dfrac{s}{\sqrt{n}} = 12.5 \pm 1.99 \dfrac{3.9}{\sqrt{81}} = (11.64, 13.36)$ using t^* for $81 - 1 = 80$ *df*.

Interpretation:

We are 95% confident that the mean number of unoccupied seats for all the airline's flights is between 11.6 and 13.4 seats.

Exam Tip: The rubrics for confidence interval questions generally have three parts. The first is *Check conditions and identify the procedure*. The conditions always include the randomization condition and the normality (large sample size) condition. (On the exam, the 10% condition has not been required. It is still important to understand, however. There have been questions asking speicifically about the need for the 10% condition.) Identifying the procedure can be done by name (for example *confidence interval for a mean*) or by formula (for example $\bar{x} \pm t^* \dfrac{s}{\sqrt{n}}$). It is a good idea to get in the habit of doing both so if you forget one, you've still done the other!

Estimating a Difference Between Groups

When we select a random sample and calculate a proportion of successes (or a mean of some measure), $\hat{p}$ (or $\bar{x}$) is a random variable. So if we independently select samples from two populations, the sample statistics are independent random variables and, as we saw in Chapter 8,

- $\mu_{X \pm Y} = \mu_X \pm \mu_Y$.

In other words, the mean of the sum (or difference) of two random variables is the sum (or difference) of the means of the individual variables:

- $\sigma^2_{X \pm Y} = \sigma^2_X + \sigma^2_Y$, if and only if X and Y are independent.
- $\sigma_{X \pm Y} = \sqrt{\sigma^2_X + \sigma^2_Y}$, if and only if X and Y are independent.

In other words, when adding or subtracting independent random variables, variances add.

This tells us what we need to create a confidence interval for the difference of two means or two proportions. The statistic will be the difference in proportions (or means). The standard error for the difference in proportions will be

$$\sqrt{\sigma^2_{\hat{p}_1} + \sigma^2_{\hat{p}_2}} = \sqrt{\frac{\hat{p}_1(1 - \hat{p}_1)}{n_1} + \frac{\hat{p}_2(1 - \hat{p}_2)}{n_2}}$$

so a confidence interval for a difference in proportions has the form

$$(\hat{p}_1 - \hat{p}_2) \pm z^* \sqrt{\frac{\hat{p}_1(1 - \hat{p}_1)}{n_1} + \frac{\hat{p}_2(1 - \hat{p}_2)}{n_2}}.$$

Similarly, the confidence interval for a difference in means is

$$(\bar{x}_1 - \bar{x}_2) \pm t^* \sqrt{\frac{s_1^2}{n_1} + \frac{s_2^2}{n_2}}.$$

There's an added complication with the difference in means, however. The adjustment needed isn't exactly t, but we can get very close if we use degrees of freedom determined by this really messy formula:

$$df = \frac{\left(\frac{s_1^2}{n_1} + \frac{s_2^2}{n_2} \right)^2}{\frac{1}{n_1 - 1}\left(\frac{s_1^2}{n_1} \right)^2 + \frac{1}{n_2 - 1}\left(\frac{s_2^2}{n_2} \right)^2}.$$

Don't worry! You don't need to know or use this formula. But your calculator will use this formula and it will often result in degrees of freedom that are not a whole number. It will be enough for you to report the *df* that your calculator gives, and not be alarmed when you see non-whole number values.

The conditions for an interval are similar to those before:

- We need *two* simple random samples independently selected from two populations.
- For *proportions*, $n_1 \hat{p}_1, n_1(1 - \hat{p}_1), n_2 \hat{p}_2, n_2(1 - \hat{p}_2)$ must all be at least 10 (or 5, according to some textbooks).

- For *means*, the populations must be approximately normal or the sample sizes must both be large enough to compensate. (If the sample size is more than 30, most textbooks consider that large enough to proceed.)
- Both samples should be less than 10% of the population size.

example: To see whether a particular candidate's popularity changes as the election approaches, a poll was taken six months before the election in which 850 randomly selected American adults were asked whether they have a positive opinion of the candidate, and another poll was taken two weeks before the election with a random sample of 750 American adults. In the first poll, 503 of those surveyed said they had a positive opinion of the candidate, and in the second poll 416 said so. Construct and interpret a 95% confidence interval for the change in the candidate's popularity over those several months.

solution:

Check conditions:

We have two samples randomly and independently selected from two populations. (The population six months before the election can be considered a different population because the opinion may have changed.)

$n_1 \hat{p}_1 = 503, n_1(1 - \hat{p}_1) = 347, n_2 \hat{p}_2 = 416, n_2(1 - \hat{p}_2) = 334$ are all at least 10.

(Subscript 1 refers to the first poll and 2 refers to the second poll.)

Both samples are all less than 10% of the population of American adults. Conditions are met for a confidence interval for a difference in proportions.

Calculations:

The 95% confidence interval is

$$(\hat{p}_2 - \hat{p}_1) \pm z^* \sqrt{\frac{\hat{p}_2(1 - \hat{p}_2)}{n_2} + \frac{\hat{p}_1(1 - \hat{p}_1)}{n_1}} = \left(\frac{416}{750} - \frac{503}{850}\right) \pm 1.96\sqrt{\frac{\frac{416}{750} \cdot \frac{334}{750}}{750} + \frac{\frac{503}{850} \cdot \frac{347}{850}}{850}} = (-0.086, 0.011)$$

Interpretation:

We are 95% confident that the candidate's popularity in the second poll was between 8.6 percentage points lower and 1.1 percentage points higher than in the first poll. Because 0 is in the confidence interval, it is plausible there was no change in the candidate's popularity.

example: A random sample of 50 male students from a large high school was selected and their temperatures were taken. Another random sample of 50 female students was selected and their temperatures were also taken. The data are summarized in the table below.

GENDER	MEAN TEMPERATURE	STANDARD DEVIATION
Female	98.4	0.75
Male	98.1	0.68

Construct a 90% confidence interval for the difference in mean temperatures for all males and all females at this school.

solution:

Check conditions and identify procedure:

- We have random samples of 50 males and 50 females from their populations at the school.
- Because both sample sizes are 50 ≥ 30, we can assume the sampling distribution of $\bar{x}$ is approximately normal.
- Because this is a large high school, there are probably more than 50 · 10 = 500 males and 500 females at the school.

Conditions are met for a confidence interval for a difference in means.

Calculations:

$$(\bar{x}_F - \bar{x}_M) \pm t^* \sqrt{\frac{s_F^2}{n_F} + \frac{s_M^2}{n_M}} \text{ (This identifies the procedure again, by formula.)}$$

$$= (98.4 - 98.1) \pm t^* \sqrt{\frac{0.75^2}{50} + \frac{0.65^2}{50}} \text{ using } t^* \text{ for 97.07 } df \text{ (as reported by my}$$

calculator—you could calculate t^* but you don't need to)

$$= (0.062, 0.538)$$

Interpretation:

We are 90% confident that the mean temperature for females at this school is between 0.062°F and 0.538°F more than the mean temperature for males at this school.

> **Examp Tip:** There are three steps to a confidence interval: Check conditions and identify the procedure, compute the interval, and interpret the interval in context. The question may not specifically ask for all three steps, but they are always required unless specifically stated otherwise.

Inference for Experiments

The examples so far have involved confidence intervals for sampling situations. Another situation in which we compare two groups is randomized experiments. If we use a completely randomized design in which each treatment is randomly assigned to half the subjects, the conditions for two-sample procedures are not met. We do not have two samples randomly selected from two populations. We have one population (the subjects in the experiment) and the groups (you could call them samples, but we do not recommend that!) are dependently selected. Once you choose one group to get one treatment, the other group is set. Also, each group is 50% of the "population"so that 10% condition is violated as well.

Both errors affect the standard deviation of the model. The problem of the groups not being independent increases the standard deviation, and the problem of the groups being more than 10% of the total number of subjects decreases the standard deviation (remember that Finite Population Correction Factor?). These two effects almost exactly cancel each other out, and we can (thank goodness!) use the same formulas.

One thing that does change is the conditions. Now there are only two:

- Treatments must be randomly assigned to subjects.
- The groups must be large enough to meet the normality conditions. So for proportions, $n_1\hat{p}_1, n_1(1 - \hat{p}_1), n_2\hat{p}_2,$ and $n_2(1 - \hat{p}_2)$ must all be at least 10 (or 5). For means, the data must be consistent with what we'd expect from an approximately normal population, or we must have large treatment groups.

• PARAMETER • ESTIMATOR	CONDITIONS	FORMULA
• Population mean: μ • Estimator: $\bar{x}$	• SRS • Normal population (or very large sample) • σ known (This is almost never the case. So you can pretty much remember "t for means.") *or* • SRS • σ unknown (This is almost always the case.) • Population approximately normal or large sample size ($n > 30$)	$\bar{x} \pm z^* \dfrac{\sigma}{\sqrt{n}}$ $\bar{x} \pm t^* \dfrac{s}{\sqrt{n}}, \mathrm{df} = n - 1$
• Population proportion: p • Estimator: $\hat{p}$	• SRS • Large population size relative to sample size • $n\hat{p} \geq 5$, $n(1 - \hat{p}) \geq 5$ (or $n\hat{p} \geq 10$, $n(1 - \hat{p}) \geq 10$)	$\hat{p} \pm z^* \sqrt{\dfrac{\hat{p}(1 - \hat{p})}{n}}$
• Difference of population means: $\mu_1 - \mu_2$ • Estimator: $\bar{x}_1 - \bar{x}_2$	• Independent SRSs • Both populations normal • σ_1, σ_2 known (Again, this is almost never the case.) *or* • Independent SRSs • σ_1, σ_2 unknown. (This is almost always the case.) • Approximately normal populations or n_1 and n_2 both "large"	$(\bar{x}_1 - \bar{x}_2) \pm z^* \sqrt{\dfrac{\sigma_1^2}{n_1} + \dfrac{\sigma_2^2}{n_2}}$ $(\bar{x}_1 - \bar{x}_2) \pm t^* \sqrt{\dfrac{s_1^2}{n_1} + \dfrac{s_2^2}{n_2}}$ df = (computed by software) df = min$\{n_1 - 1, n_2 - 1\}$ or (See explanation below.)
• Difference of population proportions: $p_1 - p_2$ • Estimator: $\hat{p}_1 - \hat{p}_2$	• SRSs from independent populations • Large population sizes relative to sample sizes • $n_1\hat{p}_1 \geq 5, n_1(1 - \hat{p}_1) \geq 5$ $n_2\hat{p}_2 \geq 5, n_2(1 - \hat{p}_2) \geq 5$	$(\hat{p}_1 - \hat{p}_2) \pm z^* \sqrt{\dfrac{\hat{p}_1(1 - \hat{p}_1)}{n_1} + \dfrac{\hat{p}_2(1 - \hat{p}_2)}{n_2}}$

Sample Size

It is always desirable to select as large a sample as possible when doing research because sample means of larger samples are less variable than sample means of small samples. However, it is often expensive or difficult to draw larger samples so that we try to find the optimum sample size: large enough to accomplish our goals, small enough that we can afford it or manage it. We will look at techniques in this section for selecting sample sizes in the case of a large sample test for a single population mean and for a single population proportion.

Sample Size for Estimating a Population Proportion

The confidence interval for a population proportion is given by

$$\hat{p} \pm z^* \sqrt{\frac{\hat{p}(1-\hat{p})}{n}}.$$

The margin of error is

$$z^* \sqrt{\frac{\hat{p}(1-\hat{p})}{n}}.$$

Let M be the desired maximum margin of error. Then,

$$M \leq z^* \sqrt{\frac{\hat{p}(1-\hat{p})}{n}}.$$

Solving for n,

$$n \geq \left(\frac{z^*}{M}\right)^2 \hat{p}(1-\hat{p}).$$

But we do not have a value of $\hat{p}$ until we collect data, so we need a way to estimate $\hat{p}$. There are two ways to choose a value of p.

1. If we have a $\hat{p}$ from a previous study, we could use that in our estimate. That can be a little risky because if we get a new value of $\hat{p}$ that is closer to 0.5 than the previous one, our standard error will be larger than we planned for, and we may not get the margin of error we want. So a safer approach would be...

2. Use 0.5 in the margin of error calculation. The standard error for a proportion (with all else equal) is largest when $p = 0.5$ so if the margin of error is small enough for $\hat{p} = 0.5$, it will be small enough for any other value of $\hat{p}$. The risk here is that you may end up using a sample size that is larger than you need, which can be more expensive.

> **example:** Historically, about 60% of a company's products are purchased by people who have purchased products from the company previously. The company is preparing to introduce a new product and wants to generate a 95% confidence interval for the proportion of its current customers who will purchase the new product. They want to be accurate within 3%. How many customers do they need to sample?

solution: Based on historical data, choose $\hat{p} = 0.6$. Then

$$n \geq \left(\frac{1.96}{0.03}\right)^2 (0.6)(0.4) = 1024.4.$$

The company needs to sample 1,025 customers. (We use 1025 since n must be an integer and greater than 1024.4.) Had it not had the historical data, it would have had to use $\hat{p} = 0.5$.

If $\hat{P} = 0.5$, $n \geq \left(\frac{1.96}{0.03}\right)^2 (0.5)(0.5) = 1067.1$. You need a sample of at least 1,068 customers.

By using $\hat{p} = 0.6$, the company was able to sample 43 fewer customers, but they run the risk of too small a sample if $\hat{p}$ turns out to be closer to 0.5. The safest bet on the AP Exam is to use 0.5.

Sample Size for Estimating a Population Mean (Large Sample)

The large sample confidence interval for a population mean is given by $\bar{x} \pm z \cdot \frac{\sigma}{\sqrt{n}}$. The margin of error is given by $z \cdot \frac{\sigma}{\sqrt{n}}$. Let M be the desired *maximum* margin of error. Then, $M \leq z \cdot \frac{\sigma}{\sqrt{n}}$. Solving for n, we have $n \geq \left(\frac{z \cdot \sigma}{M}\right)^2$. Using this "recipe," we can calculate the minimum n needed for a fixed confidence level and a fixed maximum margin of error.

One obvious problem with using this expression as a way to figure n is that we will not know σ, so we need to estimate it in some way. In an exam question, you will almost certainly be provided with an estimate of σ to use in the calculation.

In reality, researchers would probably be able to utilize some historical knowledge about the standard deviation for the type of data they are examining, as shown in the following example:

example: A machine for inflating tires, when properly calibrated, inflates tires to 32 lb, but it is known that the machine varies with a standard deviation of about 0.8 lb. How large a sample is needed in order be 99% confident that the mean inflation pressure is within a margin of error of $M = 0.10$ lb?

solution:

$$n \geq \left(\frac{2.576(0.8)}{0.10}\right)^2 = 424.49.$$ Since n must be an integer and $n \geq 424.49$, choose $n = 425$. You would need a sample of at least 425 tires.

In this course you will not need to find a sample size for constructing a confidence interval involving t. This is because you need to know the sample size before you can determine t^* since there is a different t distribution for each different number of degrees of freedom. For example, for a 95% confidence interval for the mean of a normal distribution, you know that $z^* = 1.96$ no matter what sample size you are dealing with, but you can't determine t^* without already knowing n.

❭ Rapid Review

1. True–False. A 95% confidence interval for a population proportion is given as (0.37, 0.52). This means that the probability is 0.95 that this interval contains the true proportion.

   Answer: False. Because this interval has been created, there is no repeatable random event. That's why we say, "We are 95% confident." It avoids improper use of the word probability. (The probability is 0.95 that the *process* used to generate this interval *will* capture the true proportion.)

2. You have calculated a 90% confidence interval for the mean number of hours volunteers work each week for a large charity. For each adjustment described below, if all else stayed the same, tell whether the confidence interval would become wider, narrower, or neither one and explain why.

   (a) The sample size is quadrupled.

   (b) The confidence level is changed from 90% to 95%.

   (c) The sample standard deviation was corrected to a smaller value.

   (d) The sample mean was corrected to a larger value.

   Answer:

   (a) Narrower. In the margin of error part of the calculation, you divide s by $\sqrt{n}$. Quadrupling the sample size, you would instead divide by $\sqrt{4n} = 2\sqrt{n}$. And the value of t^* gets slightly smaller, too. So the margin of error, and therefore the interval, will be narrower.

   (b) Wider. The critical value for 95% confidence is wider than for 90%. Or, more intuitively, if you want to be more confident of capturing the parameter, use a bigger net!

   (c) Narrower. The standard deviation is in the numerator of the margin of error calculation. If it gets smaller, the margin of error is smaller. Intuitively, if there is less variability in the sample, you can be more precise in your estimate.

   (d) Neither one. Changing the mean will move the center of the interval, but the margin of error will remain unchanged.

3. What is the critical value of t for a 99% confidence interval based on a sample size of 26?

   Answer: Remember that 99% refers to the *middle* 99%, so 0.5% is below that and 0.05% is above that. So you can use a calculator's inverse t function, with an area of 0.005 or 0.995 to get –2.787 or 2.787, respectively. From Table B, t Distribution Critical Values, t^* for 99% confidence and 25 df is 2.787.

4. What is the critical value of z for a 98% confidence interval for a proportion?

   Answer: Use the inverse normal command on a calculator with a tail area of 0.01 or 0.99 to get ±2.326. Or use Table A, Standard Normal Probabilities, and look for p = 0.01 or 0.99 in the body of the table. The nearest table entry to 0.99 is 0.9901, which corresponds to $z^* = 2.33$.

5. A random sample of American high school students was polled five years ago and asked whether they feel they could go one day without their smartphone. The same question was asked of a random sample of high school students this year. A 95% confidence interval was constructed for the difference in proportions of students who said they could not go a day without their smartphones. (*This year – five years ago*) and the interval is (0.024, 0.145). Does this confidence interval provide evidence that, if all American high school students had been asked, the proportion of students this year would say they could not go without their smartphones for a day is different than would have said so five years ago?

   Answer: Yes. Because 0 is not contained in the interval.

6. What is the reason we use *t* rather than z when creating a confidence interval for a mean?

   Answer: We don't know σ and must estimate with *s*. The sample standard deviation underestimates σ more often than not, making many of our intervals too narrow if we use *z*. This reduces the capture rate. We use *t* as an adjustment to get the correct capture rate.

7. You want to create a 95% confidence interval for a population proportion with a margin of error of no more than 0.05. How large a sample do you need?

   Answer: Because there is no estimate for the population proportion, we will use $p = 0.5$ to ensure that even the largest margin of error is at most 0.05.

   $$0.05 \geq 1.96 \sqrt{\frac{0.5 \cdot 0.5}{n}}$$

   Solve for *n* to get

   $$n \geq \left(\frac{1.96}{0.05} \right)^2 (0.5)(0.5) = 384.16.$$

   You would need a minimum sample size of 385 subjects. (Always round up to ensure your sample size is large enough.)

8. Why must we check that $n\hat{p}$ and $n(1 - \hat{p})$ are at least ten? And why do we care that our sample size is less than 10% of the population size?

   Answer: $n\hat{p}$ and $n(1 - \hat{p})$ must be at least 10 to ensure that a normal model will give us reasonable approximation to the sampling distribution of $\hat{p}$. Using *z* to estimate the middle 95% of a sampling distribution (and, hence, use it in the margin of error calculation for our interval) is based on a normal model being appropriate. The sample being less than 10% of the population allows us to use our standard error formulas for those sampling distributions rather than having to use the Finite Population Correction. The sample being larger than 10% of the population is not a bad thing; it just means the standard error is less than we're saying it is.

Practice Problems

Multiple-Choice

1. You are going to create a 95% confidence interval for a population proportion and want the margin of error to be no more than 0.05. Historical data indicate that the population proportion has remained constant at about 0.7. What is the minimum size random sample you need to construct this interval?

   a. 385
   b. 322
   c. 274
   d. 275
   e. 323

2. Which of the following would result in a smaller margin of error in a confidence interval for a mean?

   a. Increasing the sample size
   b. Increasing the confidence level
   c. More variability in the sample responses
   d. A smaller sample mean
   e. All of these would result in a smaller margin of error.

3. St. Norbert College in Green Bay conducts an annual survey of Wisconsin residents. In 2011 a random sample of 400 Wisconsin residents was selected and 25% of respondents said the country is headed in the right direction. In a 2016 random sample of 664 residents, 28% said they thought the country was headed in the right direction. Do these samples provide convincing evidence of a change in the proportion of adults who thought the country was headed in the right direction?

   a. No, because the 95% confidence interval for the difference contains 3%.
   b. No, because the 95% confidence interval for the difference contains 0%.
   c. No, because the 95% confidence interval for the difference does not contain 0%.
   d. Yes, because the 95% confidence interval for the difference does not contain 3%.
   e. Yes, because the 95% confidence interval for the difference contains 3%.

4. You are going to construct a 90% t confidence interval for a population mean based on a sample size of 16. What is the critical value of t (t^*) you will use in constructing this interval?

   a. 1.341
   b. 1.753
   c. 1.746
   d. 2.131
   e. 1.337

5. A 95% confidence interval for the difference between two population proportions is found to be (0.07, 0.19). Which of the following statements is (are) true?
   - I. It is unlikely that the two populations have the same proportions.
   - II. We are 95% confident that the true difference between the population proportions is between 0.07 and 0.19.
   - III. The probability is 0.95 that the true difference between the population proportions is between 0.07 and 0.19.

   a. I only
   b. II only
   c. I and II only
   d. I and III only
   e. II and III only

6. A 99% confidence interval for the true mean weight loss (in pounds) for people on the SkinnyQuick diet plan is found to be (1.3, 5.2). Which of the following is (are) correct?
   - I. The probability is 0.99 that the mean weight loss is between 1.3 lb and 5.2 lb.
   - II. The probability is 0.99 that intervals constructed by this process will capture the true population mean.
   - III. We are 99% confident that the true mean weight loss for this program is between 1.3 lb and 5.2 lb.
   - IV. This interval provides evidence that the SkinnyQuick plan is effective in reducing the mean weight of people on the plan.

   a. I and II only
   b. II only
   c. II and III only
   d. II, III, and IV only
   e. All of these statements are correct.

7. A poll was conducted in which Americans were asked whether they approve of the job the president of the United States is doing. The poll reported that 34% approve of the job the president is doing, with a margin of error of 3% and a 95% confidence level. Which of these is the best explanation of what that 3% means?

   a. There is a 3% chance that the 34% estimate is incorrect.
   b. Up to 3% of those sampled may have responded incorrectly.
   c. There is a probability of 3% that the proportion of all Americans who approve of the job the president is doing is not in the confidence interval.
   d. In about 95% of all possible samples of this size, the sample proportion will be within three percentage points of 34%.
   e. In about 95% of all possible samples of this size, the sample proportion will be within three percentage points of the proportion of all Americans who approve of the job the president is doing.

8. A 99% confidence interval for the weights of a random sample of high school wrestlers is reported as (125, 160). Which of the following statements about this interval is true?

   a. At least 99% of the weights of high school wrestlers are in the interval (125, 160).
   b. The probability is 0.99 that the true mean weight of high school wrestlers is in the interval (125, 160).

c. 99% of all samples of this size will yield a confidence interval of (125, 160).

d. The procedure used to generate this confidence interval will capture the true mean weight of high school wrestlers 99% of the time.

e. The probability is 0.99 that a randomly selected wrestler will weigh between 125 and 160 lb.

9. This year's statistics class was small (only 15 students). This group averaged 74.5 on the final exam with a sample standard deviation of 3.2. Assuming that this group is a random sample of all students who have taken statistics and the scores in the final exam for all students are approximately normally distributed, which of the following is an approximate 96% confidence interval for the true population mean of all statistics students?

a. 74.5 ± 7.245

b. 74.5 ± 7.197

c. 74.5 ± 1.871

d. 74.5 ± 1.858

e. 74.5 ± 1.772

10. A random sample of 425 residents of a city of 50,000 people was taken, and people were asked whether they consider themselves to be conservative, liberal, or moderate. A 95% confidence interval was constructed for the difference $p_C - p_L$, between the proportion of all city residents who consider themselves to be conservative and those who consider themselves to be liberal. Why is this procedure inappropriate?

a. The sample size is too small to give a good representation of 50,000 people.

b. The proportions of liberals and conservatives are not independent.

c. $n\hat{p}$ or $n(1-\hat{p})$ will almost certainly be less than 10 for one of the groups.

d. The sample size is less than 10% of the population.

e. We don't know if the population is approximately normally distributed.

Free-Response

1. You attend a large university with approximately 15,000 students. You want to construct a 90% confidence interval estimate, within 5%, for the proportion of students who favor outlawing country music. How large a sample do you need?

2. The local farmers association in Cass County wants to estimate the mean number of bushels of corn produced per acre in the county. A random sample of 13 1-acre plots produced the following results (in number of bushels per acre): 98, 103, 95, 99, 92, 106, 101, 91, 99, 101, 97, 95, 98. Construct a 95% confidence interval for the mean number of bushels per acre in the entire county. The local association has been advertising that the mean yield per acre is 100 bushels. Do you think it is justified in this claim?

3. Two groups of 40 randomly selected students were selected to be part of a study on dropout rates. Members of one group were enrolled in a counseling program designed to give them skills needed to succeed in school, and the other group received no special counseling. Fifteen of the students who received counseling dropped out of school, and 23 of the students who did not receive counseling dropped out. Construct a 90% confidence interval for the true difference between the dropout rates of the two groups. Interpret your answer in the context of the problem.

4. A hotel chain claims that the average stay for its business clients is 5 days. The manager of one hotel in the chain believes that the true stay at his hotel may actually be different from 5 days. A study conducted by the hotel of 100 randomly selected clients yields a mean of 4.55 days with a standard deviation of 3.1 days.

   (a) Construct and interpret a 95% confidence interval for the mean stay at that hotel.
   (b) Does this interval provide convincing evidence that the mean stay at this hotel is different from 5 days?

5. One researcher wants to construct a 99% confidence interval as part of a study. A colleague says such a high level isn't necessary and that a 95% confidence level will suffice. In what ways will these intervals differ?

6. A 95% confidence interval for the true difference between the mean ages of male and female statistics teachers is constructed based on a sample of 95 males and 62 females. Consider each of the following intervals that might have been constructed:
   I. (−4.5, 3.2)
   II. (2.1, 3.9)
   III. (−5.2, −1.7)

   For each of these intervals,

   (a) Interpret the interval, and
   (b) Describe the conclusion about the difference between the mean ages that might be drawn from the interval.

7. A 99% confidence interval for a population mean is to be constructed. A sample size of 20 will be used for the study. Assuming that the population from which the sample is drawn is approximately normal, what is the upper critical value needed to construct the interval?

8. Wikipedia has a "random article" link that selects an article at random. For a class project, a group of students wants to estimate the mean length of an article on Wikipedia by randomly selecting 100 articles from the more than 5 million articles currently on Wikipedia. The students will record number of words in each article and construct a confidence interval for the mean number of words. The students got into a discussion about whether this link samples with or without replacement. Explain why, for the purposes of their project, it doesn't make much difference whether it samples with or without replacement.

9. A flu vaccine is being tested for effectiveness. To test this, 350 randomly selected people are given the vaccine and observed to see if they develop the flu during the flu season. At the end of the season, 55 of the 350 did get the flu. Construct and interpret a 95% confidence interval for the true proportion of people who will get the flu despite getting the vaccine.

10. A research study gives a 95% confidence interval for the proportion of subjects helped by a new anti-inflammatory drug as (0.56, 0.65).

    (a) Interpret this interval in the context of the problem.
    (b) What is the meaning of "95% confidence interval" as stated in the problem?

11. For their statistics class, Bree and Ashling wanted to explore the effect of perceived peer pressure. They designed a paper survey in two versions. Version 1 simply asked, "On a scale of 1 (rarely) to 10 (constantly), how often do you procrastinate?" Version 2 asked the same question but students could see a list of supposed prior students' responses to the question and these prior student responses circled numbers at the higher end more often than in the middle or at the lower end. Students were randomly assigned one of the two versions of the question and their results were recorded. Summary statistics for each group are recorded in the table below:

TREATMENT	NUMBER OF SUBJECTS	MEAN	STANDARD DEVIATION
Control—Version 1	36	5.85	2.01
Primed—Version 2	32	7.45	1.76

The confidence interval for the difference in means (Version 2 – Version 1) is (0.687, 2.513). What does this Interval tell you about the effect of the "peer pressure" version of the question?

12. A study was conducted to determine if male and female 10th graders differ in performance in mathematics. Twenty-three randomly selected males and 26 randomly selected females were each given a 50-question multiple-choice test as part of the study. The scores were approximately normally distributed. The results of the study were as follows:

	MALES	FEMALES
Sample size	23	26
Mean	40.3	39.2
Std. deviation	8.3	7.6

Construct a 99% confidence interval for the true difference between the mean score for males and the mean score for females. Does the interval suggest that there is a difference between the true means for males and females?

13. A university wants to conduct a telephone poll of state residents to determine opinions about social trends. They estimate that it will cost $2.50 for each response based on the time it takes student employees to get a response to the survey. Through a private donation for this purpose, they have $1,500 to spend on gathering data for the survey.

(a) If the donor insists that they estimate all proportions to within a 2% margin of error, will they have the funds to do it?

(b) What is the *minimum* sample size needed to get the desired precision? What should the university tell the donor?

14. You want to estimate the proportion of Californians who want to outlaw cigarette smoking in all public places. Generally speaking, by how much must you increase the sample size to cut the margin of error in half?

15. The game Rock-Paper-Scissors consists of two people, at the same time, making a symbol with their hands. Each player can form a closed fist (rock), an open palm (paper), or extend two fingers opened (scissors). The idea is that "paper covers rock, scissors cuts paper, and rock smashes scissors," so that each option has one option that it beats and one option that beats it. It is considered a game of chance, but there are those who insist that there is skill involved. There are actually Rock-Paper-Scissors tournaments. One skeptic wants to see if this is true, and he watches someone who claims to be a champion player and records the proportion of games won out of 100. What value of p, the true probability of this player winning, would the skeptic expect to be in his confidence interval?

16. A pre-election poll asked likely voters, "If the election were held today, would you vote for Candidate A or Candidate B?" They randomized the order in which the candidates were listed so there would not be any bias based on the order. They reported that 51% of likely voters said they would vote for Candidate B with a margin of error of three percentage points, and referred to it as a statistical tie. What does the term "statistical tie" mean?

17. Based on the 2000 census, the population of the United States was about 281.4 million people, and the population of Nevada was about 2 million. We are interested in generating a 95% confidence interval, with a margin of error of 3%, to estimate the proportion of people who will vote in the next presidential election. How much larger a sample will we need to generate this interval for the United States than for the state of Nevada?

18. Professor Olsen has taught statistics for 41 years and has kept the scores of every test he has ever given. Every test has been worth 100 points. He is interested in the average test score over the years. He doesn't want to put all of the scores (there are thousands of them) into a computer to figure out the exact average, so he asks his daughter, Anna, to randomly select 50 of the tests and use those to come up with an estimate of the population average. Anna has been studying statistics at college and decides to create a 98% confidence interval for the true average test score. The mean test score for the 50 randomly selected tests is 73.5 with a standard deviation of 7.1. What does she tell her father?

19. A certain type of pen is claimed to operate for a mean of 190 hours. A random sample of 49 pens is tested, and the mean operating time is found to be 188 hours with a standard deviation of 6 hours. Construct a 95% confidence interval for the true mean operating time of this type of pen. Does the company's claim seem justified?

20. A young researcher thinks there is a difference between the mean ages at which males and females win Oscars for best actor or actress. The student found the mean age for all best actor winners and all best actress winners and constructed a 95% confidence interval for the mean difference between their ages. Is this an appropriate use of a confidence interval? Why or why not?

Cumulative Review Problems

1. Use a normal approximation to the binomial to determine the probability of getting between 470 and 530 heads in 1,000 flips of a fair coin.

2. A survey of the number of televisions per household found the following probability distribution:

TELEVISIONS	PROBABILITY
0	0.03
1	0.37
2	0.46
3	0.10
4	0.04

What is the mean number of television sets per household?

3. A bag of marbles contains four red marbles and five blue marbles. A marble is drawn, its color is observed, and it is returned to the bag.

(a) What is the probability that the first red marble is drawn on trial 3?
(b) What is the average wait until a red marble is drawn?

4. A study is conducted to determine which of two competing weight-loss programs is more effective. Random samples of 50 people from each program are evaluated for losing and maintaining weight loss over a 1-year period. The average number of pounds lost per person over the year is used as a basis for comparison.

(a) Why is this an observational study and not an experiment?
(b) Describe an experiment that could be used to compare the two programs. Assume that you have available 100 overweight volunteers who are not presently in any program.

5. The correlation between the first and second statistics tests in a class is 0.78.

(a) Interpret this value.
(b) What proportion of the variation in the scores on the second test can be explained by the scores on the first test?

Solutions to Practice Problems

Multiple-Choice

1. The correct answer is (e).

$p = 0.7$, $M = 0.05$, $z^* = 1.96$ (for $C = 0.95$) $\Rightarrow$

$n \geq \left(\dfrac{z^*}{M} \right)^2 (\hat{p})(1 - \hat{p}) = \left(\dfrac{1.96}{0.05} \right)^2 (0.7)(0.3) = 322.7$. You need a sample of at least

$n = 323$.

2. The correct answer is (a). Options (b) and (c) would increase the margin of error, and option (d) would not change it.

3. The correct answer is (b). It is not necessary to calculate the interval to answer this question because (b) is the only choice that is logically consistent. The question is whether 0% is in the interval, which leaves only options (b) and (c). And if 0 is in the interval, then there is not convincing evidence of a change. And the interval actually does contain 0.

4. The correct answer is (b). $n = 16 \Rightarrow df = 16 - 1 = 15$. Using a table of t distribution critical values, look in the row for 15 degrees of freedom and the column with 0.05 at the top (or 90% at the bottom). On a TI-83/84 with the `invT` function, the solution is given by `invT(0.95,15)`.

5. The correct answer is (c). Because 0 is not in the interval (0.07, 0.19), it is unlikely to be the true difference between the proportions. III is just plain wrong! We cannot make a probability statement about an interval we have already constructed. All we can say is that the process used to generate this interval has a 0.95 chance of producing an interval that does contain the true population proportion.

6. The correct answer is (d). I is not correct because you cannot make a probability statement about a found interval—the true mean is either in the interval ($P = 1$) or it isn't ($P = 0$). II is correct and is just a restatement of "Intervals constructed by this procedure will capture the true mean 99% of the time." III is true—it's our standard way of *interpreting* a confidence interval. IV is true. Since the interval constructed does not contain 0, it's unlikely that this interval came from a population whose true mean is 0. Since all the values are positive, the interval does provide statistical evidence (but not proof) that the program is effective at promoting weight loss. It does not give evidence that the amount lost is of practical importance.

7. The correct answer is (e). The margin of error is an estimate of how far off our estimate might be. We expect 95% of sample proportions to lie within three percentage points of the population proportion. So we are pretty sure the correct proportion is between 31% and 37%.

8. The correct answer is (d). A confidence level is a statement about the procedure used to generate the interval, not about any one interval. It's difficult to use the word "probability" when interpreting a confidence interval and impossible when describing an interval that has already been constructed. However, you could say, "The probability is 0.99 that an interval constructed in this manner will contain the true population proportion."

9. The correct answer is (c). For df $= 15 - 1 = 14$, $t^* = 2.264$ for a 96% confidence interval (from Table B; if you have a TI-83/84 with the `invT` function, `invT(0.98,14)=2.264`).

   The interval is $74.5 \pm (2.264)\left(\dfrac{3.2}{\sqrt{15}}\right) = 74.5 \pm 1.871$.

10. The correct answer is (b). We don't have samples independently selected from two populations. We have one sample selected from one population. If the proportion of liberals in our sample is higher than in the population, the proportion of conservatives is likely to be lower. The proportions are not independent.

Free-Response

1. $C = 0.90 \Rightarrow z^* = 1.645$, $M = 0.05$. $n \geq \left(\dfrac{1.645}{0.05} \right)^2 (0.5)(0.5) = 270.6$.

    You would need to survey at least 271 students.

2. The population standard deviation is unknown, and the sample size is small (13), so we need to use a t procedure. The problem tells us that the sample is random. A histogram of the data shows no significant departure from normality:

    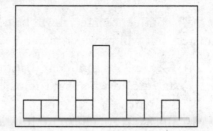

    Now, $\bar{x} = 98.1$, s = 4.21, df = 13 − 1 = 12 $\Rightarrow t^* = 2.179$. The 95% confidence interval is

    $$98.1 \pm 2.179 \left(\frac{4.21}{\sqrt{13}} \right) = (95.56, 100.64).$$

    Because 100 is contained in this interval, we do not have strong evidence that the mean number of bushels per acre differs from 100, even though the sample mean is only 98.1.

3. This is a two-proportion situation. We are told that the groups were randomly selected, but we need to check that the samples are sufficiently large:

    $$\hat{p}_1 = \frac{15}{40} = 0.375, \ \hat{p}_2 = \frac{23}{40} = 0.575.$$

    $$n_1 \hat{p}_1 = 40(0.375) = 15, \ n_1(1 - \hat{p}_1) = 40(1 - 0.375) = 25.$$

    $n_2 \hat{p}_2 = 40(0.575) = 23$, $n_2(1 - \hat{p}_2) = 40(1 - 0.575) = 17$. Since all values are greater than or equal to 5, we are justified in constructing a two-proportion z interval. For a 90% z confidence interval, $z^* = 1.645$.

    Thus, $(0.575 - 0.375) \pm 1.645 \sqrt{\dfrac{(0.375)(1 - 0.375)}{40} + \dfrac{(0.575)(1 - 0.575)}{40}} = (0.02, 0.38)$.

    We are 90% confident that the true difference between the dropout rates is between 0.02 and 0.38. Since 0 is not in this interval, we have evidence that the counseling program was effective at reducing the number of dropouts.

4. (a) Check conditions and identify procedure: We have a random sample of clients. Because $n = 100 > 30$, the sample size is large enough to use t procedures. There are probably more than $10 \cdot 100 = 1,000$ clients at this hotel. Conditions are met for a confidence interval for a mean.

Compute the interval: The 95% confidence interval is

$$\bar{x} \pm t^* \frac{s}{\sqrt{n}} = 4.55 \pm 1.98 \frac{3.1}{\sqrt{10}} = (3.93, 5.17), \text{ using } t^* \text{ for } 99 \ df.$$

Interpret the interval: We are 95% confident that the mean stay for clients at this hotel is between 3.93 days and 5.17 days.

(b) Because 5 days is in our interval, we do not have convincing evidence that the mean length of stay is different from 5 days.

5. The 99% confidence interval will be more likely to contain the population value being estimated, but will be wider than a 95% confidence interval.

6. I. (a) We are 95% confident that the true difference between the mean age of male statistics teachers and female statistics teachers is between –4.5 years and 3.2 years.
     (b) Since 0 is contained in this interval, we do not have evidence of a statistically significant difference between the mean ages of male and female statistics teachers.
   II. (a) We are 95% confident that the true difference between the mean age of male statistics teachers and female statistics teachers is between 2.1 years and 3.9 years.
     (b) Since 0 is not in this interval, we do have evidence of a real difference between the mean ages of male and female statistics teachers. In fact, since the interval contains only positive values, we have evidence that the mean age of male statistics teachers is greater than the mean age of female statistics teachers.
  III. (a) We are 95% confident that the true difference between the mean age of male statistics teachers and female statistics teachers is between –5.2 years and –1.7 years.
     (b) Since 0 is not in this interval, we have evidence of a real difference between the mean ages of male and female statistics teachers. In fact, since the interval contains only negative values, we have evidence that the mean age of male statistics teachers is less than the mean age of female statistics teachers.

7. t procedures are appropriate because the population is approximately normal. $n = 20 \Rightarrow$ df $= 20 - 1 = 19 \Rightarrow t^* = 2.861$ for $C = 0.99$.

8. Because the sample is small compared to the number of articles on Wikipedia (far less than 10% of the population size), sampling without replacement will not change the standard deviation of the sampling distribution of the sample mean enough to notice.

9. $\hat{p} = \frac{55}{350} = 0.157$, $n\hat{p} = 350(0.157) = 54.95 \geq 5$, $n(1 - \hat{p})$

$= 350(1 - 0.157) = 295.05 \geq 5$.

The conditions are present to construct a one-proportion z interval.

$$0.157 \pm 1.96 \sqrt{\frac{0.157(1 - 0.157)}{350}} = (0.119, 0.195)$$

We are 95% confident that the true proportion of people who will get the flu despite getting the vaccine is between 11.9% and 19.5%. To say that we are 95% confident means that the process used to construct this interval will, in fact, capture the true population proportion 95% of the time (that is, our "confidence" is in the procedure, not in the interval we found!).

10. (a) We are 95% confident that the true proportion of subjects helped by a new anti-inflammatory drug is (0.56, 0.65).

    (b) The process used to construct this interval will capture the true population proportion, on average, 95 times out of 100.

11. Because 0 is not in the interval and the entire interval of differences is positive, we can conclude that the "peer pressure" version of the question caused people to rate their level of procrastination as 0.687 to 2.513 points higher than the version without the peer pressure.

12. The problem states that the samples were randomly selected and that the scores were approximately normally distributed. Both samples are less than 10% of the populations of males and females in 10th grade, so we can construct a two-sample t interval.

    The interval is $(\bar{x}_M - \bar{x}_F) \pm t^* \sqrt{\dfrac{s_M^2}{n_m} + \dfrac{s_F^2}{n_F}} = (-5.04, 7.24)$, using t^* for 44.9 df. (You do not need to calculate t^* for two-sample procedures on the AP Exam. You are sometimes required to report degrees of freedom, which your calculator will give you.) We are 99% confident that male 10th graders score, on average, between 5 points lower and 7 points higher than female 10th graders on this mathematics test. Because 0 is in the interval we cannot conclude that there is a difference in performance between males and females.

13. (a) With a budget of $1,500, they can survey 600 residents. The margin of error if they survey 600 residents is, at most, $1.96\sqrt{\dfrac{0.5 \cdot 0.5}{600}} = 0.04$. They will not have enough funds to meet the donor's goal.

    (b) Solving $1.96\sqrt{\dfrac{0.5 \cdot 0.5}{n}} \le 0.02, n \ge 2,401$. The donor would need to approximately quadruple the funding for gathering the data to get the desired precision.

14. In the formula for the margin of error, n is in the denominator under the square root. So, for example, doubling the sample size will reduce the margin of error by a factor of $\sqrt{2}$. To cut the margin of error in half, you must multiply n by $2^2 = 4$.

15. The skeptic thinks this player really has a 50% chance of winning any given round, so he would expect 0.5 to be in the interval.

16. A tie would mean that the same proportion of likely voters would say they are voting for Candidate A and Candidate B. Because the two candidates are within three percentage points of each other (the margin of error) it is plausible that the same proportion of people in the population would vote for each candidate. (Note: A two-sample procedure here is inappropriate because there are not two independently selected samples from two populations.)

17. Trick question! The population size is not a part of the margin of error formula so you can use the same sample size. (Assuming the sample size is small compared to the population size!)

18. She randomly selected the sample of scores. $n = 50$ so the sample size is large enough to use t procedures. 50 is less than 10% of the thousands of test scores, so conditions are met for a confidence interval for a mean.

    The interval is $75.5 \pm 2.404 \dfrac{7.1}{\sqrt{50}} = (71.09, 75.91)$.

    She can tell her father that she is 98% confident that the mean score for all his students is between 71.09 and 75.91.

19. The problem states that we have an SRS and $n = 49$. We have a large sample size, but it is less than 10% of the population of all such pens, so we are justified in using t procedures.

    $$188 \pm 2.011 \dfrac{6}{\sqrt{49}} = (186.3, 189.7), \text{ using } t^* \text{ for 95\% confidence and 48 } df.$$

    Because 190 is not in this interval, it is not a plausible mean for the population from which we would have gotten this sample. There is some doubt about the company's claim.

20. It is not appropriate because confidence intervals use sample data to make estimates about unknown population values. In this case, the actual difference in the ages of actors and actresses is known, and the true difference can be calculated.

Solutions to Cumulative Review Problems

1. Let $X =$ the number of heads. Then X has $B(1000, 0.5)$ because the coin is fair. This binomial can be approximated by a normal distribution with mean $= 1000(0.5) = 500$ and standard deviation

   $$s = \sqrt{1000(0.5)(0.5)} = 15.81. P(470 < X < 530)$$

   $$= P\left(\frac{470 - 500}{15.81} < z < \frac{530 - 500}{15.81}\right) = P(-1.90 < z < 1.90) = 0.94. \text{ Using the TI-83/84}$$

   calculator, `normalcdf(-1.9,1.9)`.

2. $\mu_x = 0(0.3) + 1(0.37) + 2(0.46) + 3(0.10) + 4(0.04) = 1.75$.

3. (a) $P(\text{draw a red marble}) = 4/9$

   $$G(3) = \frac{4}{9}\left(1 - \frac{4}{9}\right)^{3-1} = \frac{4}{9}\left(\frac{5}{9}\right)^2 = 0.137.$$

   (b) Average wait

   $$= \frac{1}{p} = \frac{1}{4\!\!\big/\!9} = \frac{9}{4} = 2.25.$$

4. (a) It is an observational study because the researchers are simply observing the results between two different groups. To do an experiment, the researcher must manipulate the variable of interest (different weight-loss programs) in order to compare their effects.

    (b) Inform the volunteers that they are going to be enrolled in a weight-loss program and their progress monitored. As they enroll, randomly assign them to one of two groups, say Group A and Group B (without the subjects' knowledge that there are really two different programs). Group A gets one program and Group B the other. After a period of time, compare the average number of pounds lost for the two programs.

5. (a) There is a moderate to strong positive linear relationship between the scores on the first and second tests.

    (b) $0.78^2 = 0.61$. About 61% of the variation in scores on the second test can be explained by the regression on the scores from the first test.

CHAPTER 11

Inference: Testing Hypotheses

IN THIS CHAPTER

Summary: In the last chapter, we concentrated on estimating, using a confidence interval, the value of a population parameter or the difference between two population parameters. In this chapter, we test to see whether some specific hypothesized value of a population parameter, or the difference between two populations' parameters, is plausible or not. We form hypotheses and test to determine the probability of getting a particular result if the null hypothesis is true. We then make decisions about our hypothesis based on that probability.

Key Ideas
- Statistical Significance and *P*-Value
- The Logic of Hypothesis Testing
- Hypothesis-Testing Procedure
- A Significance Test for a Population Proportion
- Errors in Hypothesis Testing
- The Power of a Test
- A Significance Test for the Difference between Two Population Proportions
- *z*-Procedures versus *t*-Procedures
- A Significance Test for a Population Mean
- A Significance Test for the Difference between Two Population Means

In the last chapter we concentrated on estimating the value of a population parameter, using a confidence interval. In this chapter, we form hypotheses and perform a test to determine whether we have convincing evidence that a particular hypothesis is incorrect. To do so, we will be introduced to the logic behind significance testing, the meaning of a *P*-value, the types of errors you have to worry about when testing, and what is mean by the *power* of a test. There's a lot in this chapter, and you will need to work hard to keep it all straight.

Statistical Significance and *P*-Value

Statistical Significance

In the previous chapter, we used confidence intervals to make estimates about population values. In one of the examples, we observed that, because 0 was contained in the confidence interval, 0 was a plausible value for the difference in population proportions. In other words, we could NOT conclude that there was a difference in population proportions at all. As we progress through techniques of inference (drawing conclusions about a population from data), we often are interested in whether a sample statistic is likely under a particular assumption.

We begin by making an assumption about the population parameter. This assumption is called the **null hypothesis**. If a sample is not one we would expect based on our null hypothesis, it could be because of sampling variability, or it could be because the sample came from a different population than we thought. If the result is so far from what we expected that we think something other than chance is operating, then the result is said to be **statistically significant**.

> **example:** Todd claims that the average amount of money spent on ride tickets at a county fair is $50. If you have a random group of 50 fair goers and the average money spent on ride tickets for this group is $49.75, we have little reason to doubt Todd's claim. But if the average for this group is $38, statisticians call this result *statistically significant* meaning that we would be unlikely to have a sample average this low if Todd's claim is true.

In the above example, most people would agree that $49.75 was consistent with Todd's claim (that is, it was a likely sample mean if the population mean is $50) and that $38 is inconsistent with the claim (it is *statistically significant*). It's a bit more complicated to decide where between $38 and $49.75 the cutoff is between "reasonably likely" and "unlikely."

There are some general agreements about how unlikely a statistic needs to be, assuming the null hypothesis is true, in order to be statistically significant. Typical significance levels, symbolized by the Greek letter α, are probabilities of 0.1, 0.05, and 0.01. If a result has a lower probability of occurring than the significance level, then the result is statistically significant at that significance level.

> **example:** A statistics student determined that the probability that a random group of 50 fair goers would average $38 on county fair ride tickets when Todd's claim of an average of $50 is actually true to be less than 0.00000002. This value is so low that it seems unlikely to have occurred by chance, and so we say that the result is statistically significant. This probability is lower than any of the commonly accepted significance levels.

P-Value

We said that a result is statistically significant if it is unlikely to have occurred by chance if the null hypothesis is true. The **P-value** is what tells us just *how* unlikely a result actually is under the model based on the null hypothesis. The *P*-value is the probability, based on our model, of getting a sample statistic by chance alone as extreme, or more extreme, as the one we obtained. This requires that we have some expectation about what we ought to get. In other words, the *P*-value is the probability of getting a statistic at least as far removed from expected as we got. A decision about significance can then be made by comparing the obtained *P*-value with a stated value of α.

> **example:** Suppose it turns out that the average amount of money a *random* group of 50 people spend on ride tickets at the county fair is $47.50. What is the probability of getting an observed mean this far below the expected $50 by chance alone (that is, what is the *P*-value) if the true average is $50 (assume the population standard deviation is $8)? Is this finding significant at $\alpha = 0.05$? At $\alpha = 0.01$?
>
> **solution:** We are assuming the population is normally distributed with mean $50 and standard deviation $8. The situation is pictured below:

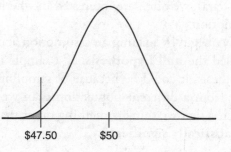

$$P(\bar{x} < 47.5) = P\left(z < \frac{47.5 - 50}{8/\sqrt{50}} = -2.21 \right) = 0.014.$$

This is the *P*-value: it's the probability of getting a sample mean as far below $50 as we did by chance alone, assuming the model based on a mean of $50 is correct. This finding is significant at the 0.05 level but not (quite) at the 0.01 level.

The Hypothesis-Testing Procedure

Now we will want to test whether the parameter has a particular value or not. More accurately, we might ask if we have convincing evidence *against* the hypothesis that $p_1 - p_2 = 0$ or if $\mu = 3$, for example. That is, we will test the hypothesis that, say, $p_1 - p_2 = 0$. In the hypothesis-testing procedure, a researcher does not look for evidence to support this hypothesis, but instead looks for evidence against this hypothesis. The process looks like this:

- *State the null and alternative hypotheses in the context of the problem.* The first hypothesis, the **null hypothesis**, is the hypothesis we are actually testing. The null hypothesis usually states that there is nothing going on: the claim is correct or that there is no distinction between groups. It is symbolized by H_0. An example of a typical null hypothesis would be $H_0: \mu_1 - \mu_2 = 0$ or $H_0: \mu_1 = \mu_2$. This is the hypothesis that μ_1 and μ_2 are the same, or that populations 1 and 2 have the same mean. Note that μ_1 and μ_2 must be identified in context (for example, μ_1 = the mean score for all people in the population before training).

  The second hypothesis, the **alternative hypothesis**, is the theory that the researcher wants to confirm by rejecting the null hypothesis. The alternative hypothesis is symbolized by H_A or H_a. There are three possible forms for the alternative hypothesis: $\neq$, $>$, or $<$. If the null is $H_0: \mu_1 - \mu_2 = 0$, then H_A could be:

$$H_A: \mu_1 - \mu_2 \neq 0 \text{ (this is called a \textbf{two-sided alternative})}$$

or

$$H_A: \mu_1 - \mu_2 > 0 \text{ (this is a \textbf{one-sided alternative})}$$

or

$$H_A: \mu_1 - \mu_2 < 0 \text{ (also a \textbf{one-sided alternative}).}$$

(In the case of the one-sided alternative $H_A: \mu_1 - \mu_2 > 0$, the null hypothesis is sometimes written: $H_0: \mu_1 - \mu_2 \leq 0$. This actually makes pretty good sense: if the researcher is wrong in a belief that the difference is greater than 0, then any finding less than or equal to 0 fails to provide evidence in favor of the alternative.)

- *Identify which procedure you intend to use and show that the conditions for its use are present.* We identified the conditions for constructing a confidence interval in the previous chapter. Now, we will want to identify the conditions needed to do hypothesis testing. For the most part, they are similar to those you have already studied. One exception is this: when we are conducting a test for a proportion, we have an assumption about the value of p. This value is referred to as p_0. Because we are creating a model based on p_0, we will use p_0 when checking conditions. Specifically, we check that np_0 and $n(1 - p_0)$ are greater than 10. For a test for a difference in proportions, out null hypothesis is that $p_1 = p_2$. We don't have an estimate for the *value* of p_1 and p_2, so the best we can do is treat the two samples as one big sample to estimate that value. We define $\hat{p}$ as the *pooled proportion* and we calculate it as the total number of successes over the total sample size. Symbolically:

$$\hat{p} = \frac{x_1 + x_2}{n_1 + n_2} = \frac{n_1 \hat{p}_1 + n_2 \hat{p}_2}{n_1 + n_2}.$$ The condition we must check to ensure that the shape of our sampling distribution model for $\hat{p}_1 - \hat{p}_2$ is approximately normal is that $n_1 \hat{p}$, $n_1(1 - \hat{p})$, $n_2 \hat{p}$, and $n_2(1 - \hat{p})$, are all at least 5 (or 10). If you are going to state a significance level α it can be done here.

- *Compute the value of the test statistic and the P-value.*
- *Give a conclusion, linked to your computations, in the context of the problem.*

Exam Tip: The four steps above have been incorporated into AP exam scoring for any question involving a hypothesis test. Note that for the third item (compute the value of the test statistic and the P-value), the mechanics in the problem are only one part of a complete solution. *All* four steps must be present in order to receive a 4 ("complete response") on the problem.

If you stated a significance level in the second step of the process, the conclusion can be based on a comparison of the P-value with α. If you didn't state a significance level, you can argue your conclusion based on the value of the P-value alone: if it is small, you have evidence against the null; if it is not small, you do not have evidence against the null.

Many statisticians will argue that you are better off to argue directly from the P-value and not use a significance level. One reason for this is the arbitrariness of the P-value. That is, if $\alpha = 0.05$, you would reject the null hypothesis for a P-value of 0.04999 but not for a P-value of 0.05001 when, in reality, there is no practical difference between them.

The conclusion can be (1) that we reject H_0 (because of a sufficiently small P-value) or (2) that we do not reject H_0 (because the P-value is too large). We do **NOT** accept the null: we either reject it or fail to reject it. If we reject H_0, we can say in context that we have convincing evidence in favor of H_A.

> **example:** Consider, one last time, Todd and his claim that the average money spent on fair rides is $50. A sample was taken and we got $47.50; we assumed the population standard deviation was $8. A test of the hypothesis that the average money spent was $50 on average against the alternative that Todd is incorrect might look something like the following (we will fill in many of the details, especially those in the third part of the process, in the following chapters):

- Let μ be the true average money spent by Todd's fair goers. H_0: μ, = $50 (or H_0: μ, ≥ $50, because the alternative is one sided) and H_A: μ, < $50.
- Because we know the population standard deviation σ, we will use a z-test. We assume the 50 people is an SRS of all the people attending the fair, and the central limit theorem tells us that the sampling distribution of $\bar{x}$ is approximately normal. We will use a significance level of $\alpha = 0.05$.
- In the previous section, we determined that the P-value for this situation (the probability of getting an average as far away from our expected value as we got) is 0.014.
- Because the P-value < α (0.014 < 0.05), we can reject H_0. We have good evidence that the true mean amount of money spent on ride tickets is actually less than $50 (note that we *aren't* claiming anything about exactly how much they do spend on average, just that it's likely to be less than $50).

Recap of Hypothesis (Significance) Testing

Here's a summary of the logic of hypothesis testing:

I. State hypotheses in the context of the problem. The first hypothesis, the **null hypothesis**, is the hypothesis we are actually testing. The null hypothesis usually states that there is nothing going on: the claim is correct or there is no distinction between groups. It is symbolized by H_0.

The second hypothesis, the **alternative hypothesis**, is the theory that the researcher wants to confirm by rejecting the null hypothesis. The alternative hypothesis is symbolized by H_A. There are three forms for the alternative hypothesis: ≠, >, or <. That is, if the null is H_0: $\mu_1 - \mu_2 = 0$, then H_A could be:

$$H_A: \mu_1 - \mu_2 = 0 \text{ (a two-sided alternative)}$$

$$H_A: \mu_1 - \mu_2 > 0 \text{ (a one-sided alternative)}$$

$$H_A: \mu_1 - \mu_2 < 0 \text{ (a one-sided alternative)}.$$

(In the case of the one-sided alternative H_A: $\mu_1 - \mu_2 > 0$, the null hypothesis is sometimes written H_0: $\mu_1 - \mu_2 \leq 0$.)

II. Identify which procedure you intend to use and show that the conditions for its use are satisfied. If you are going to state a significance level, α, it can be done here.

III. Compute the value of the test statistic and the P-value.

IV. Using the value of the test statistic and/or the P-value, give a conclusion in the context of the problem.

If you stated a significance level, the conclusion can be based on a comparison of the P-value with α. If you didn't state a significance level, you can argue your conclusion based on the relative size of the P-value alone: if it is small, you have evidence against the null hypothesis; if it is not small, you do not have evidence against the null hypothesis.

The conclusion can be (1) that we reject H_0 (because of a sufficiently small P-value) or (2) that we do not reject H_0 (because the P-value is too large). We *never* accept the null hypothesis: we either reject it or fail to reject it. If we reject H_0, we can say that we have convincing evidence in favor of H_A.

Significance testing involves making a decision about whether or not an observed result is statistically significant. That is, is the result sufficiently unlikely, if the null hypothesis were true, so as to provide good evidence for rejecting the null hypothesis in favor of the alternative? The four steps in the hypothesis testing process outlined above are the four steps that are required on the AP exam when doing inference problems. In brief, every test of a hypothesis should have the following four steps:

I. State the null and alternative hypotheses in the context of the problem, defining all symbols. (Make sure that you are using parameters in words and/or symbols!)

II. Identify the appropriate test and check that the conditions for its use are present.

III. Do the correct mechanics, including calculating the value of the test statistic and the P-value.

IV. State a correct conclusion in the context of the problem.

> **Exam Tip:** You are not required to *number* the four steps on the exam, but it is a good idea to do so—then you are sure you have done everything required.

Inference for a Single Population Proportion

We are usually more interested in estimating a population proportion with a confidence interval than we are in testing that a population proportion has a particular value. However, significance testing techniques for a particular population proportion exist and follow a pattern similar to those of the previous two sections. The main difference is that the only test statistic is z. The logic is based on using a normal approximation to the binomial as discussed in Chapter 9.

• **Hypothesis** • **Estimator** • **Standard Error**	**Conditions**	**Test Statistic**
• Null hypothesis $H_0: p = p_0$ • Estimator:	• SRS • $np_0 \geq 5$, $n(1 - p_0) \geq 5$ (or $np_0 \geq 10$, $n(1 - p_0) \geq 10$)	$z = \dfrac{\hat{p} - p_0}{\sqrt{\dfrac{p_0(1 - p_0)}{n}}}$

$\hat{p} = \dfrac{X}{n}$ where X is the count of successes

• Standard error:

$$s_{\hat{p}} = \sqrt{\frac{p_0(1 - p_0)}{n}}$$

Notes on the preceding table:

- The standard error for a hypothesis test of a single population proportion is different from that for a confidence interval for a population proportion. The standard error for a confidence interval,

$$s_{\hat{p}} = \sqrt{\frac{\hat{p}(1-\hat{p})}{n}},$$

is a function of $\hat{p}$, the sample proportion, whereas the standard error for a hypothesis test,

$$s_{\hat{p}} = \sqrt{\frac{p_0(1-p_0)}{n}},$$

is a function of the hypothesized value of p. This is because a test assumes, for the sake of our model, that $p = p_0$. So we use that hypothesized value in the calculation of the standard error. An interval makes no such assumption.

- Like the conditions for the use of a z-interval, we require that the np_0 and $n(1 - p_0)$ be large enough to justify the normal approximation. As with determining the standard error, we use the hypothesized value of p rather than the sample value. "Large enough" means either $np_0 \geq 5$ and $n(1 - p_0) \geq 5$, or $np_0 \geq 10$ and $n(1 - p_0) \geq 10$ (it varies by textbook, but either 5 or 10 is acceptable).

example: Consider a screening test for prostate cancer that its maker claims will detect the cancer in 85% of the men who actually have the disease. One hundred seventy-five random men who have been previously diagnosed with prostate cancer are given the screening test, and 141 of the men are identified as having the disease. Does this finding provide evidence that the screening test detects the cancer at a rate different from the 85% rate claimed by its manufacturer?

solution:

I. Let p = the true proportion of men with prostate cancer who test positive.

H_0: $p = 0.85$.
H_A: $p \neq 0.85$

II. We want to use a one-proportion z-test. We were told that the men were **random**. $175(0.85) = 148.75 > 5$ and $175(1 - 0.85) = 26.25 > 5$ (remember either 5 or 10 can be used to check the sample size condition), so the conditions are present to use this test.

III. $\hat{p} = \dfrac{141}{175} = 0.806.$

$z = \dfrac{0.806 - 0.85}{\sqrt{\dfrac{0.85(1 - 0.85)}{175}}} = 1.64 \Rightarrow P\text{-value} = 0.10.$

Using the TI-83/84, the same answer is obtained by using 1-PropZTest in the STAT TESTS menu. On the AP exam, we recommend that you simply name the test as in part II, then list the z-value and the P-value reported by your calculator.

IV. Because the P-value is reasonably large, we do not have sufficient evidence to reject the null hypothesis. There is not convincing evidence to believe that the screening test detects the cancer at a rate different from the 85% rate claimed by its manufacturer.

example: Maria has a quarter that she suspects is out of balance. In fact, she thinks it turns up heads more often than it would if it were fair. Being a senior, she has lots of time on her hands, so she decides to flip the coin 300 times and count the number of heads. There are 165 heads in the 300 flips. Does this provide evidence at the 0.05 level that the coin is biased in favor of heads? At the 0.01 level?

solution:

> **I.** Let p = the true proportion of heads in 300 flips of a fair coin.
>
> H_0: $p = 0.50$ (or H_0: $p \leq 0.50$).
> H_A: $p > 0.50$.
>
> **II.** We will use a one-proportion z-test.
>
> We will assume that the 300 flips are a random sample of all flips. $300(0.50) = 150 > 5$ and $300(1 - 0.50) = 150 > 5$, so the conditions are present for a one-proportion z-test.
>
> **III.** $\hat{p} = \dfrac{165}{300} = 0.55$, $z = \dfrac{0.55 - 0.50}{\sqrt{\dfrac{0.50(1 - 0.50)}{300}}} = 1.73 \Rightarrow P\text{-value} = 0.042$.
>
> (This can be done using `1-PropZTest` in the `STAT TESTS` menu of the TI-83/84. We will repeat this many times—on the AP exam, you SHOULD do the mechanics using the test menu from your brand of calculator. Just identify the procedure and list the values reported. Whether you identify the procedure by name or formula, the components of the name of the procedure are all important: one proportion, z, and test must be clear.)
>
> **IV.** The question asked about both $\alpha = 0.05$ and $\alpha = 0.01$. We note that this finding would have been significant at $\alpha = 0.05$ but not at $\alpha = 0.01$. For the two significance levels, the solutions would look something like this:
>
> (1) Because the P-value (0.042) $< \alpha$ (0.05), we reject the null hypothesis. We have evidence that the true proportion of heads is greater than 0.50.
>
> (2) Because the P-value (0.042) $> \alpha$ (0.01), we fail to reject the null hypothesis. We do not have convincing evidence that the true proportion of heads is greater than 0.50.

Type I and Type II Errors and the Power of a Test

When we do a hypothesis test we never really know if we have made the correct decision or not. We can try to minimize our chances of being wrong, but there are trade-offs involved. If we are given a hypothesis, it may be true or it may be false. We can decide to reject the hypothesis or not to reject it. This leads to four possible outcomes:

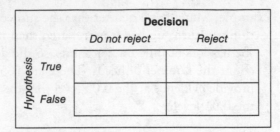

		Decision	
		Do not reject	Reject
Hypothesis	True		
	False		

Two of the cells in the table are errors and two are not. Filling those in, we have

		Decision	
		Do not reject	Reject
Hypothesis	True	OK	Error
	False	Error	OK

Note that, in statistics, the term "error" does not mean somebody did something wrong. It refers to variability. An "error" occurs because the sampling variability caused a sample statistic that led us to the wrong decision. The errors have names that are rather unspectacular: If the (null) hypothesis is true, and we mistakenly reject it, it is a **Type I error**. If the hypothesis is false, and we mistakenly fail to reject it, it is a **Type II error**. We note that the probability of a Type I error is equal to α, the significance level. (This is because a *P*-value ***less than*** α causes us to reject H_0. If H_0 is true, and we decide to reject it because we got an unusual sample, we have made a Type I error.) We call the probability of a Type II error β. Filling the table with this information, we have:

		Decision	
		Do not reject	Reject
Hypothesis	True	OK	Type I Error $P(Type\ I) = \alpha$
	False	Type II Error $P(Type\ II) = \beta$	OK

The cell in the lower right-hand corner is important. The probability of correctly rejecting a false hypothesis (in favor of the alternative) is called the **power of the test**. The power of the test equals $1 - \beta$. Finally, then, our decision table looks like this:

		Decision	
		Do not reject	Reject
Hypothesis	True	OK	Type I Error $P(Type\ I) = \alpha$
	False	Type II Error $P(Type\ II) = \beta$	OK [power = $1 - \beta$]

Exam Tip: You will not need to know how to actually calculate *P*(*Type II error*) or the power of the rest on the AP exam. You *will* need to understand the concept of each, what affects each, and how they are related.

example: Sticky Fingers is arrested for shoplifting. The judge, in her instructions to the jury, says that Sticky is innocent until proven guilty. That is, the jury's hypothesis is that Sticky is innocent. Identify Type I and Type II errors in this situation and explain the consequence of each.

solution: Our hypothesis is that Sticky is innocent. A Type I error involves mistakenly rejecting a true hypothesis. In this case, Sticky *is* innocent, but because we reject innocence, he is found guilty. The risk in a Type I error is that Sticky goes to jail for a crime he didn't commit.

A Type II error involves failing to reject a false hypothesis. If the hypothesis is false, then Sticky is guilty, but because we think he's innocent, we acquit him. The risk in Type II error is that Sticky goes free even though he is guilty.

In life we often have to choose between possible errors. In the example above, the choice was between sending an innocent person to jail (a Type I error) or setting a criminal free (a Type II error). Which of these is the more serious error is not a statistical question—it's a social one.

We can decrease the chance of Type I error by adjusting α. By making α very small, we could virtually ensure that we would never mistakenly reject a true hypothesis. However, this would result in a large Type II error because we are making it hard to reject the null under any circumstance, even when it is false.

The probability of making a Type II error is smaller and, hence, the power of the test greater if:

- The sample size is increased.
- The standard deviation of our sample data is decreased (this is not always under the control of the researcher but, for example, if more precise measurements are possible the variability in the data could be reduced).
- We increase the significance level (α). (This could be seen as dishonest—manipulating the significance level to get the result you want.)
- The effect size (the difference between the hypothesized parameter and the true value) is larger. A bigger difference is easier to detect!

example: A pharmaceutical company created a new skin cream to help sunburn heal faster. They are going to conduct a clinical trial (a randomized experiment) to determine if the cream decreases the time for sunburn to heal. Suppose that, in fact, this cream does work. Which of the following is equivalent to the power of the test?

(a) The decrease in the amount of time it takes for the cream to heal the sunburn
(b) The probability that the researchers conclude from the experiment that the cream is ineffective
(c) The probability that the researchers conclude from the experiment that there is no conclusive evidence
(d) The probability that the researchers conclude from the experiment that the cream works
(e) The probability that the null hypothesis is false

solution: The power of the test is the probability of rejecting a false null hypothesis in favor of an alternative—in this case, the hypothesis that the cream doesn't work. If we correctly reject this hypothesis, the researchers will conclude that the cream does, in fact, work. Hence, (d) is the correct answer.

Inference for the Difference Between Two Population Proportions

The logic behind inference for two population proportions is the same as for the others we have studied. As with the one-sample case, there are some differences between z-intervals and z-tests in terms of the computation of the standard error. The following table gives the essential details.

• **Hypothesis** • **Estimator** • **Standard Error**	**Conditions**	**Test Statistic**
• Null hypothesis $H_0: p_1 - p_2 = 0$ (or $H_0: p_1 = p_2$) • Estimator: $\hat{p}_1 - \hat{p}_2$	• SRSs from independent populations	$z = \dfrac{\hat{p}_1 - \hat{p}_2}{\sqrt{\hat{p}(1-\hat{p})\left(\dfrac{1}{n_1}+\dfrac{1}{n_2}\right)}}$ where

where $\hat{p}_1 = \dfrac{X_1}{n_1}$ $\qquad$ $\hat{p} = \dfrac{n_1\hat{p}_1 + n_2\hat{p}_2}{n_1 + n_2} \geq 5 \,(\text{or } 10)$ $\qquad$ $\hat{p} = \dfrac{X_1 + X_2}{n_1 + n_2}$

$\hat{p}_2 = \dfrac{X_2}{n_2}$

• Standard error:

$s_{\hat{p}_1 - \hat{p}_2} = \sqrt{\hat{p}(1-\hat{p})\left(\dfrac{1}{n_1}+\dfrac{1}{n_2}\right)}$

where $\hat{p} = \dfrac{X_1 + X_2}{n_1 + n_2}$

example: Two concentrations of a new vaccine designed to prevent infection are developed and a study is conducted to determine if they differ in their effectiveness. Participants were randomly placed into treatment groups, one group receiving Vaccine A and the other Vaccine B. The results from the study are given in the following table.

	Vaccine A	**Vaccine B**
Infected	102	95
Not infected	123	190
Total	225	285

Does this study provide statistical evidence at the $\alpha = 0.01$ level that the vaccines differ in their effectiveness?

solution:

I. Let p_1 = the population proportion infected after receiving Vaccine A.

Let p_2 = the population proportion infected after receiving Vaccine B.

H_0: $p_1 - p_2 = 0$.
H_A: $p_1 - p_2 \neq 0$.

II. We will use a two-proportion z-test at $\alpha = 0.01$.

$$\hat{p}_1 = \frac{102}{225} = 0.453, \quad \hat{p}_2 = \frac{95}{285} = 0.333,$$

$$\hat{p} = \frac{102 + 95}{225 + 285} \approx 0.387$$

$$n_1 \hat{p} = 225 \cdot 0.387 \approx 87.1, \quad n_1(1 - \hat{p}) = 225 \cdot 0.613 \approx 137.9,$$
$$n_2 \hat{p} = 285 \cdot 0.387 \approx 110.3, \quad n_2(1 - \hat{p}) = 285 \cdot 0.613 \approx 174.7$$

Patients were randomly assigned treatments. All values are larger than 5 and we assume that treatments of vaccine have been randomly applied, so the conditions necessary are present for the two-proportion z-test.

III. $z = 2.76$ and P-value = 0.0057
On the AP exam, you should simply name the procedure. The name from your calculator is probably sufficient. You can just report the values listed as we did here—especially in the case of a two-sample test such as this.

IV. Because $P = 0.0057 < 0.01$, we reject the null hypothesis. We have convincing evidence that the two vaccines differ in their effectiveness.

---

z-Procedures versus t-Procedures

We are going to move from procedures that involve counting and producing proportions to those in which we are measuring and producing means. With proportions, assuming the proper conditions are met, we deal only with large samples—that is, with z-procedures. When doing inference for a population mean, or for the difference between two population means, we typically use t-procedures because we do not know the population standard deviation. We use t-procedures when doing inference for a population mean or for the difference between two populations' means when the following are true about each sample:

(a) The sample is a simple random sample from the population.
and

(b) The sample size is large (rule of thumb: $n \geq 30$) or the *population* from which the sample is drawn is approximately normally distributed (or, at least, does not depart dramatically from normal).

You ***could*** use z-procedures when doing inference for a population mean or for the difference between two populations' means when the following are true about each sample:

(a) The sample is a simple random sample from the population.
and

(b) The population from which the sample is drawn is normally distributed (in this case, the sampling distribution of $\bar{x}$ or $\bar{x}_1 - \bar{x}_2$ will also be normally distributed).
and

(c) The population standard deviation is known. (Remember Todd and his claim about county fair ride tickets?) *This would be a rare situation and is unlikely to be seen on the AP exam.*

Historically, many texts allowed you to use z-procedures when doing inference for means if your sample size was large enough to argue, based on the central limit theorem, that the sampling distribution of $\bar{x}$ or $\bar{x}_1 - \bar{x}_2$ is approximately normal. The basic assumption is that, for large samples, the sample standard deviation s is a reasonable estimate of the population standard deviation σ. Today, most statisticians would tell you that it's better practice to *always* use t-procedures when doing inference for a population mean or for the difference between two population means. You can receive credit on the AP exam for doing a large sample problem for means using z-procedures if you specify that you are doing so because the sample size is large, but it's definitely better practice to use t-procedures.

When using t-procedures, it is important to check in step II of the hypothesis test procedure that the data could plausibly have come from an approximately normal population.

A stemplot, boxplot, dotplot, or normal probability plot can be used to show that there are no outliers or extreme skewness in the data. (For the exam itself, our preference is a dotplot because they are quick to produce.) t-procedures are **robust** against these assumptions, which means that the procedures still work reasonably well even with some violation of the condition of normality, provided there is not much skewness. Some texts use the following guidelines for sample size when deciding whether or not to use t-procedures:

- $n < 15$. Use t-procedures if the data show no outliers and no skewness.
- $15 < n < 40$. Use t-procedures unless there are outliers or marked skewness.
- $n > 40$. Use t-procedures for any distribution.

(For the two-sample case discussed later, these guidelines can still be used if you replace n with n_1 and n_2.)

Inference for a Single Population Mean

In step II of the hypothesis-testing procedure, we need to identify the test to be used and justify the conditions needed. The test can be identified by name or by formula. For example, you could say "I will do a one-sample t-test for a population mean" to identify the procedure in words. To identify the procedure with a formula, write the formula for the **test statistic,** which will usually have the following form:

$$\text{Test Statistic} = \frac{\text{estimator} - \text{hypothesized value}}{\text{standard error}}.$$

When doing inference for a single mean, the *estimator* is $\bar{x}$, the hypothesized value is μ_0 in the null hypothesis $H_0: \mu = \mu_0$, and the standard error is the estimate of the standard deviation of $\bar{x}$, which is

$$s_{\bar{x}} = \frac{s}{\sqrt{n}} \ (\text{df} = n-1).$$

• Hypothesis • Estimator • Standard Error	Conditions	Test Statistic
• Null hypothesis H_0: $\mu = \mu_0$ • Estimator: $\bar{x}$ • Standard error: $s_{\bar{x}} = \dfrac{s}{\sqrt{n}}$	• SRS • σ known • Population normal or large sample size	$z = \dfrac{\bar{x} - \mu_o}{\sigma / \sqrt{n}}$
	• SRS • Large sample **or** population approximately normal	$t = \dfrac{\bar{x} - \mu_o}{s / \sqrt{n}}$, df $= n - 1$

Note: **Paired samples** (dependent samples) are a special case of one-sample statistics (H_0: $\mu_d = 0$).

example: A study was done to determine if 12- to 15-year-old girls who want to be engineers differ in IQ from the average of all girls. The mean IQ of all girls in this age range is known to be about 100 with a standard deviation of 15 (In fact, the test is designed that way!). A random sample of 49 girls who state that they want to be engineers is selected and their IQ is measured. The mean IQ of the girls in the sample is 104.5. Does this finding provide evidence, at the 0.05 level of significance, that the mean IQ of 12- to 15-year-old girls who want to be engineers differs from the average? This is an extremely rare situation in which we know the populaiton standard deviation ($\sigma = 15$), so we can use z rather than t.

solution 1 (test statistic approach): The solution to this problem will be put into a form that emphasizes the format required when writing out solutions on the AP exam.

I. Let μ = the mean IQ of all girls who want to be engineers.

H_0: $\mu = 100$.
H_A: $\mu \neq 100$.
(The alternative is two-sided because the problem wants to know if the mean IQ "differs" from 100. It would have been one-sided if it had asked whether the mean IQ of girls who want to be engineers is higher than average.)

II. Since σ is known, we will use a one-sample z-test at $\alpha = 0.05$.

Conditions:

- The problem states that we have a random sample.

- Sample size n is large (greater than 30).

- σ is known (the rare case!).

III. $z = \dfrac{104.5 - 100}{15 / \sqrt{49}} = 2.10 \Rightarrow P\text{-value} = 2(1 - 0.9821) = 0.0358$ (from Table A).

IV. Because $P = 0.0358 < \alpha = 0.05$, we reject H_0. We have convincing evidence that the mean IQ for all girls who want to be engineers differs from the mean IQ of all girls in this age range.

Notes on the above solution:

- Had the alternative hypothesis been one-sided, the P-value would have been $1 - 0.9821 = 0.0179$. We multiplied by 2 in step III because we needed to consider the area in *both* tails.
- The problem told us that the significance level was 0.05. Had it not mentioned a significance level, we could have chosen one, or we could have argued a conclusion based only on the derived P-value without a significance level.
- The linkage between the P-value and the significance level must be made explicit in part IV. Some sort of statement, such as "Since $P = 0.0358 < \alpha = 0.05$ . . ." or, if no significance level is stated, "Since the P-value is low . . ." will indicate that your conclusion is based on the P-value determined in step III.

Using Confidence Intervals for Two-Sided Alternatives

Consider a two-sided significance test at, say, $\alpha = 0.05$ and a confidence interval with $C = 0.95$. A sample statistic that would result in a significant result at the 0.05 level would also generate a 95% confidence interval that does not contain the hypothesized value. Confidence intervals for two-sided hypothesis tests could then be used in place of generating a test statistic and finding a P-value. If the sample value generates a C-level confidence interval that does not contain the hypothesized value of the parameter, then a significance test based on the same sample value would reject the null hypothesis at $\alpha = 1 - C$. Questions sometimes ask for such a decision to be made based on a confidence interval. Beware, however. If a question simply asks if there is statistical evidence to support a hypothesis, the question is asking for a **test**, not an interval. You should *never* use confidence intervals for hypothesis tests involving one-sided alternative hypotheses. (One-sided confidence intervals are not a part of this course.)

solution 2 (confidence interval approach—in this case an appropriate approach but only because H_A is two-sided):

I. Let $\mu =$ the true mean IQ of girls who want to be engineers.

$H_0: \mu = 100$
$H_A: \mu \neq 100$

II. We will use a 95% z-confidence interval ($C = 0.95$).

Conditions:

- The sample is random.
- We have a large sample size ($n = 49$).
- The standard deviation is known ($\sigma = 15$).

III. $\bar{x} = 104.5, z^* = 1.96 \rightarrow 104.5 \pm 1.96\left(\dfrac{15}{\sqrt{49}}\right) = (100.3, 108.7)$

We are 95% confident that the mean IQ of all girls who want to be engineers is in the interval (100.3, 108.7).

IV. Because 100 is not in the 95% confidence interval for μ, we reject H_0. We have strong evidence that the true mean IQ for girls who want to be engineers differs from the mean IQ of all girls in this age range.

Inference for the Difference Between Two Population Means

The two-sample case for the difference in the means of two independent samples is more complicated than the one-sample case. The hypothesis testing logic is identical, however, so the contrasts are in the mechanics needed to do the problems, not in the process. For hypotheses about the differences between two means, the procedures are summarized in the following table.

• Hypothesis • Estimator • Standard Error	Conditions	Test Statistic
• Null hypothesis: $H_0: \mu_1 - \mu_2 = 0$ OR $H_0 : \mu_1 = \mu_2$ • Estimator: $\bar{x}_1 - \bar{x}_2$	• σ_1^2 and σ_2^2 known– Use z-procedures but if unknown use t-procedures	$z = \dfrac{\bar{x}_1 - \bar{x}_2}{\sqrt{\dfrac{\sigma_1^2}{n_1} + \dfrac{\sigma_2^2}{n_2}}}$ (rarely used since σ_1^2 and σ_2^2 are rarely known)
• Standard error: $s_{\bar{x}_1 - \bar{x}_2} = \sqrt{\dfrac{s_1^2}{n_1} + \dfrac{s_2^2}{n_2}}$	• Two independent random samples from approximately normal populations *or* large sample sizes ($n_1 \geq 30$ and $n_2 \geq 30$)	$t = \dfrac{\bar{x}_1 - \bar{x}_2}{\sqrt{\dfrac{s_1^2}{n_1} + \dfrac{s_2^2}{n_2}}}$ df = (the often non-integer value given by your calculator) or df = min$\{ n_1 - 1, n_2 - 1\}$

Note on the above table:

- The "non-integer value" is based on the formula given in Chapter 10.

example: A statistics teacher, Mr. Gosset, gave a quiz to his 8:00 am class and to his 9:00 am class. There were 50 points possible on the quiz. The data for the two classes are in this table:

	n	$\bar{x}$	s
Group 1 (8:00 AM)	34	40.2	9.57
Group 2 (9:00 AM)	31	44.9	4.86

Before the quiz, some members of the 9:00 am class had been bragging that later classes do better in statistics. Considering these two classes as random samples from the populations of 8:00 am and 9:00 am classes, do these data provide evidence at the 0.01 level of significance that students in 9:00 am classes do better than those in 8:00 am classes?

solution:

> **I.** Let μ_1 = the true mean score for 8:00 am classes.
> Let μ_2 = the true mean score for 9:00 am classes.
> $H_0: \mu_1 - \mu_2 = 0$
> $H_A: \mu_1 - \mu_2 < 0$
>
> **II.** We will use a *two-sample t*-test at $\alpha = 0.01$. We assume these samples are random samples from independent populations. *t*-procedures are justified because both sample sizes are larger than 30.
>
> (Note: If the sample sizes were not large enough, we would need reasonable evidence that the samples were drawn from populations that are approximately normally distributed. To that end, we would examine graphs of the sample data.)
>
> **III.** Using the test menu of our calculator we activate the command for a `2-Sample t Test`: $t = -2.53$, df = 49.92, and *P*-value = 0.007.
>
> **IV.** Because *P*-value (0.007) < α (0.01), we reject the null hypothesis. We have convincing evidence that the true mean for 9:00 am classes is higher than the true mean for 8:00 am classes.

Watch out for Matched Pairs!

There are problems that appear to need the two-sample procedures as described in the example above. However, the samples are not drawn from independent populations—the data are paired in some way. When you have paired data you should run a one-sample hypothesis test on the *differences*!

example: A company president believes that there are more absences on Monday than on other days of the week. The company has 45 workers. The following table gives the number of worker absences on Mondays and Wednesdays for a random sample of eight weeks from the past two years. Do the data provide evidence that there are more absences on Mondays?

	Week 1	Week 2	Week 3	Week 4	Week 5	Week 6	Week 7	Week 8
Monday	5	9	2	3	2	6	4	1
Wednesday	2	5	4	0	3	1	2	0

solution: Because the data are paired on a weekly basis, the data we use for this problem are the *differences* between the days of the week for each of the eight weeks. Adding a row to the table that gives the differences (absences on Monday minus absences on Wednesday), we have:

	Week 1	Week 2	Week 3	Week 4	Week 5	Week 6	Week 7	Week 8
Monday	5	9	2	3	2	6	4	1
Wednesday	2	5	4	0	3	1	2	1
Difference	3	4	−2	3	−1	5	2	0

I. Let μ_d = the true mean difference between number of absences on Monday and absences on Wednesday where d = Mon – Wed:

H_0: $\mu_d = 0$ (or H_0: $\mu_d \leq 0$).
H_A: $\mu_d > 0$.

II. We will use a one-sample t-test for the difference scores.

Conditions:

A boxplot of the data shows no significant departures from normality (no outliers or severe skewness). Also, we are told that the eight weeks are randomly selected.

III. Using the t Test function on your calculator, you should find that $t =$ 1.99 with df = 7 and P-value = 0.044.

IV. The P-value is small. This provides us with convincing evidence that, on average over the past two years, there have been more absences on Mondays than on Wednesdays.

Calculator Tip: The TI-83/84 and other brands can do each of the significance tests described in this chapter, as well as those in Chapters 12 and 13. Graphing calculators will also do all of the confidence intervals we considered in Chapter 10. On the TI-83/84 you get to the tests by entering STAT TESTS. Note that Z-Test and T-Test are what are often referred to as "one-sample tests." The two-sample tests are identified as 2-SampZTest and 2-SampTTest. Most of the tests will give you a choice between Data and Stats. Data means that you have your data in a list, and Stats means that you have the summary statistics of your data. Once you find the test you are interested in, just follow the calculator prompts and enter the values requested.

› Rapid Review

1. A researcher reports that a finding of $x = 3.1$ is significant at the 0.05 level of significance. What is the meaning of this statement?

   Answer: Under the assumption that the null hypothesis is true, the probability of getting a value at least as extreme as the one obtained is less than 0.05. It was unlikely to have occurred by chance.

2. Let μ_1 = the mean score on a test of agility using a new training method and let μ_2 = the mean score on the test using the traditional method. Consider a test of H_0: $\mu_1 - \mu_2 = 0$. A large sample significance test finds $P = 0.04$. What conclusion, in the context of the problem, do you report if

   (a) $\alpha = 0.05$?

   (b) $\alpha = 0.01$?

Answer:

(a) Because the *P*-value is less than 0.05, we reject the null hypothesis at the 5% significance level. We have evidence that there is a non-zero difference between the traditional and new training methods.

(b) Because the *P*-value is greater than 0.01, we do not have sufficient evidence to reject the null hypothesis at the 1% significance level. We do not have strong support for the hypothesis that the training methods differ in effectiveness.

3. True–False: In a hypothesis test concerning a single mean, we can use either *z*-procedures or *t*-procedures as long as the sample size is at least 20.

   Answer: False. With a sample size of only 20, we cannot use *z*-procedures unless we know that the population from which the sample was drawn is approximately normal *and* σ is known. We can use *t*-procedures if the data do not have outliers or severe skewness; that is, if it is reasonable to believe based on a graph of the data that the population from which the sample was drawn is approximately normal.

4. We are going to conduct a two-sided significance test for a population proportion. The null hypothesis is H_0: $p = 0.3$. The simple random sample of 225 subjects yields $\hat{p} = 0.35$. What is the standard error, $s_{\hat{p}}$, involved in this procedure if

   (a) you are constructing a confidence interval for the true population proportion?

   (b) you are doing a significance test for the null hypothesis?

   Answer:

   (a) For a confidence interval, you use the value of $\hat{p}$ in the standard error. Hence, $s_{\hat{p}} = \sqrt{\dfrac{(0.35)(1-0.35)}{225}} = 0.0318$.

   (b) For a significance test, you use the hypothesized value of p. Hence,

   $$s_{\hat{p}} = \sqrt{\dfrac{(0.3)(1-0.3)}{225}} = 0.03055.$$

5. For the following data,

   (a) Assuming treatments were randomly assigned, justify the use of a two-proportion *z*-test for H_0: $p_1 - p_2 = 0$.

   (b) What is the value of the test statistic for H_0: $p_1 - p_2 = 0$?

   (c) What is the *P*-value of the test statistic for the two-sided alternative?

	n	*x*	$\hat{p}$
Group 1	40	12	0.3
Group 2	35	14	0.4

   Answer:

   (a) $\hat{p} = \dfrac{12+14}{40+35} = 0.35$

$n_1 \hat{p} = 40 \cdot 0.35 = 14$, $n_1(1 - \hat{p}) = 40 \cdot 0.65 = 26$

$n_2 \hat{p} = 35 \cdot 0.35 = 12.25$, $n_2(1 - \hat{p}) = 35 \cdot 0.65 = 22.75$ are all greater than 5.

Since all values are at least 5, the conditions are present for a two-proportion z-test.

(b) $z = \dfrac{0.3 - 0.4}{\sqrt{0.35(1 - 0.35)\left(\dfrac{1}{40} + \dfrac{1}{35}\right)}} = -0.91$

(c) $z = -0.91$, P-value $= 2(0.18) = 0.36$ (from Table A). But of course, you ought to just use your calculator to get these values. On the TI-83/84, you would simply activate the 2-proportion z-test, then list the z- and P-values.

6. You want to conduct a one-sample test (t-test) for a population mean. Your random sample size of 10 yields the following data: 26, 27, 34, 29, 38, 30, 28, 30, 30, 23. Should you proceed with your test? Explain.

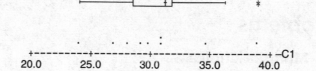

Answer: A boxplot of the data shows that the 38 is an outlier. Further, the dotplot of the data casts doubt on the approximate normality of the population from which this sample was drawn. Hence, you should *not* use a t-test on these data.

7. A randomized experiment was conducted to see if adding compost to a potted plant will help the plant to grow taller. Unknown to the researchers, this compost actually does improve the height of plants. But in their experiment, they got a P-value of 0.15 and, thus, failed to reject their null hypothesis. They concluded there was not convincing evidence that the compost helped. What type of error did they make?

Answer: The null hypothesis is false but the researchers failed to reject it. That is a Type II error.

8. Mary comes running into your office and excitedly tells you that she got a statistically significant finding from the data on her most recent research project. What is she talking about?

Answer: Mary means that the finding she got had such a small probability of occurring by chance that she has concluded it probably wasn't just chance variation but a real difference from expected.

9. Which of the following statements is correct?

   I.   The t distribution has more area in its tails than the z distribution (normal).

   II. When constructing a confidence interval for a population mean, you would always use z^* rather than t^* if you have a sample size of at least 30 ($n > 30$).

   III. When constructing a two-sample t interval, the "conservative" method of choosing degrees of freedom (df $= \min\{n_1 - 1, n_2 - 1\}$) will result in a wider confidence interval than other methods.

Answer:

• I is correct. A t distribution, because it must estimate the population standard deviation, has more variability than the normal distribution.

- II is not correct. It is definitely not correct that you would always use z^* rather than t^* in this situation. A more interesting question is *could* you use z^* rather than t^*? The answer to that question is a qualified "yes." The difference between z^* and t^* is small for large sample sizes (e.g., for a 95% confidence interval based on a sample size of 50, $z^* = 1.96$ and $t^* = 2.01$) and, while a t interval would have a somewhat larger margin of error, the intervals constructed would capture roughly the same range of values. In fact, many traditional statistics books teach this as the proper method. Now, having said that, the best advice is to *always* use t when dealing with a one-sample situation when s is unknown (confidence interval or hypothesis test), and use z only when you know, or have a very good estimate of, the population standard deviation.

- III is correct. The conservative method ($df = \min\{n_1 - 1, n_2 - 1\}$) will give a larger value of t^*, which, in turn, will create a larger margin of error, which will result in a wider confidence interval than other methods for a given confidence level.

Practice Problems

Multiple-Choice

1. Which of the following will increase the power of a test?
   a. Increase n.
   b. Increase a.
   c. Reduce the amount of variability in the sample.
   d. Consider an alternative hypothesis further from the null.
   e. All of these will increase the power of a test.

2. Under a null hypothesis, a sample value yields a P-value of 0.015. Which of the following statements is (are) true?
   I. This finding is statistically significant at the 0.05 level of significance.
   II. This finding is statistically significant at the 0.01 level of significance.
   III. The probability of getting a sample value as extreme as this one by chance alone if the null hypothesis is true is 0.015.

   a. I and III only
   b. I only
   c. III only
   d. II and III only
   e. I, II, and III

3. In a test of the null hypothesis $H_0: p = 0.35$ with $a = 0.01$, against the alternative hypothesis $H_A : p < 0.35$, a large random sample produced a z-score of -2.05. Based on this, which of the following conclusions can be drawn?
   a. It is likely that $p < 0.35$.
   b. $p < 0.35$ only 2% of the time.
   c. If the z-score were positive instead of negative, we would be able to reject the null hypothesis.
   d. We do not have sufficient evidence to claim that $p < 0.35$.
   e. 1% of the time we will reject the alternative hypothesis in error.

4. A paint manufacturer advertises that one gallon of its paint will cover 400 square feet of interior wall. Some local painters suspect the average coverage is considerably less and decide to conduct an experiment to find out. If μ represents the true average number of square feet covered by the paint, which of the following are the correct null and alternative hypotheses to be tested?

   a. $H_0: \mu = 400$, $H_A: \mu > 400$
   b. $H_0: \mu \geq 400$, $H_A: \mu \neq 400$
   c $H_0: \mu = 400$, $H_A: \mu \neq 400$
   d. $H_0: \mu \neq 400$, $H_A: \mu < 400$
   e. $H_0: \mu \geq 400$, $H_A: \mu < 400$

5. A school district claims that the average teacher in the district earns $48,000 per year. The teachers' organization argues that the average salary is less. A random sample of 25 teachers yields a mean salary of $47,500 with a sample standard deviation of $2,000. Assuming that the distribution of all teachers' salaries is approximately normally distributed, what is the value of the t-test statistic and the P-value for a test of the hypothesis $H_0: \mu = 48,000$ against $H_A: \mu < 48,000$? (Remember, df $= 25 - 1 = 24$.)

   a. $t = 1.25$, $0.10 < P < 0.15$
   b. $t = -1.25$, $0.20 < P < 0.30$
   c. $t = 1.25$, $0.20 < P < 0.30$
   d. $t = -1.25$, $0.10 < P < 0.15$
   e. $t = -1.25$, $P > 0.25$

6. Which of the following conditions is (are) necessary to justify the use of z-procedures in a significance test about a population proportion?
   I. The samples must be drawn from a normal population.
   II. The population must be much larger (10–20 times) than the sample.
   III. $np_0 \geq 5$ and $n(1 - p_0) \geq 5$.

   a. I only
   b. I and II only
   c. II and III only
   d. III only
   e. I, II, and III

7. A minister claims that more than 70% of the adult population attends a religious service at least once a month. Let $p =$ the proportion of adults who attend church. The null and alternative hypotheses you would use to test this claim would be:

   a. $H_0: p \leq 0.7$, $H_A: p > 0.7$
   b. $H_0: m \leq 0.7$, $H_A: m > 0.7$
   c. $H_0: p = 0.7$, $H_A: p \neq 0.7$
   d. $H_0: p \leq 0.7$, $H_A: p < 0.7$
   e. $H_0: p \geq 0.7$, $H_A: p < 0.7$

8. When is it OK to use a confidence interval instead of computing a P-value in a hypothesis test?

   a. In any significance test
   b. In any hypothesis test with a two-sided alternative hypothesis
   c. Only when the hypothesized value of the parameter is *not* in the confidence interval
   d. Only when you are conducting a hypothesis test with a one-sided alternative
   e. Only when doing a test for a single population mean or a single population proportion

9. Which of the following is *not* a required step for a significance test?

   a. State null and alternative hypotheses in the context of the problem.
   b. Identify the test to be used and justify the conditions for using it.
   c. State the significance level for which you will decide to reject the null hypothesis.
   d. Compute the value of the test statistic and the *P*-value.
   e. State a correct conclusion in the context of the problem.

10. Which of the following best describes what we mean when say that *t*-procedures are *robust*?

   a. The *t*-procedures work well with almost any distribution.
   b. The numerical value of *t* is not affected by outliers.
   c. The *t*-procedures will still work reasonably well even if the assumption of normality is violated.
   d. *t*-procedures can be used as long as the sample size is at least 40.
   e. *t*-procedures are as accurate as *z*-procedures.

11. For a hypothesis test of $H_0: \mu = \mu_0$ against the alternative $H_A: \mu < \mu_0$, the *z*-test statistic is found to be 2.00. This finding is

   a. significant at the 0.05 level but not at the 0.01 level.
   b. significant at the 0.01 level but not at the 0.05 level.
   c. significant at both the 0.01 and the 0.05 levels.
   d. significant at neither the 0.01 nor the 0.05 levels.
   e. not large enough to be considered significant.

12. Two types of tennis balls were tested to determine which one goes faster on a serve. Eight different players served one of each type of ball and the results were recorded:

SERVER	TYPE A	TYPE B
Raphael	120	115
Roger	125	122
Serena	119	114
Venus	110	114
Andy	118	115
Maria	82	91
Lleyton	115	110
Ana	105	106

Assuming that the speeds are approximately normally distributed, how many degrees of freedom will there be in the appropriate *t*-test used to determine which type of tennis ball travels faster?

   a. 6
   b. 7
   c. 16
   d. 15
   e. 14

13. Two statistics teachers want to compare their teaching methods. They decide to give the same final exam and use the scores on the exam as a basis for comparison. They decide that the value of interest to them will be the proportion of students in each class who score above 80% on the final. One class has 32 students, and one has 27 students. Which of the following would be the most appropriate test for this situation?

   a. Two proportion z-test
   b. Two-sample t-test
   c. Chi-square goodness-of-fit test
   d. One-sample z-test
   e. Chi-square test for independence

Free-Response

1. A university is worried that it might not have sufficient housing for its students for the next academic year. It's very expensive to build additional housing, so it is operating under the assumption (hypothesis) that the housing it has is sufficient, and it will spend the money to build additional housing only if it is convinced it is necessary (that is, it rejects its hypothesis).
   (a) For the university's assumption, what is the risk involved in making a Type I error?
   (b) For the university's assumption, what is the risk involved in making a Type II error?

2. A large high school has been waging a campaign against smoking. Before the campaign began in 2004, a random sample of 100 students from the junior and senior classes found 27 who admitted to smoking (we understand that some students who do so would be reluctant to admit it on a survey). To assess the success of their program, in early 2007 they surveyed a random sample of 175 juniors and seniors, and 30 responded that they smoked. Is this good evidence that the use of cigarettes has been reduced at the school?

3. Twenty-six pairs of identical twins are enrolled in a study to determine the impact of training on ability to memorize a string of letters. Two programs (A and B) are being studied. One member of each pair is randomly assigned to one of the two groups and the other twin goes into the other group. Each group undergoes the appropriate training program, and then the scores for pairs of twins are compared. The means and standard deviations for groups A and B are determined as well as the mean and standard deviation for the difference between each twin's score. Is this study a *one-sample* or *two-sample* situation, and how many degrees of freedom are involved in determining the t-value?

4. Which of the following statements is (are) correct? Explain.
   I. A confidence interval can be used instead of a test statistic in any hypothesis test involving means or proportions.
   II. A confidence interval can be used instead of a test statistic in a two-sided hypothesis test involving means or proportions.
   III. The standard error for constructing a confidence interval for a population proportion and the standard error for a significance test for a population proportion are the same.
   IV. The standard error for constructing a confidence interval for a population mean and the standard error for a significance test for a population mean are the same.

5. The average math SAT score at Centennial High School over the years is 520. The mathematics faculty believes that this year's class of seniors is the best the school has ever had in mathematics. One hundred seventy-five seniors take the exam and achieve an average score of 531 with a sample standard deviation of 96. Does this performance provide good statistical evidence that this year's class is, in fact, superior?

6. An avid reader, Booker Worm, claims that he reads books that average more than 375 pages in length. A random sample of 13 books on his shelf had the following number of pages: 595, 353, 434, 382, 420, 225, 408, 422, 315, 502, 503, 384, 420. Do the data support Booker's claim? Test at the 0.05 level of significance.

7. The statistics teacher, Dr. Tukey, gave a 50-point quiz to his class of 10 students and they didn't do very well, at least by Dr. Tukey's standards (which are quite high). Rather than continuing to the next chapter, he spent some time reviewing the material and then gave another quiz. The quizzes were comparable in length and difficulty. The results of the two quizzes were as follows.

Student	1	2	3	4	5	6	7	8	9	10
Quiz 1	42	38	34	37	36	26	44	32	38	31
Quiz 2	45	40	36	38	34	28	44	35	42	30

Do the data indicate that the review was successful, at the 0.05 level, at improving the performance of the students on this material? Give statistical evidence for your conclusion.

8. The new reality TV show, "I Want to Marry a Statistician," has been showing on Monday evenings, and ratings show that it has been watched by 55% of the viewing audience each week. The producers are moving the show to Wednesday night but are concerned that such a move might reduce the percentage of the viewing public watching the show. After the show has been moved, a random sample of 500 people who are watching television on Wednesday night is surveyed and asked what show they are watching. Two hundred fifty-five respond that they are watching "I Want to Marry a Statistician." Does this finding provide evidence at the 0.01 level of significance that the percentage of the viewing public watching "I Want to Marry a Statistician" has declined?

9. Harvey is running for student body president. A survey is conducted by the AP Statistics class in an attempt to predict the outcome of the election. They randomly sample 30 students, 16 of whom say they plan to vote for Harvey. Harvey figures (correctly) that 53.3% of students in the sample intend to vote for him and is overjoyed at his soon-to-be-celebrated victory. Explain carefully why Harvey should not get too excited until the votes are counted.

10. A company uses two different models, call them model A and model B, of a machine to produce electronic locks for hotels. The company has several hundred of each machine in use in its various factories. The machines are not perfect, and the company would like to phase out of service the one that produces more defects in the locks. A random sample of 13 model A machines and 11 model B machines is tested and the data for the average number of defects per week are given in the following table.

	n	$\bar{x}$	s
Model A	13	11.5	2.3
Model B	11	13.1	2.9

Dotplots of the data indicate that there are no outliers or strong skewness in the data and that there are no strong departures from normal. Do these data provide statistically convincing evidence that the two machines differ in terms of the number of defects produced?

11. Take another look at the preceding problem. Suppose there were 30 of each model machine that were sampled. Assuming that the sample means and standard deviations are the same as given in problem #9, how might this have affected the hypothesis test you performed in that problem?

12. The directors of a large metropolitan airport claim that security procedures are 98% accurate in detecting banned metal objects that passengers may try to carry onto a plane. The local agency charged with enforcing security thinks the security procedures are not as good as claimed. A study of 250 passengers showed that screeners missed nine banned carry-on items. What is the P-value for this test and what conclusion would you draw based on it?

13. A random sample of 175 married couples is selected to see if women have a stronger reaction than men to videos that contain violent material. At the conclusion of the study, each couple is given a questionnaire designed to measure the intensity of their reaction. Higher values indicate a stronger reaction. The means and standard deviations for all men, all women, and the differences between husbands and wives are as follows:

	$\bar{x}$	s
Men	8.56	1.42
Women	8.97	1.84
Difference (Husband–Wife)	−0.38	1.77

Do the data give strong statistical evidence that wives have a stronger reaction to violence in videos than do their husbands?

14. An election is bitterly contested between two rivals. In a poll of 750 potential voters taken 4 weeks before the election, 420 indicated a preference for candidate Grumpy over candidate Dopey. Two weeks later, a new poll of 900 randomly selected potential voters found 465 who plan to vote for Grumpy. Dopey immediately began advertising that support for Grumpy was slipping dramatically and that he was going to win the election. Statistically speaking (say at the 0.05 level), how happy should Dopey be (i.e., how sure is he that support for Grumpy has dropped)?

15. Consider, once again, the situation of problem #8 above. In that problem, a one-sided, two-proportion z-test was conducted to determine if there had been a drop in the proportion of people who watch the show "I Want to Marry a Statistician" when it was moved from Monday to Wednesday evenings.

Suppose instead that the producers were interested in whether the popularity ratings for the show had changed in *any* direction since the move. Over the seasons the show had been broadcast on Mondays, the popularity rating for the show (10 high, 1 low) had averaged 7.3. After moving the show to the new time, ratings were taken for 12 consecutive weeks. The average rating was determined to be 6.1 with a sample standard deviation of 2.7. Does this provide evidence, at the 0.05 level of significance, that the ratings for the show have changed? Use a confidence interval, rather than a *t*-test, as part of your argument. A dotplot of the data indicates that the ratings are approximately normally distributed.

Cumulative Review Problems

1. How large a sample is needed to estimate a population proportion within 2.5% at the 99% level of confidence if
    a. you have no reasonable estimate of the population proportion?
    b. you have data that show the population proportion should be about 0.7?

2. Let X be a binomial random variable with $n = 250$ and $p = 0.6$. Use a normal approximation to the binomial to approximate $P(X > 160)$.

3. Write the mathematical expression you would use to evaluate $P(X > 2)$ for a binomial random variable X that has $B(5, 0.3)$ (that is, X is a binomial random variable equal to the number of successes out of 5 trials of an event that occurs with probability of success $p = 0.3$). Do not evaluate.

4. An individual is picked at random from a group of 55 office workers. Thirty of the workers are female, and 25 are male. Six of the women are administrators. Given that the individual picked is female, what is the probability she is an administrator?

5. A random sample of 25 cigarettes of a certain brand were tested for nicotine content, and the mean was found to be 1.85 mg with a standard deviation of 0.75 mg. Find a 90% confidence interval for the mean number of mg of nicotine in this type of cigarette. Assume that the amount of nicotine in cigarettes is approximately normally distributed. Interpret the interval in the context of the problem.

Solutions to Practice Problems

Multiple-Choice

1. The correct answer is (e).

2. The correct answer is (a). It is not significant at the 0.01 level because 0.015 is greater than 0.01.

3. The correct answer is (d). To reject the null at the 0.01 level of significance, we would need to have $z < -2.33$.

4. The correct answer is (e). Because we are concerned that the actual amount of coverage might be less than 400 square feet, the only options for the alternative hypothesis are (d) and (e) (the alternative hypothesis in (a) is in the wrong direction, and the alternatives in (b) and (c) are two-sided). The null hypothesis given in (d) is not a form we would use for a null (the only choices are =, ≤, or ≥). We might see $H_0 : \mu = 400$ rather than $H_0 : \mu \geq 400$. Both are correct statements of a null hypothesis against the alternative $H_A : \mu < 400$.

5. The correct answer is (d).

$$t = \frac{47500 - 48000}{2000 \big/ \sqrt{25}} = -1.25 \Rightarrow 0.10 < P < 0.15$$

for the one-sided alternative. The calculator answer is $P = 0.112$. Had the alternative been two-sided, the correct answer would have been (b).

6. The correct answer is (c). If the sample size conditions are met, it is not necessary that the samples be drawn from a normal population.

7. The correct answer is (a). Often you will see the null written as H_0: $p = 0.7$ rather than H_0: $p \leq 0.7$. Either is correct.

8. The correct answer is (b). In AP Statistics, we consider confidence intervals to be two-sided. A two-sided α-level significance test will reject the null hypothesis whenever the hypothesized value of the parameter is not contained in the $C - 1 = \alpha$ level confidence interval.

9. (c) is not one of the required steps. You can state a significance level that you will later compare to the P-value, but it is not required. You can simply argue the strength of the evidence against the null hypothesis based on the P-value alone—small values of P provide evidence against the null.

10. (c) is the most correct response. (a) is incorrect because t-procedures do not work well with, for example, small samples that come from non-normal populations. (b) is false because the numerical value of t is, like z, affected by outliers. t-procedures are generally OK to use for samples of size 40 or larger, but this is not what is meant by *robust*, so (d) is incorrect. (e) is not correct since the t-distribution is more variable than the standard normal. It becomes closer to z as sample size increases but is "as accurate" only in the limiting case.

11. The correct answer is (d). The alternative hypothesis, H_0: $\mu = \mu_0$, would require a negative value of z to be evidence against the null. Because the given value is positive, we conclude that the finding is in the wrong direction to support the alternative and, hence, is not going to be significant at *any* level.

12. The correct answer is (b). Because the data are paired, the appropriate t-test is a one-sample test for the mean of the difference in scores. In this case, df $= n - 1 = 8 - 1 = 7$.

13. The correct answer is (a). The problem states that the teachers will record for comparison the number of students in each class who score above 80%. Because the enrollments differ in the two classes, we need to compare the proportion of students who score above 80% in each class. Thus the appropriate test is a two-proportion z-test. Note that, although it is not one of the choices, a chi-square test for homogeneity of proportions could also be used, since we are interested in whether the proportions of those getting above 80% are the same across the two populations.

Free-Response

1. (a) A Type I error is made when we mistakenly reject a true null hypothesis. In this situation, that means that we would mistakenly reject the true hypothesis that the available housing is sufficient. The risk would be that a lot of money would be spent on building additional housing when it wasn't necessary.

(b) A Type II error is made when we mistakenly fail to reject a false hypothesis. In this situation that means we would fail to reject the false hypothesis that the available housing is sufficient. The risk is that the university would have insufficient housing for its students.

2.

> **I.** Let p_1 = the true proportion of students who admit to smoking in 2004.
>
> Let p_2 = the true proportion of students who admit to smoking in 2007.
>
> $H_0: p_1 = p_2$ (or $H_0: p_1 - p_2 = 0$; or $H_0: p_1 \leq p_2$; or $H_0: p_1 - p_2 > 0$).
> $H_A: p_1 > p_2$ (or $H_0: p_1 > p_2$).
>
> **II.** We will use a two-proportion z-test. The survey involved drawing random samples from independent populations.
>
> $$\hat{p} = \frac{27 + 30}{100 + 175} \approx 0.21$$
>
> $n_1 \hat{p} = 100 \cdot 0.21 = 21$, $n_1(1 - \hat{p}) = 100 \cdot 0.79 = 79$
>
> $n_2 \hat{p} = 175 \cdot 0.21 = 36.75$, $n_2(1 - \hat{p}) = 175 \cdot 0.79 = 138.25$, are all greater than 5 (or 10) so conditions are met for a two-proportion z test.
>
> **III.** Using the 2-prop z-test from the test menu on your calculator we find that $z = 1.949$ and the P-value = 0.026.
>
> **IV.** Since the P-value is small (a difference in sample proportions this extreme would occur only about 2.6% of the time if there had been no decrease in usage), we have evidence that the rate of smoking among students (at least among juniors and seniors) has decreased.

3. This is a paired study because the scores for each pair of twins are compared. Hence, it is a one-sample situation, and there are 26 pieces of data to be analyzed, which are the 26 difference scores between the twins. Hence, df = 26 − 1 = 25.

4. • I is not correct. A confidence interval, at least in AP Statistics, cannot be used in any one-sided hypothesis test—only two-sided tests.

   • II is correct. A confidence interval constructed from a random sample that does not contain the hypothesized value of the parameter can be considered statistically significant evidence against the null hypothesis.

   • III is not correct. The standard error for a confidence interval based on the sample proportions is

$$s_{\hat{p}} = \sqrt{\frac{\hat{p}(1-\hat{p})}{n}}.$$

The standard error for a significance test based on the hypothesized population value is

$$s_{\hat{p}} = \sqrt{\frac{p_0(1-p_0)}{n}}.$$

   • IV is correct.

5.

> **I.** Let μ = the true mean score for all students taking the exam.
>
> H_0: μ = 520 (or, H_0: $\mu \leq$ 520)
> H_A: μ > 520
>
> **II.** We will use a one-sample *t*-test. We consider the 175 students to be a random sample of students taking the exam. The sample size is larger than 40, so the conditions for inference are present. (*Note*: Due to the very large sample size, it is reasonable that the sample standard deviation is a good estimate of the population standard deviation. This means that you would receive credit on the AP exam for doing this problem as a *z*-test although a *t*-test is preferable. But you must explicitly appeal to the very large sample size if you use a *z*-test.)
>
> **III.** Using the one-sample *t*-test from the test menu on your calculator we find that with $\bar{x}$ = 531 and s_x = 96, *t* = 152, df = 174, and *P*-value = 0.0657.
>
> **IV.** The *P*-value, which is greater than 0.05, is not small enough to provide convincing evidence that the current class is superior in math ability as measured by the SAT.

6.

> **I.** Let μ = the true average number of pages in the books Booker reads.
>
> H_0: $\mu \leq$ 375
> H_A: μ > 375
>
> **II.** We are going to use a one-sample *t*-test test at α = 0.05. The problem states that the sample is a random sample. A dotplot of the data shows good symmetry and no significant departures from normality (while the data do spread out quite a bit from the mean, there are no outliers):
>
>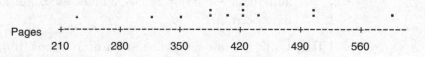
>
> The conditions for the *t*-test are present.
>
> **III.** Using the one-sample *t*-test from the test menu on your calculator we find that with $\bar{x}$ = 412.5 and s_x = 91.35, *t* = 1.48, df = 12, and *P*-value = 0.082.
>
> **IV.** Because *P* > 0.05, we cannot reject H_0. We do not have convincing evidence to back up Booker's claim that the books he reads actually average more than 375 pages in length.

7. The data are paired by individual students, so we need to test the difference scores for the students rather than the means for each quiz. The differences are given in the following table.

Student	1	2	3	4	5	6	7	8	9	10
Quiz 1	42	38	34	37	36	26	44	32	38	31
Quiz 2	45	40	36	38	34	28	44	35	42	30
Difference (Q2 – Q1)	3	2	2	1	–2	2	0	3	4	–1

I. Let μ_d = the mean of the differences between the scores of students on Quiz 2 and Quiz 1.

H_0: $\mu_d = 0$
H_A: $\mu_d > 0$

II. This is a matched pairs t-test. That is, it is a one-sample t-test for a population mean. We assume that these are random samples from the populations of all students who took both quizzes. The significance level is $\alpha = 0.05$.

A boxplot of the differences in scores shows no significant departures from normality, so the conditions to use the one-sample t-test have been met.

III. Using the one-sample t-test from the test menu on your calculator we find that with $\bar{x}_d = 1.4$ and $s_d = 1.9$, $t = 2.33$, df = 9, and P-value = 0.022.

IV. Because $P < 0.05$, we reject the null hypothesis. The data provide convincing evidence at the 0.05 level that the review was successful at improving student performance on the material.

8.

I. Let p = the true proportion of Wednesday night television viewers who are watching "I Want to Marry a Statistician."

H_0: $p = 0.55$
H_A: $p < 0.55$

II. We want to use a one-proportion z-test at $\alpha = 0.01$. $500(0.55) = 275 > 5$ and $500(1 - 0.55) = 225 > 5$. Thus, the conditions needed for this test have been met.

III. Using the one-sample z-test for proportions from the test menu on your calculator (`STAT TESTS 1-PropZTest`) with $\hat{p} = 255/500 \Rightarrow x = 255$, $n = 500$ the calculator shows $z = -1.80$ and P-value = 0.036.

IV. Because $P > 0.01$, we do not have sufficient evidence to reject the null hypothesis. The evidence is not convincing evidence to conclude at the 0.01 level that the proportion of viewers has dropped since the program was moved to a different night.

9. Let's suppose that the Stats class constructed a 95% confidence interval for the true proportion of students who plan to vote for Harvey (we are assuming that this a random sample from the population of interest, and we note that both $n\hat{p}$ and $n(1 - \hat{p})$ are greater than 10). $\hat{p} = \dfrac{16}{30} = 0.533$ (as Harvey figured). Then a 95% confidence interval for the true proportion of votes Harvey can expect to get is $0.533 \pm 1.96 \sqrt{\dfrac{(0.533)(1-0.533)}{30}} = (0.355, 0.712)$. That is, we are 95% confident that between 35.5% and 71.2% of students plan to vote for Harvey. It is certainly possible

for him to have a majority, but there are plausible values for the true proportion that are less than a majority as well. Harvey shouldn't get too complacent! (The argument is similar with a 90% CI: (0.384, 0.683); or with a 99% CI: (0.299, 0.768).)

10.

> **I.** Let μ_1 = the true mean number of defects produced by machine A.
> Let μ_2 = the true mean number of defects produced by machine B.
>
> H_0: $\mu_1 - \mu_2 = 0$
> H_A: $\mu_1 - \mu_2 \neq 0$
>
> **II.** We use a two-sample t-test for the difference between means. The conditions that need to be present for this procedure are given in the problem: both samples are simple random samples from independent, approximately normal populations.
>
> **III.** Using the two-sample t-test (2SampTTest) from the test menu on your calculator we find that with $\bar{x}_1 = 11.5$, $s_1 = 2.3$, $n_1 = 13$ $\bar{x}_2 = 13.1$, $s_2 = 2.9$, $n_2 = 11$, $t = -1.478$, df = 18.99, and P-value = 0.1557.
>
> **IV.** The P-value is too large to reject the null. We do not have convincing evidence that the types of machines actually differ in the number of defects produced.

11. Using a two-sample t-test and increasing the sample sizes to 30 each, steps I and II would not change. Step III would change to

$$t = -2.37 \text{ df} = 55.14 \Rightarrow P\text{-}value = 0.0214.$$

In step IV, a student could argue that we could reject the null because the P-value is small. This is an example of how larger sample sizes make it easier to detect statistically significant differences.

12. H_0: $p \geq 0.98$, H_A: $p < 0.98$, $\hat{p} = \dfrac{241}{250} = 0.964$.

$$z = \frac{0.964 - 0.98}{\sqrt{\dfrac{(0.98)(0.02)}{250}}} = -1.81, \ P\text{-value} = 0.035.$$

This P-value is lower than an alpha level of 0.05. This provides evidence against the null and in favor of the alternative that security procedures actually detect less than the claimed percentage of banned objects. (However, a different student could argue that this P-value is higher than an alpha level of 0.01 and therefore does NOT provide convincing evidence that the security procedures detect less than 98% of all banned objects.)

13.

> **I.** Let μ_d = the mean of the differences between the scores of husbands and wives.
>
> H_0: $\mu_d = 0$
> H_A: $\mu_d < 0$
>
> **II.** This is a matched-pairs situation and we will use a one-sample t-test for a population mean. We are told that this is a random sample of married couples. Because $n = 175 > 30$, t procedures are appropriate.
>
> **III.** Using the one-sample t-test on your calculator, we find that with $\bar{x}_d = -0.38$ and $s_d = 1.77$, $t = -2.84$, df $= 174$, and P-value $= 0.0023$.
>
> **IV.** Because P is very small (smaller than and reasonable value of alpha), we reject H_0. The data provide convincing evidence that women have a stronger reaction to violence in videos than do men.

14.

> **I.** Let p_1 = the true proportion of voters who plan to vote for Grumpy 4 weeks before the election.
>
> Let p_2 = the true proportion of voters who plan to vote for Grumpy 2 weeks before the election.
>
> H_0: $p_1 - p_2 = 0$
> H_A: $p_1 - p_2 > 0$
>
> **II.** We will use a two-proportion z-test for the difference between two population proportions. Both samples are random samples of the voting populations at the time. And,
>
> $$\hat{p} = \frac{420 + 425}{750 + 900} \approx 0.536$$
>
> $n_1 \hat{p} = 420 \cdot 0.536 = 225.12$, $n_1(1 - \hat{p}) = 420 \cdot 0.464 = 194.88$
>
> $n_2 \hat{p} = 900 \cdot 0.536 = 482.4$, $n_2(1 - \hat{p}) = 900 \cdot 0.464 = 417.6$
>
> All values are larger than 5, so the conditions needed for the two-proportion z-test are present.
>
> **III.** From the test menu 2-PropZTest yields $z = 1.75$ and a P-value $= 0.039$.
>
> **IV.** Because $P < 0.05$, we can reject the null hypothesis. Candidate Dopey may have cause for celebration—there is convincing evidence that support for candidate Grumpy is dropping.

15.

> **I.** Let μ = the true mean popularity rating for "I Want to Marry a Statistician."
>
> H_0: $\mu = 7.3$
> H_A: $\mu \neq 7.3$
>
> **II.** We will use a one-sample t confidence interval (at the direction of the problem—otherwise we would most likely have chosen to do a one-sample t significance test) at $\alpha = 0.05$, which, for the two-sided test, is equivalent to a confidence level of 0.95 ($C = 0.95$). We will assume that the ratings are a random sample of the population of all ratings. The sample size is small, but we are told that the ratings are approximately normally distributed, so that the conditions necessary for the inference are present.
>
> **III.** $\bar{x} = 6.1$, $s = 2.7$. For df $= 12 - 1 = 11$ and a 95% confidence interval, $t^* = 2.201$ A 95% t confidence interval from your calculator gives (4.384, 7.816)
>
> **IV.** Since 7.3 is in the interval (4.384, 7.816) and is therefore a plausible value for μ, we cannot reject H_0 at the 0.05 level of significance. We do not have convincing evidence that there has been a significant change in the popularity rating of the show after its move to Wednesday night.
>
> (*Note*: A one-sample t-test for these data yields P-value $= 0.15$.)

Solutions to Cumulative Review Problems

1. a. $n \geq \left(\dfrac{2.576}{2(0.025)} \right)^2 = 2654.3$; use a sample size ≥ 2655

    b. $n \geq (0.7)(1 - 0.7)\left(\dfrac{2.576}{0.025} \right)^2 = 2229.6$; use a sample size ≥ 2230.

2. $\mu_X = (250)(0.6) = 150$, $\sigma_X = \sqrt{(250)(0.6)(0.4)} = 7.75$

    Using the exact binomial given by `1-binomcdf(250,0.6,160)=0.087`, or by using the normal approximation to the binomial:

    $$P(X > 160) = P\left(z > \frac{160 - 150}{7.75} \right) = P(z > 1.29) = 0.099.$$

3. You can use the following calculation:

    $$\binom{5}{3}(0.3)^3(0.7)^2 + \binom{5}{4}(0.3)^4(0.7)^1 + \binom{5}{5}(0.3)^5(0.7)^0$$

4. P (the worker is an administrator | the worker is female)

$$= \frac{P(\text{A and F})}{P(\text{F})} = \frac{6/55}{30/55} = \frac{6}{30} = 0.20.$$

5. A confidence interval is justified because we are dealing with a random sample from an approximately normally distributed population. df = 25 − 1 = 24 with $t* = 1.711$. Using the `Tinterval` from the test menu of our calculator we get (1.5934, 2.1066). Therefore we are 90% confident that the true mean mg per cigarette for this type of cigarette is between 1.59 mg and 2.11 mg. (Notice the three parts are included: checked conditions, named the procedure with appropriate calculations, and interpreted the results in context.)

CHAPTER 12

Inference for Regression

IN THIS CHAPTER

Summary: In the last two chapters, we've considered inference for population means and proportions and for the difference between two population means or two population proportions. In this chapter, we extend the study of linear regression begun in Chapter 6 to include inference for the slope of a regression line, including both confidence intervals and significance testing. Finally, we will look at the use of technology when doing inference for regression.

Key Ideas
✪ Simple Linear Regression (Review)
✪ Significance Test for the Slope of a Regression Line
✪ Confidence Interval for the Slope of a Regression Line
✪ Inference for Regression Using Technology

Simple Linear Regression

When we studied data analysis earlier in this text, we distinguished between *statistics* and *parameters*. Statistics are measurements or values that describe samples, and parameters are measurements that describe populations. We have also seen that statistics can be used to estimate parameters. Thus, we have used $\bar{x}$ to estimate the population mean μ, s to estimate the population standard deviation σ, etc. In Chapter 6, we introduced the least-squares regression line ($\hat{y} = a + bx$), which was based on a set of ordered pairs. $\hat{y}$ is actually a statistic because it is based on sample data. In this chapter, we study the parameter, μ_y, which is estimated by $\hat{y}$.

Before we look at the model for linear regression, let's consider an example to remind us of what we did in Chapter 6:

example: The following data are pulse rates and heights for a group of 10 female statistics students (The scatterplot of the data and a residual plot indicate that a linear model is appropriate):

Height	70	60	70	63	59	55	64	64	72	66
Pulse	78	70	65	62	63	68	76	58	73	53

a. What is the least-squares regression line for predicting pulse rate from height?
b. What is the correlation coefficient between height and pulse rate? Interpret the correlation coefficient in the context of the problem.
c. What is the predicted pulse rate of a 67″ tall student?
d. Interpret the slope of the regression line in the context of the problem.

solution:

a. $\widehat{Pulse\ rate} = 47.17 + 0.302\ (Height)$. (Done on the TI-83/84 with *Height* in L1 and *Pulse* in L2, the LSRL can be found STAT CALC LinReg(a+bx) L1,L2,Y1.)
b. $r = 0.21$. There is a weak, positive, linear relationship between Height and Pulse rate.
c. $\widehat{Pulse\ rate} = 47.17 + 0.302(67) = 67.4$. (On the Ti-83/84: Y1(67) = 67.42. Remember that you can paste Y1 to the home screen by entering VARS Y-VARS Function Y1.)
d. For each increase in height of one inch, the pulse rate is predicted to increase by about 0.302 beats per minute.

When doing inference for regression, we use $\hat{y} = a + bx$ to estimate the true population regression line. Similar to what we have done with other statistics used for inference, we use a and b as estimators of population parameters α and β, the intercept and slope of the population regression, respectively. The conditions necessary for doing inference for regression are:

* For each given value of x, the values of the response variable y-values are independent and normally distributed.
* For each given value of x, the standard deviation, σ, of y-values is the same.
* The mean response of the y-values for the fixed values of x are linearly related by the equation $\mu_y = \alpha + \beta x$.

example: Consider a situation in which we are interested in how well a person scores on an agility test after a fixed number of 3-oz. glasses of wine. Let x be the number of glasses consumed. Let x take on the values 1, 2, 3, 4, 5, and 6. Let y be the score on the agility test (scale: 1–100). Then for any given value x_i, there will be a distribution of y-values with mean μ_{y_i}. The conditions for inference for regression are that (i) each of these distributions of y-values is normally distributed, (ii) each of these distributions of y-values has the same standard deviation σ, and (iii) each of the μ_{y_i} lies on a line.

Remember that a *residual* was the error involved when making a prediction from a regression equation (residual = actual value of y − predicted value of $y = y_i - \hat{y}_i$). Not surprisingly, the standard error of the predictions is a function of the squared residuals:

$$s = \sqrt{\frac{SS_{RES}}{n-2}} = \sqrt{\frac{\sum (y_i - \hat{y}_i)^2}{n-2}}.$$

s is an estimator of σ, the standard deviation of the residuals. Thus, there are actually three parameters to worry about in regression: α, β, and σ, which are estimated by a, b, and s, respectively.

The final statistic we need to do inference for regression is the standard error of the slope of the regression line given by the following equation. You will not need to use this formula on the exam:

$$s_b = \frac{s}{\sqrt{\sum (x_i - \overline{x})^2}} = \frac{\sqrt{\dfrac{\sum (y_i - \hat{y}_i)^2}{n-2}}}{\sqrt{\sum (x_i - \overline{x})^2}}$$

In summary, inference for regression depends upon estimating $\mu_y = \alpha + \beta x$ with $\hat{y} = a + bx$. For each x, the response values of y are independent and follow a normal distribution, each distribution having the same standard deviation. Inference for regression depends on the following statistics:

- a, the estimate of the y intercept, α, of μ_y
- b, the estimate of the slope, β, of μ_y
- s, the standard error of the residuals
- s_b, the standard error of the slope of the regression line

In the section that follows, we explore inference for the slope of a regression line in terms of a significance test and a confidence interval for the slope.

Inference for the Slope of a Regression Line

Inference for regression consists of either a significance test or a confidence interval for the slope of a regression line. The null hypothesis in a significance test is usually H_0: $\beta = 0$, although it is possible to test H_0: $\beta = \beta_0$. Our interest is the extent to which a least-squares regression line is a good model for the data. That is, the significance test is a test of a linear model for the data.

We note that in theory we could test whether the slope of the regression line is equal to any specific value. However, the usual test is whether the slope of the regression line is zero or not. If the slope of the line is zero, then there is no linear relationship between the x and y variables (remember: $b = r\dfrac{s_y}{s_x}$; if $r = 0$, then $b = 0$).

The alternative hypothesis is often two sided (i.e., H_A: $\beta \neq 0$). We could do a one-sided test if we believed that the data were positively or negatively related.

Significance Test for the Slope of a Regression Line

The basic details of a significance test for the slope of a regression line are given in the following table:

• HYPOTHESIS		
• ESTIMATOR		
• STANDARD ERROR	CONDITIONS	TEST STATISTIC
• Null hypothesis H_0: $\beta = \beta_0$ (most often: H_0: $\beta = 0$) • Estimator: b (from: $\hat{y} = a + bx$) • Standard error of the residuals:	• For each given value of x, the values of the response variable y are independent and normally distributed.	

Continued

$$s = \sqrt{\frac{SS_{RES}}{n-2}} = \sqrt{\frac{\sum(y_i - \hat{y}_i)^2}{n-2}}.$$

(Gives the variability of the vertical distances of the y-values from the regression line)

- Standard error of the slope:

$$s_b = \frac{s}{\sqrt{\sum(x_i - \overline{x})^2}}$$

(Gives the variability of the estimates of the slope of the regression line)

- For each given value of x, the standard deviation of y is the same.

- The mean response of the y-values for the fixed values of x are linearly related by the equation
$\mu_y = \alpha + \beta x$.

$$t = \frac{b - \beta_o}{s_b}$$
$$= \frac{b}{s_b} \text{ (if } \beta_o = 0),$$
$$df = n - 2$$

example: The data in the following table give the top 15 states in terms of per-pupil expenditure in 1985 and the average teacher salary in the state for that year.

STATE/SALARY		PER-PUPIL EXPENDITURE
MN	27360	3982
CO	25892	4042
OR	25788	4123
PA	25853	4168
WI	26525	4247
MD	27186	4349
DE	24624	4517
MA	26800	4642
RI	29470	4669
CT	26610	4888
DC	33990	5020
WY	27224	5440
NJ	27170	5536
NY	30678	5710
AK	41480	8349

Test the hypothesis, at the 0.01 level of significance, that there is no straight-line relationship between per-pupil expenditure and teacher salary. Assume that the conditions necessary for inference for linear regression are present.

solution:

I. Let $\beta =$ the true slope of the regression line for predicting salary from per-pupil expenditure.

$$H_0: \beta = 0.$$
$$H_A: \beta \neq 0.$$

II. We will use the t-test for the slope of the regression line. The problem states that the conditions necessary for linear regression are present.

III. The regression equation is

$$\widehat{Salary} = 12027 + 3.34 \ PPE$$
$$(s = 2281, s_b = 0.5536)$$

$$t = \frac{3.34 - 0}{0.5536} = 6.04, \qquad \text{df} = 15 - 2$$
$$= 13 \Rightarrow P - \text{value} = 0.0000.$$

(To do this significance test for the slope of a regression line on the TI-83/84, first enter *Per-Pupil Expenditure* (the explanatory variable) in L1 and *Salary* (the response variable) in L2. Then go to STAT TESTS LinRegTTest and enter the information requested. The calculator will return the values of t, p (the P-value), df, a, b, s, r^2, and r. Minitab, and some other computer software packages, will not give the the value of r—you'll have to take the appropriate square root of r^2—but will give you the value of s_b. If you need s_b for some reason—such as constructing a confidence interval for the slope of the regression line—and only have access to a calculator, you can find it by noting that, since $t = \frac{b}{s_b}$, then $s_b = \frac{b}{t}$. Note that Minitab reports the P-value as 0.0000.)

IV. Because $P < \alpha$, we reject H_0. We have evidence that the true slope of the regression line is not zero. We have evidence that there is a linear relationship between amount of per-pupil expenditure and teacher salary.

A significance test that the slope of a regression line equals zero is closely related to a test that there is no correlation between the variables. That is, if ρ is the population correlation coefficient, then the test statistic for $H_0: \beta = 0$ is equal to the test statistic for $H_0: \rho = 0$. You aren't required to know it for the AP exam, but the t-test statistic for $H_0: \rho = 0$, where r is the sample correlation coefficient, is

$$t = r\sqrt{\frac{n-2}{1-r^2}}, \ \text{df} = n - 2.$$

Because this and the test for a nonzero slope are equivalent, it should come as no surprise that

$$r\sqrt{\frac{n-2}{1-r^2}} = \frac{b}{s_b}.$$

Confidence Interval for the Slope of a Regression Line

In addition to doing hypothesis tests on $H_0: \beta = \beta_0$, we can construct a confidence interval for the true slope of a regression line. The details follow:

• PARAMETER • ESTIMATOR	CONDITIONS	FORMULA
• Population slope: β • Estimator: b (from: $\hat{y} = a + bx$) • Standard error of the residuals. • Standard error of the slope.	• For each given value of x, the values of the response variable y are independent and normally distributed. • For each given value of x, the standard deviation of y is the same. • The mean response of the y-values for the fixed values of x are linearly related by the equation $\mu_y = \alpha + \beta x$.	$b \pm t^* s_b$, df $= n - 2$ (where t^* is the upper critical value of t for a C-level confidence interval)

example: Consider once again the earlier example on predicting teacher salary from per-pupil expenditure. Construct a 95% confidence interval for the slope of the population regression line.

solution: When we were doing a test of $H_0: \beta = 0$ for that problem, we found that $\widehat{Salary} = 12027 + 3.34\ PPE$. The slope of the regression line for the 15 points, and hence our estimate of β, is $b = 3.34$. We also had $t = 6.04$.

Our confidence interval is of the form $b \pm t^* s_b$. We need to find t^* and s_b. For $C = 0.95$, df $= 15 - 2 = 13$, we have $t^* = 2.160$ (from Table B; if you have a TI-83/84 with the invT function, use invT(0.975,13)). Now, as mentioned earlier, $s_b = \dfrac{b}{t} = \dfrac{3.34}{6.04} = 0.5530$.

Hence, $b \pm t^* s_b = 3.34 \pm 2.160(0.5530) = (2.15, 4.53)$. We are 95% confident that the true slope of the regression line is between 2.15 and 4.53. Note that, since 0 is *not* in this interval, this finding is consistent with our earlier rejection of the hypothesis that the slope equals 0. This is another way of saying that we have statistically significant evidence of a predictive linear relationship between *PPE* and *Salary*.

Inference for Regression Using Technology

If you had to do them from the raw data, the computations involved in doing inference for the slope of a regression line would be daunting.

For example, would you want to compute $s_b = \dfrac{s}{\sqrt{\sum(x_i - \overline{x})^2}} = \dfrac{\sqrt{\dfrac{\sum(y_i - \hat{y}_i)^2}{n-2}}}{\sqrt{\sum(x_i - \overline{x})^2}}$ by hand?

Fortunately, you probably will never have to do this by hand, but instead can rely on computer output you are given, or you will be able to use your calculator to do the computations.

Consider the following data that were gathered by counting the number of cricket chirps in 15 seconds and noting the temperature.

Number of Chirps	22	27	35	15	28	30	39	23	25	18	35	29
Temperature (°F)	64	68	78	60	72	76	82	66	70	62	80	74

We want to use technology to test the hypothesis that the slope of the regression line is 0 and to construct a confidence interval for the true slope of the regression line.

First let us look at the Minitab regression output for these data.

The regression equation is
$\widehat{Temp} = 44.0 + 0.993$ Number

Predictor	Coef	St Dev	t ratio	P
Constant	44.013	1.827	24.09	0.000
Number	0.99340	0.06523	15.23	0.000
$s = 1.538$	R-sq = 95.9%		R-sq(adj) = 95.5%	

You should be able to read most of this table, but you are not responsible for all of it. You see the following table entries:

- The regression equation, $\widehat{Temp} = 44.0 + 0.993$ Number, is the least squares regression line (LSRL) for predicting temperature from the number of cricket chirps.
- Under "Predictor" are the y-intercept and explanatory variable of the regression equation, called "Constant" and "Number" (for number of chirps) in this example.
- Under "Coef" are the values of the "Constant" (which equals the y-intercept, the a in $\hat{y} = a + bx$; here, $a = 44.013$) and the slope of the regression line (which is the coefficient of "Number" in this example, the b in $\hat{y} = a + bx$; here, $b = 0.99340$).
- For the purposes of this book, we are not concerned with the "Stdev," "t-ratio," or "P" for "Constant" therefore only the "44.013" is meaningful for us.
- "Stdev" of "Number" is the standard error of the slope (what we have called s_b, the variability of the estimates of the slope of the regression line, which *equals* here

$\dfrac{s}{\sqrt{\sum(x_i - \overline{x})^2}}$ $s_b = 0.06523$); "t-ratio" is the value of the t-test statistic ($t = \dfrac{b}{s_b}$,

$df = n - 2$; here, $t = \dfrac{0.99340}{0.06523} = 15.23$); and P is the P-value associated with the test statistic assuming a two-sided test (here, $P = 0.000$; if you were doing a *one*-sided test, you would need to divide the given P-value by 2).

- s is the standard error of the residuals, which is the variability of the vertical distances of the y-values from the regression line; $s = \sqrt{\dfrac{\sum(y_i - \hat{y}_i)^2}{n-2}}$; (here, $s = 1.538$).

- "R-sq" is the coefficient of determination (or, r^2; here R-sq = 95.9% ⇒ 95.9% of the variation in temperature that is explained by the regression on the number of chirps in 15 seconds; note that, here, $r = \sqrt{0.959} = 0.979$—it's positive since $b = 0.9934$ is positive). You don't need to worry about "R-sq(adj)."

Thankfully, all of the mechanics needed to do a t-test for the slope of a regression line are contained in this printout. You need only to quote the appropriate values in your write-up. Thus, for the problem given above, we see that $t = 15.23 \Rightarrow P$-value = 0.000.

Exam Tip: You may be given a problem that has both the raw data and the computer printout based on the data. If so, there is *no* advantage to doing the computations all over again because they have already been done for you.

A confidence interval for the slope of a regression line follows the same pattern as all confidence intervals (estimate ± (critical value) × (standard error)): $b \pm t^*s_b$, based on $n - 2$ degrees of freedom. A 99% confidence interval for the slope in this situation (df = $10 \Rightarrow t^* = 3.169$ from Table B) is $0.9934 \pm 3.169(0.06523) = (0.787, 1.200)$.

If you *have* to do a confidence interval using the calculator and do not have a TI-83/84 with the `LinRegTInt` function, you first need to determine s_b. Because you know that $t = \dfrac{b}{s_b} \Rightarrow s_b = \dfrac{b}{t}$, it follows that $s_b = \dfrac{0.9934}{15.2295} = 0.0652$, which agrees with the standard error of the slope ("St Dev" of "Number") given in the computer printout.

A 95% confidence interval for the slope of the regression line for predicting temperature from the number of chirps per minute is then given by $0.9934 \pm 2.228(0.0652) = (0.848, 1.139)$. $t^* = 2.228$ is based on $C = 0.95$ and df = $12 - 2 = 10$. Using `LinRegTInt`, if you have it, results in the following (note that the "s" given in the printout is the standard error of the residuals, not the standard error of the slope).

❯ Rapid Review

1. The regression equation for predicting grade point average from number of hours studied is determined to be $\widehat{GPA} = 1.95 + 0.05(Hours)$. Interpret the slope of the regression line.

   Answer: For each additional hour studied, the GPA is predicted to increase by 0.05 points.

2. Which of the following is *not* a necessary condition for doing inference for the slope of a regression line?

   a For each given value of the independent variable, the response variable is normally distributed.

   b. The values of the predictor and response variables are independent.

   c. For each given value of the independent variable, the distribution of the response variable has the same standard deviation.

   d. The mean response values lie on a line.

Answer: (b) is not a condition for doing inference for the slope of a regression line. In fact, we are trying to find out the degree to which they are not independent.

3. True–False: Significance tests for the slope of a regression line are always based on the hypothesis $H_0: \beta = 0$ versus the alternative $H_A: \beta \neq 0$.

Answer: False. While the stated null and alternative may be the usual hypotheses in a test about the slope of the regression line, it is possible to test that the slope has some particular nonzero value or that the alternative can be one sided ($H_A: B > 0$ or $H_A: \beta < 0$). Note that most computer programs will test only the two-sided alternative by default. The TI-83/84 will test either a one- or two-sided alternative.

4. Consider the following Minitab printout:

The regression equation is
$\hat{y} = 282 + 0.634\,x$

Predictor	Coef	St Dev	t ratio	P
Constant	282.459	3.928	71.91	0.000
x	0.63383	0.07039	9.00	0.000
s = 9.282	R-sq = 81.0%	R-sq(adj) = 80.0%		

   a. What is the slope of the regression line?

   b. What is the standard error of the residuals?

   c. What is the standard error of the slope?

   d. Do the data indicate a predictive linear relationship between x and y?

Answer:

   a. 0.634

   b. 9.282

   c. 0.07039

   d. Yes, the t-test statistic = 9.00 $\Rightarrow$ P-value = .000. That is, the probability is close to zero of getting a slope of 0.634 if, in fact, the true slope was zero.

5. A t-test for the slope of a regression line is to be conducted at the 0.02 level of significance based on 18 data values. As usual, the test is two sided. What is the upper critical value for this test (that is, find the minimum positive value of t^* for which a finding would be considered significant)?

Answer: There are $18 - 2 = 16$ degrees of freedom. Since the alternative is two sided, the rejection region has 0.01 in each tail. Using Table B, we find the value at the intersection of the df = 16 row and the 0.01 column: $t^* = 2.583$. If you have a TI-83/84 with the invT function, invT(0.99,16)=2.583. This is, of course, the same value of t^* you would use to construct a 98% confidence interval for the slope of the regression line.

6. In the printout from question #4, we were given the regression equation $\hat{y} = 282 + 0.634x$. The t-test for H_0: $\beta = 0$ yielded a P-value of 0.000. What is the conclusion you would arrive at based on these data?

Answer: Because P is very small, we would reject the null hypothesis that the slope of the regression line is 0. We have strong evidence of a predictive linear relationship between x and y.

7. Suppose the computer output for regression reports $P = 0.036$. What is the P-value for H_A: $\beta > 0$ (assuming the test was in the correct direction for the data)?

Answer: 0.018. Computer output for regression assumes the alternative is two sided (H_A: $\beta \neq 0$). Hence the P-value reported assumes the finding could have been in either tail of the t-distribution. The correct P-value for the one-sided test is one-half of this value.

Practice Problems

Multiple-Choice

1. Which of the following statements is (are) true?
   I. In the computer output for regression, s is the estimator of σ, the standard deviation of the residuals.
   II. The t-test statistic for the H_0: $\beta = 0$ has the same value as the t-test statistic for H_0: $\rho = 0$.
   III. The t-test for the slope of a regression line is always two sided (H_A: $\beta \neq 0$).

   a. I only
   b. II only
   c. III only
   d. I and II only
   e. I and III only

Use the following output in answering questions 2–4:

A study attempted to establish a linear relationship between IQ score and musical aptitude. The following table is a partial printout of the regression analysis and is based on a sample of 20 individuals.

The regression equation is				
$\widehat{MusApt} = -22.3 + 0.493$ IQ					
Predictor	Coef	St Dev	t ratio	P
Constant	−22.26	12.94	−1.72	.102
IQ	0.4925	0.1215		
$s = 6.143$	R-sq = 47.7%	R-sq(adj) = 44.8%		

2. The value of the t-test statistic for H_0: $\beta = 0$ is
   a. 4.05
   b. −1.72
   c. 0.4925
   d. 6.143
   e. 0.0802

3. A 99% confidence interval for the slope of the regression line is
    a. $0.4925 \pm 2.878(6.143)$
    b. $0.4925 \pm 2.861(0.1215)$
    c. $0.4925 \pm 2.861(6.143)$
    d. $0.4925 \pm 2.845(0.1215)$
    e. $0.4925 \pm 2.878(0.1215)$

4. Which of the following best interprets the slope of the regression line?
    a. A student with an IQ one point above another student has a Musical Aptitude score 0.4925 points higher.
    b. As IQ score increases, so does the Musical Aptitude score.
    c. A student with an IQ one point above another student is predicted to have a Musical Aptitude score 0.4925 points higher.
    d. For each additional point of Musical Aptitude, IQ is predicted to increase by 0.4925 points.
    e. There is a strong predictive linear relationship between IQ score and Musical Aptitude.

5. A group of 12 students take both the SAT Math and the SAT Verbal. The least-squares regression line for predicting Verbal Score from Math Score is determined to be $\overline{Verbal\ Score} = 106.56 + 0.74(Math\ Score)$. Further, $s_b = 0.11$. Which of the following is a 95% confidence interval for the slope of the regression line?
    a. 0.74 ± 0.245
    b. 0.74 ± 0.242
    c. 0.74 ± 0.240
    d. 0.74 ± 0.071
    e. 0.74 ± 0.199

Free-Response

1–5. The following table gives the ages in months of a sample of children and their mean height (in inches) at that age.

Age	18	19	20	21	22	23	24	25	26	27	28
Height	30.0	30.7	30.7	30.8	31.0	31.4	31.5	31.9	32.0	32.6	32.9

1. Find the correlation coefficient and the least-squares regression line for predicting height (in inches) from age (in months).

2. Draw a scatterplot of the data and the LSRL on the plot. Does the line appear to be a good model for the data?

3. Construct a residual plot for the data. Does the line still appear to be a good model for the data?

4. Use your LSRL to predict the height of a child of 35 months. How confident should you be in this prediction?

5. Interpret the slope of the regression line found in question #1 in the context of the problem.

6. In 2002, there were 23 states in which more than 50% of high school graduates took the SAT test. The following printout gives the regression analysis for predicting SAT Math from SAT Verbal from these 23 states.

The regression equation is

Predictor	Coef	St Dev	t ratio	P
Constant	185.77	71.45	2.60	0.017
Verbal	0.6419	0.1420	4.52	0.000
s = 7.457	R-sq = 49.3%		R-sq(adj) = 46.9%	

a. What is the equation of the least-squares regression line for predicting Math SAT score from Verbal SAT score?
b. Determine the slope of the regression line and interpret in the context of the problem.
c. Identify the standard error of the slope of the regression line and interpret it in the context of the problem.
d. Identify the standard error of the residuals and interpret it in the context of the problem.
e. Assuming that the conditions needed for doing inference for regression are present, what are the hypotheses being tested in this problem, what test statistic is used in the analysis, what is its value, and what conclusion would you make concerning the hypothesis?

7. For the regression analysis of question #6:
a. Construct and interpret a 95% confidence interval for the true slope of the regression line.
b. Explain what is meant by "95% confidence interval" in the context of the problem.

8. It has been argued that the average score on the SAT test drops as more students take the test (nationally, about 46% of graduating students took the SAT). The following data are the Minitab output for predicting SAT Math score from the percentage taking the test (PCT) for each of the 50 states. Assuming that the conditions for doing inference for regression are met, test the hypothesis that scores decline as the proportion of students taking the test rises. That is, test to determine if the slope of the regression line is negative. Test at the 0.01 level of significance.

The regression equation is $\widehat{SAT\ Math} = 574 - 99.5\ PCT$

Predictor	Coef	St Dev	t ratio	P
Constant	574.179	4.123	139.25	0.000
PCT	−99.516	8.832	−11.27	0.000
s = 17.45	R-sq = 72.6%	R-sq(adj) = 72.0%		

9. Some bored researchers got the idea that they could predict a person's pulse rate from his or her height (earlier studies had shown a very weak linear relationship between pulse rate and weight). They collected data on 20 college-age women. The following table is part of the Minitab output of their findings.

The regression equation is

Pulse = [   ] Height

Predictor	Coef	St Dev	t ratio	P
Constant	52. 00	37.24	1.40	0.180
Height	0.2647	0.5687	[ ]	
$s = 10.25$	R-sq = 1.2%	R-sq(adj) = 0.0%		

a. Determine the t-ratio and the P-value for the test.
b. Construct a 99% confidence interval for the slope of the regression line used to predict pulse rate from height.
c. Do you think there is a predictive linear relationship between height and pulse rate? Explain.
d. Suppose the researchers were hoping to show that there was a positive linear relationship between pulse rate and height. Are the t-ratio and P-value the same as in part (a)? If not, what are they?

Cumulative Review Problems

1. You are testing the hypothesis H_0: $p = 0.6$. You sample 75 people as part of your study and calculate that $\hat{p} = 0.7$.

   a. What is $s_{\hat{p}}$ for a significance test for p?
   b. What is $s_{\hat{p}}$ for a confidence interval for p?

2. A manufacturer of lightbulbs claims a mean life of 1500 hours. A mean of 1450 hours would represent a significant departure from this claim. Suppose, in fact, the mean life of bulbs is only 1450 hours. In this context, what is meant by the power of the test (no calculation is required)?

3. Complete the following table by filling in the shape of the sampling distribution of $\bar{x}$ for each situation.

SITUATION	SHAPE OF SAMPLING DISTRIBUTION
• Shape of parent population: normal • Sample size: $n = 8$	
• Shape of parent population: normal • Sample size: $n = 35$	
• Shape of parent population: strongly skewed to the left • Sample size: $n = 8$	
• Shape of parent population: strongly skewed to the left • Sample size: $n = 35$	
• Shape of parent population: unknown • Sample size: $n = 8$	
• Shape of parent population: unknown • Sample size: $n = 50$	

4. The following is most of a probability distribution for a discrete random variable.

X	2	6	7	9
p(x)	0.15	0.25		0.40

Find the mean and standard deviation of this distribution.

5. Consider the following scatterplot and regression line.

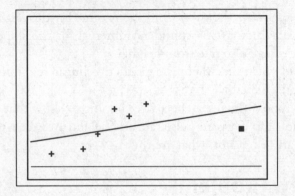

a. Would you describe the point marked with a box as an outlier, influential point, neither, or both?
b. What would be the effect on the correlation coefficient of removing the box-point?
c. What would be the effect on the slope of the regression line of removing the box-point?

Solutions to Practice Problems

Multiple-Choice

1. The correct answer is (d). II is true since it can be shown that $t = \dfrac{b}{s_b} = r\sqrt{\dfrac{n-2}{1-r^2}}$.

   III is not true since, although we often use the alternative $H_A: \beta \neq 0$, we can certainly test a null with an alternative that states that there is a positive or a negative association between the variables.

2. The correct answer is (a). $t = \dfrac{b}{s_b} = \dfrac{0.4925}{0.1215} = 4.05$.

3. The correct answer is (e). For $n = 20$, df $= 20 - 2 = 18 \Rightarrow t^* = 2.878$ for $C = 0.99$.

4. The correct answer is (c). Note that (a) is not correct since it doesn't have "predicted" or "on average" to qualify the increase. (b) is a true statement but is not the best interpretation of the slope. (d) has mixed up the response and explanatory variables. (e) is also true ($t = 4.05 \Rightarrow P$-value $= 0.0008$) but is not an interpretation of the slope.

5. The correct answer is (a). A 95% confidence interval at $12 - 2 = 10$ degrees of freedom has a critical value of $t^* = 2.228$ (from Table B; if you have a TI-83/84 with the invT function, invT(0.975,10) = 2.228). The required interval is $0.74 \pm (2.228)(0.11) = 0.74 \pm 0.245$.

Free-Response

1. $r = 0.9817$, height $= 25.41 + 0.261$(age)

    (Assuming that you have put the age data in `L1` and the height data in `L2`, remember that this can be done on the TI-83/84 as follows: `STAT CALC LinReg(a+bx) L1,L2,Y1`.)

2. The line does appear to be a good model for the data.

    (After the regression equation was calculated on the TI-83/84 and the LSRL stored in `Y1`, this was constructed in `STAT PLOT` by drawing a scatterplot with `Xlist:L1` and `Ylist:L2`.)

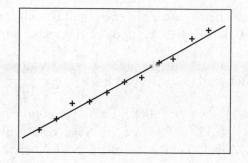

3. The residual pattern seems quite random. A line still appears to be a good model for the data.

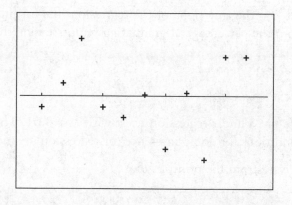

(This scatterplot was constructed on the TI-83/84 using `STAT PLOT` with `Xlist:L1` and `Ylist:RESID`. Remember that the list of residuals for the most recent regression is saved in a list named `RESID`.)

4. $\widehat{Height} = 25.41 + 0.261(35) = 34.545$ (`Y1(35) = 34.54`). You probably shouldn't be too confident in this prediction. 35 is well outside of the data on which the LSRL was constructed and, even though the line appears to be a good fit for the data, there is no reason to believe that a linear pattern is going to continue indefinitely. (If it did, a 25-year-old would have a predicted height of $25.41 + 0.261(12 \times 25) = 103.71''$, or 8.64 feet!)

5. The slope of the regression line is 0.261. This means that, for an increase in age of 1 month, height is predicted to increase by 0.261 inches. You could also say, that, for an increase in age of 1 month, height will increase on average by 0.261 inches.

6. a. $\widehat{Math} = 185.77 + 0.6419(Verbal)$.

    b. $b = 0.6419$. For each additional point scored on the SAT Verbal test, the score on the SAT Math test is predicted to increase by 0.6419 points (or: will increase *on average* by 0.6419 points). (Very important on the AP exam: be very sure to say "is predicted" or "on average" if you'd like maximum credit for the problem!)

    c. The standard error of the slope is $s_b = 0.1420$. This is an estimate of the variability of the standard deviation of the estimated slope for predicting SAT Verbal from SAT Math.

    d. The standard error of the residuals is $s = 7.457$. This value is a measure of variation in SAT Verbal for a fixed value of SAT Math.

    e. • The hypotheses being tested are $H_0: \beta = 0$ (which is equivalent to $H_0: \rho = 0$) and $H_A \beta \neq 0$, where β is the slope of the regression line for predicting SAT Verbal from SAT Math.

       • The test statistic used in the analysis is $t = \dfrac{b}{s_b} = \dfrac{0.6419}{0.1420} = 4.52$, df $= 23 - 2 = 21$.

7. a. df $= 23 - 2 = 21 \Rightarrow t^* = 2.080$. The 95% confidence interval is: $0.6419 \pm 2.080(0.1420) = (0.35, 0.94)$. We are 95% confident that, for each 1 point increase in SAT Verbal, the true increase in SAT Math is between 0.35 points and 0.94 points.

    b. The procedure used to generate the confidence interval would produce intervals that contain the true slope of the regression line, on average, 0.95 of the time.

8.   I. Let $\beta =$ the true slope of the regression line for predicting SAT Math score from the percentage of graduating seniors taking the test.

$$H_0: \beta = 0.$$

$$H_A: \beta < 0.$$

       II. We use a linear regression t test with $\alpha = 0.01$. The problem states that the conditions for doing inference for regression are met.

       III. We see from the printout that

$$t = \frac{b}{s_b} = \frac{-99.516}{8.832} = -11.27$$

       based on $50 - 2 = 48$ degrees of freedom. The P-value is 0.000. (Note: The P-value in the printout is for a two-sided test. However, since the P-value for a one-sided test would only be half as large, it is still 0.000.)

       IV. Because $P < 0.01$, we reject the null hypothesis. We have very strong evidence that there is a negative linear relationship between the proportion of students taking SAT Math and the average score on the test.

9. a. $t = \dfrac{b}{s_b} = \dfrac{0.2647}{0.5687} = 0.47$, df $= 20 - 2 = 18 \Rightarrow P$-value $= 0.644$.

    b. df $= 18 \Rightarrow t^* = 2.878$; $0.2647 \pm 2.878(0.5687) = (-1.37, 1.90)$.

c. No. The P-value is very large, giving no grounds to reject the null hypothesis that the slope of the regression line is 0. Furthermore, the correlation coefficient is only $r = \sqrt{0.012} = 0.11$, which is very close to 0. Finally, the confidence interval constructed in part (b) contains the value 0 as a likely value of the slope of the population regression line.

d. The t-ratio would still be 0.47. The P-value, however, would be half of the 0.644, or 0.322 because the computer output assumes a two-sided test. This is a lower P-value but is still much too large to infer any significant linear relationship between pulse rate and height.

Solutions to Cumulative Review Problems

1. a. $s_{\hat{p}} = \sqrt{\dfrac{(0.6)(0.4)}{75}} = 0.057$.

    b. $s_{\hat{p}} = \sqrt{\dfrac{(0.7)(0.3)}{75}} = 0.053$.

2. The *power of the test* is the probability of correctly rejecting a false hypothesis against a particular alternative. In other words, the *power* of this test is the probability of rejecting the claim that the true mean is 1500 hours against the alternative that the true mean is only 1450 hours.

SITUATION	SHAPE OF SAMPLING DISTRIBUTION
• Shape of parent population: normal • Sample size: $n = 8$	**Normal**
• Shape of parent population: normal • Sample size: $n = 35$	**Normal**
• Shape of parent population: strongly skewed to the left • Sample size: $n = 8$	**Skewed somewhat to the left**
• Shape of parent population: strongly skewed to the left • Sample size: $n = 35$	**Approximately normal (central limit theorem)**
• Shape of parent population: unknown • Sample size: $n = 8$	**Similar to parent population**
• Shape of parent population: unknown • Sample size: $n = 50$	**Approximately normal (central limit theorem)**

4. $P(7) = 1 - (0.15 + 0.25 + 0.40) = 0.20$.

$\mu_x = 2(0.15) + 6(0.25) + 7(0.20) + 9(0.40) = 6.8$.

$\sigma_X = \sqrt{(2-6.8)^2(0.15)+(6-6.8)^2(0.25)+(7-6.8)^2(0.20)+(9-6.8)^2(0.40)} = 2.36$.

(Remember that this can be done by putting the X-values in L1, the $p(x)$-values in L2, and doing STAT CALC 1-Var Stats L1,L2.)

5. a. The point is both an outlier and an influential point. It is an outlier because it is removed from the general pattern of the data. It is an influential observation because it is an outlier in the x direction and its removal would have an impact on the slope of the regression line.

   b. Removing the point would increase the correlation coefficient. That is, the remaining data are better modeled by a line without the box-point than with it.

   c. Removing the point would make the slope of the regression line more positive (steeper) than it is already.

CHAPTER 13

Inference for Categorical Data: Chi-Squared Tests

IN THIS CHAPTER

Summary: In this final chapter, we will look at inference for categorical variables. Up until now, we have studied primarily data analysis and inference only for one or two numerical variables (those whose outcomes we can measure), and proportions, which are categorical variable with only two possible values (success and failure). The material in this chapter opens up for us a wide range of research topics, allowing us to compare categorical variables across several values. For example, we will ask new questions like, "Is there an association between gender and political party preference?"

Key Ideas
- ✪ Chi-Square Goodness-of-Fit Test
- ✪ Chi-Square Test for Independence
- ✪ Chi-Square Test for Homogeneity of Proportions (Populations)
- ✪ χ^2 versus z^2

Chi-Square Goodness-of-Fit Test

The following are the approximate percentages for the different blood types among white Americans: A: 40%; B: 11%; AB: 4%; O: 45%. A random sample of 1000 black Americans yielded the following blood type data: A: 270; B: 200; AB: 40; O: 490. Does this sample provide evidence that the distribution of blood types among black Americans differs from that of white Americans, or could the sample values simply be due to sampling variation? This is the kind of question we can answer with the chi-square **goodness-of-fit test**. ("Chi" is the Greek letter χ; chi-square is, logically enough, χ^2.) With the chi-square goodness-of-fit test, we note that there is one categorical variable (blood type) and one population (black Americans). In this chapter we will also encounter a situation in which there is one categorical variable measured across two or more populations (called a chi-square test for homogeneity of proportions) and a situation in which there are two categorical variables measured across a single population (called a chi-square test for independence).

To answer this question, we need to compare the **observed values** in the sample with the **expected values** we would get *if* the sample of black Americans really had the same distribution of blood types as white Americans. The values we need for this are summarized in the following table:

BLOOD TYPE	PROPORTION OF POPULATION	EXPECTED VALUES	OBSERVED VALUES
A	0.40	$(0.4)(1000) = 400$	270
B	0.11	$(0.11)(1000) = 110$	200
AB	0.04	$(0.04)(1000) = 40$	40
O	0.45	$(0.45)(1000) = 450$	490

It appears that the numbers vary noticeably for types A and B, but not as much for types AB and O. The table can be rewritten as follows:

BLOOD TYPE	OBSERVED VALUES	EXPECTED VALUES
A	270	400
B	200	110
AB	40	40
O	490	450

Before working through this problem, a note on symbols. Often in this book, and in statistics in general, we use English letters for statistics (measurements from data) and Greek letters for parameters (population values). Hence, $\bar{x}$ is a sample mean and μ is a population mean; s is a sample standard deviation and σ is a population standard deviation, etc. We follow this same convention in this chapter: we will use χ^2 when referring to a population value or to the name of a test and use X^2 when referring to the chi-square statistic.

The chi-square statistic (X^2) calculates the squared difference between the observed and expected values relative to the expected value for each category. The X^2 statistic is computed as follows:

$$X^2 = \sum \frac{(\text{Observed} - \text{Expected})^2}{\text{Expected}} = \sum \frac{(O-E)^2}{E}.$$

The chi-square distribution is based on the number of degrees of freedom, which equals, for the goodness-of-fit test, the number of categories minus 1 (df = $c - 1$). The X^2 statistic follows approximately a unique chi-square distribution, assuming a random sample and a large enough sample, for each different number of degrees of freedom. The probability that a sample has a X^2 value as large as it does can be read from a table of X^2 critical values or determined from a calculator. There is a X^2 table in the back of this book, and you will be supplied a table like this on the AP exam. We will demonstrate both the use of tables and the calculator in the examples and problems that follow.

A hypothesis test for χ^2 goodness-of-fit follows the by now familiar pattern. The essential parts of the test are summarized in the following table.

HYPOTHESES	CONDITIONS	TEST STATISTIC
• Null hypothesis: H_0: p_1 = population proportion for category 1 p_2 = population proportion for category 2 . . . p_n = population proportion for category n This is usually stated in words: "The proportions of the population in each category are the same as those given in the table above" or something like that. • Alternative hypothesis: H_A: at least one of the proportions in H_0 is not equal to the claimed population proportion.	• Observations are based on a random sample. • The number of each expected count is at least 5. (Some texts use the following condition for expected counts: at least 80% of the counts are greater than 5 and none is less than 1. It is acceptable for you to use either of these.)	$X^2 = \sum \dfrac{(O-E)^2}{E}$ df = $c - 1$

Now, let's answer the question posed in the opening paragraph of this chapter. We will use the by now familiar four-step hypothesis-testing procedure introduced in Chapter 11.

example: The following are the approximate percentages for the different blood types among white Americans: A: 40%; B: 11%; AB: 4%; O: 45%. A random sample of 1000 black Americans yielded the following blood type data: A: 270; B: 200; AB: 40; O: 490. Does this sample indicate that the distribution of blood types among black Americans differs from that of white Americans?

solution:

I. Let p_A = proportion of black Americans with type A blood; p_B = proportion with type B blood; p_{AB} = proportion with type AB blood; p_O = proportion with type O blood. H_0: $p_A = 0.40$, $p_B = 0.11$, $p_{AB} = 0.04$, $p_0 = 0.45$.

H_A: at least one of these proportions is not as stated.

II. We will use the χ^2 goodness-of-fit test. The problem states that the sample is a random sample. The expected values are type A: 400; type B: 110; type AB: 40; type O: 450. Each of these is greater than 5. The conditions needed for the test are satisfied.

III. The data are summarized in the table:

BLOOD TYPE	OBSERVED VALUES	EXPECTED VALUES
A	270	400
B	200	110
AB	40	40
O	490	450

$$X^2 = \sum \frac{(O-E)^2}{E} = \frac{(270-400)^2}{400} + \frac{(200-110)^2}{110}$$
$$+ \frac{(40-40)^2}{40} + \frac{(490-450)^2}{450} = 119.44,$$
$$\text{df} = 4 - 1 = 3.$$

From the X^2 table (Table C, which is read very much like Table B), we see that 119.44 is much larger than any value for df = 3. Hence, the P-value < 0.0005. (A calculator gives a P-value of 1.02×10^{-25}—more about how to do this coming up next.)

IV. Because the P-value is so small, we reject the null hypothesis. We have very strong evidence that the proportions of the various blood types among black Americans differ from the proportions among white Americans.

Calculator Tip: All newer graphing calculators have two χ^2 tests in the test menu. The goodness-of-fit test will require the observed and expected frequencies to be entered into lists. Older calculators that do not have the goodness-of-fit test put you at a slight disadvantage, but the computation is simple enough with list operations. If you put the observed frequencies in the first list, and the expected fewquencies in the second, you can create a third list with the formula $\dfrac{(List1 - List2)^2}{List2}$, then calculate the sum of $List2$ to get the χ^2 statistic. Then use the χ^2 cdf function to calculate the area above the χ^2 statistic.

example: The statistics teacher, Mr. Hinders, used his calculator to simulate rolling a die 96 times and storing the results in a list L1. He did this by entering MATH PRB randInt(1,6,96)→(L1). Next he sorted the list

(STAT SortA(L1)). He then counted the number of each face value. The results were as follows (this is called a one-way table):

FACE VALUE	OBSERVED
1	19
2	15
3	10
4	14
5	17
6	21

Does it appear that the teacher's calculator is simulating a fair die? (That is, are the observations consistent with what you would expect to get *if* the die were fair?)

solution:

I. Let p_1, p_2,..., p_6 be the population proportion for each face of the die to appear on repeated trials of a roll of a fair die.

H_0: $p_1 = p_2 = p_3 = p_4 = p_5 = p_6 = \frac{1}{6}$.

H_A: At least one of the proportions is not equal to 1/6.

II. We will use a χ^2 goodness-of-fit test. If the die is fair, we would expect to get

$$\left(\frac{1}{6}\right)(96) = 16$$

of each face. Because all expected values are greater than 5, the conditions are met for this test.

III. FACE VALUE	OBSERVED	EXPECTED
1	19	16
2	15	16
3	10	16
4	14	16
5	17	16
6	21	16

$X^2 = 4.75$ (calculator result), df $= 6 - 1 = 5$ P-value > 0.25 (Table C) or P-value $= 0.45$ (calculator).

(*Remember*: To get X^2 on the calculator, put the Observed values in L1, the Expected values in L2, let L3 = (L1-L2)²/L2, then LIST MATH SUM (L3) will be X^2. The corresponding probability is then found by DISTR χ^2 cdf (4.75,100,5). This can also be done on a TI-83/84 that has the χ^2 GOF-Test.)

IV. Because $P > 0.25$, we fail to reject the null hypothesis. We do not have convincing evidence that the calculator is failing to simulate a fair die.

Inference for Two-Way Tables

Two-Way Tables (Contingency Tables) Defined

A **two-way table**, or **contingency table**, for categorical data is simply a rectangular array of cells. Each cell contains the frequencies for the joint values of the row and column variables. If the row variable has r values, then there will be r rows of data in the table. If the column variable has c values, then there will be c columns of data in the table. Thus, there are $r \times c$ cells in the table. (The **dimension** of the table is $r \times c$.) The **marginal totals** are the sums of the observations for each row and each column.

example: A class of 36 students is polled concerning political party preference. The results are presented in the following *two-way* table.

Political Party Preference

		Democrat	Republican	Independent	Total
Gender	**Male**	11	7	2	20
	Female	7	8	1	16
	Total	18	15	3	36

The values of the row variable (Gender) are "Male" and "Female." The values of the column variable (Political Party Preference) are "Democrat," "Republican," and "Independent." There are $r = 2$ rows and $c = 3$ columns. We refer to this as a 2×3 table (the number of rows always comes first). The row marginal totals are 20 and 16; the column marginal totals are 18, 15, and 3. Note that the sum of the row and column marginal totals must both add to the total number in the sample.

In the example above, we had one population of 36 students and two categorical variables (gender and party preference). In this type of situation, we are interested in whether or not the variables are independent in the population. That is, does knowledge of one variable provide you with information about the other variable? Another study might have drawn a simple random sample of 20 males from, say, the senior class and another simple random sample of 16 females. Now we have two populations rather than one, but only one categorical variable. Now we might ask if the proportions of Democrats, Republicans, and Independents in each population are the same. Either way we do it, we end up with the same contingency table given in the example. We will look at how these differences in design play out in the next couple of sections.

Chi-Square Test for Independence

A random sample of 400 residents of a large western city are polled to determine their attitudes concerning the affirmative action admissions policy of the local university. The residents are classified according to ethnicity (white, black, Asian) and whether or not they favor the affirmative action policy. The results are presented in the following table.

Attitude Toward Affirmative Action

		Favor	Do Not Favor	Total
	White	130	120	250
Ethnicity	**Black**	75	35	110
	Asian	28	12	40
	Total	233	167	400

We are interested in whether or not, in the population of this large city, ethnicity and attitude toward affirmative action are associated (note that, in this situation, we have one population and two categorical variables). That is, does knowledge of a person's ethnicity give us information about that person's attitude toward affirmative action? Another way of asking this is, "Are the variables *ethnicity* and *attitude toward affirmative action* independent in the population?" As part of a hypothesis test, the null hypothesis is that the two variables are independent, and the alternative is that they are not: H_0: the variables *ethnicity* and *attitude toward affirmative action* are independent among all residents of this city *vs.* H_A: the variables are not independent among all residents of this city. Alternatively, we could say H_0: the variables *ethnicity* and *attitude toward affirmative action* are not associated among all residents of this city *vs.* H_A: the variables *ethnicity* and *attitude toward affirmative action* are associated among all residents of this city.

The test statistic for the independence hypothesis is the same chi-square statistic we saw for the goodness-of-fit test:

$$X^2 = \sum \frac{(O-E)^2}{E}.$$

For a two-way table, the number of degrees of freedom is calculated as (*number of rows* − 1)(*number of columns* − 1) = $(r-1)(c-1)$. As with the goodness-of-fit test, we require that we are dealing with a random sample and that the number of expected values in each cell be at least 5 (or some texts say there are no empty cells and at least 80% of the cells have more than 5 expected values).

Calculation of the expected frequencies for chi-square can be labor intensive if there are many cells, but it is *usually* done by technology (see the next Calculator Tip for details). However, you should know how expected frequencies are arrived at.

> **example** (calculation of expected frequency): Suppose we are testing for independence of the variables (ethnicity and opinion) in the previous example. For the two-way table with the given marginal values, find the expected frequency for the cell marked "Exp."

		Favor	Do Not Favor	Total
Ethnicity	**White**			250
	Black		Exp	110
	Asian			40
	Total	233	167	400

> **solution:** There are two ways to approach finding an expected value, but they are numerically equivalent and you can use either. The first way is to find the probability of being in the desired location by chance and then multiplying that value times the total in the table (as we found an expected value with discrete random variables). The probability of being in the "Black" row is $\frac{110}{400}$ and the probability of being in the "Do Not Favor" column is $\frac{167}{400}$.
>
> Assuming independence, the probability of being in "Exp" by chance is then $\left(\frac{110}{400}\right)\left(\frac{167}{400}\right)$. Thus, Exp $= \left(\frac{110}{400}\right)\left(\frac{167}{400}\right)(400) = 45.925$.

The second way is to argue, under the assumption that there is no relation between ethnicity and opinion, that we'd expect each cell in the "Do Not Favor" column to show the same proportion of outcomes. In this case, each row of the "Do Not Favor" column would contain $\frac{167}{400}$ of the row total. Thus, Exp $= \left(\frac{167}{400}\right)(110) = 45.925$. Most of you will probably find using the calculator easier.

Calculator Tip: The easiest way to obtain the expected values is to use your calculator. To do this, let's use the data from the previous examples:

130	120
75	35
28	12

In mathematics, a rectangular array of numbers such as this is called a matrix. Matrix algebra is a separate field of study, but we are only concerned with using the matrix function on our calculator to find a set of expected values (which we'll need to check the conditions for doing a hypothesis test using the chi-square statistics).

On a TI-84, go to MATRIX EDIT [A]. Note that our data matrix has three rows and two columns, so make the dimension of the matrix (the numbers right after MATRIX [A]) read 3×2. The calculator expects you enter the data by rows, so just enter 130, 120, 75, 35, 28, 12 in order and the matrix will be correct. Now, QUIT the MATRIX menu and go to STAT TESTS χ^2-Test (Note: Technically we don't yet know that we have the conditions present to do a χ^2-test, but this is the way we'll find our expected values.) Enter [A] for Observed and [B] for Expected (you can paste "[A]" by entering MATRIX NAMES [A]). Then choose Calculate. The calculator will return some things we don't care about yet ($X^2 = 10.748$, $p = 0.0046$, and df = 2). Now return to the MATRIX menu and select NAMES [B] and press ENTER. You should get the following matrix of expected values:

$$\begin{bmatrix} 145.625 & 104.375 \\ 64.075 & 45.925 \\ 23.3 & 16.7 \end{bmatrix}$$

Note that the entry in the second row and second column, 45.925, agrees with our hand calculation for "Exp" in the previous example.

The χ^2-test for independence can be summarized as follows:

HYPOTHESES	CONDITIONS	TEST STATISTIC
• Null hypothesis: H_0: The row and column variables are independent (or: they are not associated). • Alternative hypothesis: H_0: The row and column variables are not independent (or: they are associated).	• Observations are based on a random sample. • The number of each expected count is at least 5. (Some texts use the following condition: all expected counts are greater than 1, and at least 80% of the expected counts are greater than 5.)	$X^2 = \sum \dfrac{(O-E)^2}{E}$ df $= (r-1)(c-1)$

example: A study of 150 randomly selected cities was conducted to determine if crime rate is associated with outdoor temperature. The results of the study are summarized in the following table:

Crime Rate

		BELOW	NORMAL	ABOVE
Temperature	**Below**	12	8	5
	Normal	35	41	24
	Above	4	7	14

Do these data provide evidence, at the 0.02 level of significance, that the crime rate is associated with the temperature at the time of the crime?

solution:

I.

H_0: The crime rate is independent of temperature (or, H_0: Crime Rate and Temperature are not associated).

H_A: The crime rate is not independent of temperature (or, H_A: Crime Rate and Temperature are associated).

II. We will use a chi-square test for independence.
The cities were randomly selected.
A matrix of expected values (using the TI-83/84 as explained in the previous Calculator Tip) is found to be:

$\begin{bmatrix} 8.5 & 9.33 & 7.17 \\ 34 & 37.33 & 28.67 \\ 8.5 & 9.33 & 7.17 \end{bmatrix}$. Since all expected values are greater than 5, the conditions

are present for a chi-square test.

III. $X^2 = \sum \dfrac{(O-E)^2}{E} = \dfrac{(12-8.5)^2}{8.5} + \cdots + \dfrac{(14-7.17)^2}{7.17} = 12.92, \text{df} = (3-1)(3-1) = 4$

$\Rightarrow 0.01 < P\text{-value} < 0.02$ (from Table C; or P-value = DISTR χ^2cdf $(12.92, 1000, 410) = 0.012$).
(Note that the entire problem could be done by entering the observed values in MATRIX [A] and using STAT TESTS χ^2-Test.]

IV. Since $P < 0.02$, we reject H_0. We have strong evidence that the number of crimes committed is related to the temperature at the time of the crime.

Chi-Square Test for Homogeneity of Proportions (or Homogeneity of Populations)

In the previous section, we tested for the independence of two categorical variables measured on a single population. In this section we again use the chi-square statistic but will investigate whether or not the values of a single categorical variable are proportional among two or more populations. In the previous section, we considered a situation in which a sample of 36 students was selected and were then categorized according to gender and political party preference. We then asked if gender and party preference are independent in the population. Now suppose instead that we had selected a random sample of 20 males from the population of males in the school and another, independent, random sample of 16 females from the population of females in the school. Within each sample we classify the students as Democrat, Republican, or Independent. The results are presented in the following table, which you should notice is *exactly* the same table we presented earlier when gender was a category.

Political Party Preference

		DEMOCRAT	REPUBLICAN	INDEPENDENT	TOTAL
Gender	**Male**	11	7	2	20
	Female	7	8	1	16
	Total	18	15	3	36

Because "Male" and "Female" are now considered separate populations, we do not ask if gender and political party preference are independent in the population of students. We ask instead if the proportions of Democrats, Republicans, and Independents are the *same* within the populations of Males and Females. This is the test for homogeneity of proportions (or homogeneity of populations). Let the proportion of Male Democrats be p_1; the proportion of Female Democrats be p_2; the proportion of Male Republicans be p_3; the proportion of Female Republicans be p_4; the proportion of Independent Males be p_5; and the proportion of Independent Females be p_6. Our null and alternative hypotheses are then

H_0: $p_1 = p_2$, $p_3 = p_4$, $p_5 = p_6$.
H_A: Not all of the proportions stated in the null hypothesis are true.

It works just as well, and might be a bit easier, to state the hypotheses as follows.

H_0: The proportions of Democrats, Republicans, and Independents are the same among Male and Female students.
H_A: Not all of the proportions stated in the null hypothesis are equal.

For a given two-way table the expected values are the same under a hypothesis of homogeneity or independence.

example: A university dean suspects that there is a difference between how tenured and nontenured professors view a proposed salary increase. She randomly selects 20 nontenured instructors and 25 tenured staff to see if there is a difference. She gets the following results.

	FAVOR PLAN	DO NOT FAVOR PLAN
Tenured	15	10
Nontenured	8	12

Do these data provide good statistical evidence that tenured and nontenured faculty differ in their attitudes toward the proposed salary increase?

solution:

> I. Let p_1 = the proportion of tenured faculty who favor the plan and let p_2 = the proportion of nontenured faculty who favor the plan.
> H_0: $p_1 = p_2$.
> H_A: $p_1 \neq p_2$.
>
> II. We will use a chi-square test for homogeneity of proportions. The samples of tenured and nontenured instructors are given as random. We determine that the expected values are given by the matrix $\begin{bmatrix} 12.78 & 12.22 \\ 10.22 & 9.78 \end{bmatrix}$. Because all expected values are greater than 5, the conditions for the test are present.
>
> III. $X^2 = 1.78$, df $= (2-1)(2-1) = 1 \Rightarrow 0.15 < P$-value < 0.20 (from Table C; on the TI-83/84: $P = \chi^2 \texttt{cdf}(1.78, 1000, 1) = 0.182$).
>
> IV. The P-value is not small enough to reject the null hypothesis. These data do not provide strong statistical evidence that tenured and nontenured faculty differ in their attitudes toward the proposed salary plan.

$X^2 = z^2$

The example just completed could have been done as a two-proportion z-test where p_1 and p_2 are defined the same way as in the example (that is, the proportions of tenured and nontenured staff that favor the new plan). Then

$$\hat{p}_1 = \frac{15}{25}$$

and

$$\hat{p}_2 = \frac{8}{20}.$$

Computation of the z-test statistics for the two-proportion z-test yields $z = 1.333$. Now, $z^2 = 1.78$. Because the chi-square test and the two-proportion z-test are testing the same thing, it should come as little surprise that z^2 equals the obtained value of X^2 in the example. For a 2×2 table, the X^2 test statistic and the value of z^2 obtained for the same data in a two-proportion z-test are the same. Note that the z-test is somewhat more flexible in the case that a one-sided test allows us to consider a particular direction of difference, whereas the chi-square test has only one tail and does not allow us to test for a particular direction of the difference. However, this advantage *only* holds for a 2×2 table since there is no way to use a simple z-test for situations with more than two rows and two columns.

> **Digression:** You aren't required to know these, but you might be interested in the following (unexpected) facts about a χ^2 distribution with k degrees of freedom:
>
> - The *mean* of the χ^2 distribution $= k$.
> - The *median* of the χ^2 distribution ($k > 2$) $\approx k - \frac{2}{3}$.
> - The *mode* of the χ^2 distribution $= k - 2$.
> - The *variance* of the χ^2 distribution $= 2k$.

› Rapid Review

1. A study yields a chi-square statistic value of 20 ($X^2 = 20$). What is the P-value of the test if

   a. the study was a goodness-of-fit test with $n = 12$?
   b. the study was a test of independence between two categorical variables, the row variable with 3 values and the column variable with 4 values?

   Answer:

   a. $n = 12 \Rightarrow df = 12 - 1 = 11 \Rightarrow 0.025 < P < 0.05$.
   (Using the TI-83/84: χ^2cdf (20,1000,11) = 0.045.)
   b. $r = 3$, $c = 4 \Rightarrow df. = (3 - 1)(4 - 1) = 6 \Rightarrow 0.0025 < P < 0.005$.
   (Using the TI-83/84: χ^2 cdf (20,1000,6) = 0.0028.)

2–4. The following data were collected while conducting a chi-square test for independence:

Preference

	BRAND A	BRAND B	BRAND C
Male	16	22	15
Female	18 (X)	30	28

2. What null and alternative hypotheses are being tested?

   Answer:

   H_0: Gender and Preference are independent (or: H_0: Gender and Preference are not associated).
   H_A: Gender and Preference are not independent (H_A: Gender and Preference are associated).

3. What is the expected frequency of the cell marked with the X?

   Answer: Identifying the marginals on the table we have

16	22	15	53
18 (X)	30	28	76
34	52	43	129

   Since there are 34 values in the column with the X, we expect to find $\dfrac{34}{129}$ of each row total in the cells of the first column. Hence, the expected value for the cell containing X is $\left(\dfrac{34}{129}\right)(76) = 20.03$.

4. How many degrees of freedom are involved in the test?

   Answer: df $= (2 - 1)(3 - 1) = 2$.

5. The null hypothesis for a chi-square goodness-of-fit test is given as:

   H_0: $p_1 = 0.2$, $p_2 = 0.3$, $p_3 = 0.4$, $p_4 = 0.1$. Which of the following is an appropriate alternative hypothesis?

   a. H_A: $p_1 \neq 0.2$, $p_2 \neq 0.3$, $p_3 \neq 0.4$, $p_4 \neq 0.1$.
   b. H_A: $p_1 = p_2 = p_3 = p_4$.
   c. H_A: Not all of the proportions stated in H_0 are correct.
   d. H_A: $p_1 \neq p_2 \neq p_3 \neq p_4$.

   Answer: c

Practice Problems

Multiple-Choice

1. Find the expected frequency of the cell marked with the "****" in the following 3×2 table (the bold face values are the marginal totals):

observation	observation	**19**
observation	***	**31**
observation	observation	**27**
45	**32**	**77**

   a. 74.60
   b. 18.12
   c. 12.88
   d. 19.65
   e. 18.70

2. A χ^2 goodness-of-fit test is performed on a random sample of 360 individuals to see if the number of birthdays each month is proportional to the number of days in the month. X^2 is determined to be 23.5. The P-value for this test is

   a. $0.001 < P < 0.005$
   b. $0.02 < P < 0.025$
   c. $0.025 < P < 0.05$
   d. $0.01 < P < 0.02$
   e. $0.05 < P < 0.10$

3. Two random samples, one of high school teachers and one of college teachers, are selected and each sample is asked about their job satisfaction. Which of the following are appropriate null and alternative hypotheses for this situation?

   a. H_0: The proportion of each level of job satisfaction is the same for high school teachers and college teachers.
      H_A: The proportions of teachers highly satisfied with their jobs is higher for college teachers.
   b. H_0: Teaching level and job satisfaction are independent.
      H_A: Teaching level and job satisfaction are not independent.
   c. H_0: Teaching level and job satisfaction are associated.
      H_A: Teaching level and job satisfaction are not associated.

d. H_0: The proportion of each level of job satisfaction is the same for high school teachers and college teachers.

H_A: Not all of the proportions of each level of job satisfaction are the same for high school teachers and college teachers.

e. H_0: Teaching level and job satisfaction are independent.

H_A: Teaching level and job satisfaction are not associated.

4. A group separated into men and women are asked their preference toward certain types of television shows. The following table gives the results.

	Program Type A	Program Type B
Men	5	20
Women	3	12

Which of the following statements is (are) true?

I.   The variables gender and program preference are independent.
II. For these data, $X^2 = 0$.
III. The variables gender and program preference are related.

a.   I only
b.   I and II only
c.   II only
d.   III only
e.   II and III only

5. For the following two-way table, compute the value of X^2.

	C	D
A	15	25
B	10	30

a. 2.63
b. 1.22
c. 1.89
d. 2.04
e. 1.45

6. The main difference between a χ^2 test for independence and a χ^2 test for homogeneity of proportions is which of the following?

a. They are based on a different number of degrees of freedom.
b. One of the tests is for a two-sided alternative and the other is for a one-sided alternative.
c. In one case, two variables are compared within a single population. In the other case, two populations are compared in terms of a single variable.
d. For a given value of X^2, they have different P-values.
e. There are no differences between the tests. They measure exactly the same thing.

7. A study is to be conducted to help determine if ethnicity is related to blood type. Ethnic groups are identified as White, African-American, Asian, Latino, or Other. Blood types are A, B, O, and AB. How many degrees of freedom are there for a chi-square test of independence between Ethnicity and Blood Type?

a. $5 \times 4 = 20$
b. $5 \times 3 = 15$
c. $4 \times 4 = 16$
d. $5 + 4 - 2 = 7$
e. $4 \times 3 = 12$

8. Which of the following statements is (are) correct?

   I. A condition for using a χ^2 test is that most expected frequencies must be at least 5 and that all must be at least 1.

   II. A χ^2 test for goodness of fit tests the degree to which a categorical variable has a specific distribution.

   III. Expected cell frequencies are computed in the same way for goodness of fit tests and tests of independence.

   a.   I only
   b.   II only
   c.   I and II only
   d.   II and III only
   e.   I, II and III

Free-Response

1. An AP Statistics student noted that the probability distribution for a binomial random variable with $n = 4$ and $p = 0.3$ is approximately given by

n	P
0	0.240
1	0.412
2	0.265
3	0.076
4	0.008

   (*Note:* $\Sigma p = 1.001$ rather than 1 due to rounding.)

   The student decides to test the randBin function on her TI-83/84 by putting 500 values into a list using this function (randBin(4,0.3,500) → L1) and counting the number of each outcome. (Can you think of an efficient way to count each outcome?) She obtained

n	Observed
0	110
1	190
2	160
3	36
4	4

   Do these data provide evidence that the randBin function on the calculator is correctly generating values from this distribution?

Calculator Tip: It's a bit of a digression, but if you actually wanted to do the experiment in question 1, you would need to have an efficient way of counting the number of each outcome. You certainly don't want to simply scroll through all 500 entries and tally each one. Even sorting them first and then counting would be tedious (more so if n were bigger than 4). One easy way is to draw a histogram of the data and then `TRACE` to get the totals. Once you have your 500 values from `randBin` in `L1`, go to `STAT PLOTS` and set up a histogram for `L1`. Choose a WINDOW something like `[-0.5,4.5,1,-1,300,1,1]`. Be sure that `Xscl` is set to 1. You may need to adjust the `Ymax` from `300` to get a nice picture on your screen. Then simply `TRACE` across the bars of the histogram and read the value of n for each outcome off of the screen. The reason for having x go from –0.5 to 4.5 is so that the (integer) outcomes will be in the middle of each bar of the histogram.

2. A chi-square test for the homogeneity of proportions is conducted on three populations and one categorical variable that has four values. Computation of the chi-square statistic yields $\chi^2 = 17.2$. Is this result significant at the 0.01 level of significance?

3. Which of the following best describes the difference between a test for independence and a test for homogeneity of proportions? Discuss the correctness of each answer.

   a. There is no difference because they both produce the same value of the chi-square test statistic.

   b. A test for independence has one population and two categorical variables, whereas a test for homogeneity of proportions has more than one population and only one categorical variable.

   c. A test for homogeneity of proportions has one population and two categorical variables, whereas a test for independence has more than one population and only one categorical variable.

   d. A test for independence uses count data when calculating chi-square and a test for homogeneity uses percentages or proportions when calculating chi-square.

4. Compute the expected frequency for the cell that contains the frog. You are given the marginal distribution.

	D	E	F	G	Total
A					94
B					96
C					119
Total	74	69	128	38	309

5. Restaurants in two parts of a major city were compared on customer satisfaction to see if location influences customer satisfaction. A random sample of 38 patrons from

the Big Steak Restaurant in the eastern part of town and another random sample of 36 patrons from the Big Steak Restaurant on the western side of town were interviewed for the study. The restaurants are under the same management, and the researcher established that they are virtually identical in terms of decor, service, menu, and food quality. The results are presented in the following table.

Patrons' Ratings of Restaurants

	Excellent	Good	Fair	Poor
Eastern	10	12	11	5
Western	6	15	7	8

Do these data provide good evidence that location influences customer satisfaction?

6. A chi-square test for goodness of fit is done on a variable with 15 categories. What is the minimum value of χ^2 necessary to reject the null hypothesis at the 0.02 level of significance?

7. The number of defects from a manufacturing process by day of the week are as follows:

	Monday	Tuesday	Wednesday	Thursday	Friday
Number:	36	23	26	25	40

The manufacturer is concerned that the number of defects is greater on Monday and Friday. Test, at the 0.05 level of significance, the claim that the proportion of defects is the same each day of the week.

8. A study was done on opinions concerning the legalization of marijuana at Mile High College. One hundred fifty-seven respondents were randomly selected from a large pool of faculty, students, and parents at the college. Respondents were given a choice of favoring the legalization of marijuana, opposing the legalization of marijuana, or favoring making marijuana a legal but controlled substance. The results of the survey were as follows.

	FAVOR LEGALIZATION	OPPOSE LEGALIZATION	FAVOR LEGALIZATION WITH CONTROL
Students	17	9	6
Faculty	33	40	27
Parents	5	8	12

Do these data support, at the 0.05 level, the contention that the type of respondent (student, faculty, or parent) is related to the opinion toward legalization? Is this a test of independence or a test of homogeneity of proportions?

Cumulative Review Problems

Use the computer output given below to answer Questions 1 and 2.

The regression equation is
$y = -136 + 3.98\ X$

Predictor	Coef	St Dev	t ratio	P
Constant	−136.10	42.47	−3.20	.024
X	3.9795	0.6529	6.09	.002

$s = 8.434$ R-sq = 88.1%   R-sq(adj) = 85.8%

1. Based on the computer output above, what conclusion would you draw concerning $H_0: \beta = 0$? Justify your answer.

2. Use the computer output above to construct a 99% confidence interval for the slope of the regression line ($n = 8$). Interpret your interval.

3. If you roll two dice, the probability that you roll a sum of 10 is approximately 0.083.

   a. What is the probability that you first roll a 10 on the 10th roll?
   b. What is the average number of rolls until you first roll a 10?

4. An experiment is conducted by taking measurements of a personality trait on identical twins and then comparing the results for each set of twins. Would the analysis of the results assume two independent populations or proceed as though there were a single population? Explain.

5. The lengths and widths of a certain type of fish were measured and the least-squares regression line for predicting width from length was found to be: $\widehat{width} = -0.826 + 0.193\ (length)$. The graph that follows is a residual plot for the data:

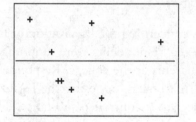

   a. The fish whose width is 3.8 had a length of 25.5. What is the residual for this point?

   b. Based on the residual plot, does a line seem like a good model for the data?

Solutions to Practice Problems

Multiple-Choice

1. The correct answer is (c). The expected value for that cell can be found as follows:

$$\left(\frac{32}{77}\right)(31) = 12.88\ .$$

2. The correct answer is (d). Since there are 12 months in a year, we have $n = 12$ and df $= 12 − 1 = 11$. Reading from Table C, $0.01 < P\text{-value} < 0.02$. The exact value is given by the TI-83/84 as χ^2 cdf (23.5,1000,11) = 0.015.

3. The correct answer is (d). Because we have independent random samples of teachers from each of the two levels, this is a test of homogeneity of proportions. It would have been a test of independence if there was only one sample broken into categories (correct answer would be (b) in that case).

4. The correct answer is (b). The expected values for the cells are exactly equal to the observed values; e.g., for the 1st row, 1st column, $\text{Exp} = \left(\dfrac{25}{40}\right)(8) = 5$, so X^2 must equal $0 \Rightarrow$ the variables are independent, and are not related.

5. The correct answer is (e). The expected values for this two-way table are given by the matrix: $\begin{bmatrix} 12.5 & 27.5 \\ 12.5 & 27.5 \end{bmatrix}$.

    Then,

$$X^2 = \frac{(15-12.5)^2}{12.5} + \frac{(25-27.5)^2}{27.5} + \frac{(10-12.5)^2}{12.5}$$
$$+ \frac{(30-27.5)^2}{27.5} = 1.45.$$

6. The correct answer is (c). In a χ^2 test for independence, we are interested in whether or not two categorical variables, measured on a single population, are related. In a χ^2 test for homogeneity of proportions, we are interested in whether two or more populations have the same proportions for each of the values of a single categorical variable.

7. The correct answer is (e). Using Ethnicity as the row variable, there are five rows ($r = 5$) and four columns ($c = 4$). The number of degrees of freedom for an $r \times c$ table is $(r-1)(c-1)$. In this question, $(5-1)(4-1) = 4 \times 3 = 12$.

8. The correct answer is (c). In III, the expected count for each category in a goodness-of-fit test is found by multiplying the proportion of the distribution of each category by the sample size. The expected count for a test of independence is found by multiplying the row total by the column total and then dividing by n.

Free-Response

1.

I. Let $p_1 =$ the proportion of 0s, $p_2 =$ the proportion of 1s, $p_3 =$ the proportion of 2s, $p_4 =$ the proportion of 3s, and $p_5 =$ the proportion of 4s.

H_0: $p_1 = 0.24$, $p_2 = 0.41$, $p_3 = 0.27$, $p_4 = 0.07$, $p_5 = 0.01$.

H_A: Not all of the proportions stated in H_0 are correct.

II. We will use a chi-square goodness-of-fit test. The observed and expected values for the 500 trials are shown in the following table:

N	OBSERVED	EXPECTED
0	110	$(0.24)(500) = 120$
1	190	$(0.41)(500) = 205$
2	160	$(0.27)(500) = 135$
3	36	$(0.07)(500) = 35$
4	4	$(0.01)(500) = 5$

Continued

We note that all expected values are at least 5, so the conditions necessary for the chi-square test are present.

III. $X^2 = \sum \dfrac{(O-E)^2}{E} = \dfrac{(110-120)^2}{120} + \cdots + \dfrac{(4-5)^2}{5} = 6.79$, df $= 4 \Rightarrow 0.10 <$ P-value

< 0.15 (from Table C). Using the TI-83/84, χ^2 cdf $(6.79,1000,40) = 0.147$.

IV. The P-value is greater than any commonly accepted significance level. Hence, we do not reject H_0 and conclude that we do not have good evidence that the calculator is not correctly generating values from $B(4, 0.3)$.

2. For a 3×4 two-way table, df $= (3 - 1)(4 - 1) = 6 \Rightarrow 0.005 < P$-value < 0.01 (from Table C). The finding is significant at the 0.01 level of significance. Using the TI-83/84, P-value $= \chi^2$ cdf $(17.2,1000,6) = 0.009$.

3. (a) is not correct. For a given set of observations, they both do produce the same value of chi-square. However, they differ in that they are different ways to design a study. (b) is correct. A test of independence hypothesizes that two categorical variables are independent within a given population. A test for homogeneity of proportions hypothesizes that the proportions of the values of a single categorical variable are the same for more than one population. (c) is incorrect. It is a reversal of the actual difference between the two designs. (d) is incorrect. You always use count data when computing chi-square.

4. The expected value of the cell with the frog is $\dfrac{128}{309}$ (96) = 39.77.

5. Based on the design of the study this is a test of homogeneity of proportions.

I. H_0: The proportions of patrons who rate the restaurant Excellent, Good, Fair, and Poor are the same for the Eastern and Western sides of town.

H_A: Not all the proportions are the same.

II. We will use the chi-square test for homogeneity of proportions.

Calculation of expected values (using the TI-83/84). yields the following results:

$\begin{bmatrix} 8.22 & 13.86 & 9.24 & 6.68 \\ 7.78 & 13.14 & 8.76 & 6.32 \end{bmatrix}$. Since all expected values are at least 5, the conditions

necessary for this test are present.

III. $X^2 = \sum \dfrac{(O-E)^2}{E} = \dfrac{(10-8.22)^2}{8.22} + \cdots + \dfrac{(8-6.32)^2}{6.32} = 2.86$, df $= (2-1)(4-1) = 3 \Rightarrow$

P-value > 0.25 (from Table C). χ^2 cdf $(2.86,1000,3) = 0.414$.

IV. The P-value is larger than any commonly accepted significance level. Thus, we cannot reject H_0. We do not have evidence that location influences customer satisfaction.

6. If $n = 15$, then df $= 15 - 1 = 14$. In the table we find the entry in the column for tail probability of 0.02 and the row with 14 degrees of freedom. That value is 26.87. Any value of X^2 larger than 26.87 will yield a P-value less than 0.02.

7.

> **I.** Let p_1 be the true proportion of defects produced on Monday.
>
> Let p_2 be the true proportion of defects produced on Tuesday.
>
> Let p_3 be the true proportion of defects produced on Wednesday.
>
> Let p_4 be the true proportion of defects produced on Thursday.
>
> Let p_5 be the true proportion of defects produced on Friday.
>
> H_0: $p_1 = p_2 = p_3 = p_4 = p_5$ (the proportion of defects is the same each day of the week.)
>
> H_A: At least one proportion does not equal the others (the proportion of defects is not the same each day of the week).
>
> **II.** We will use a chi-square goodness-of-fit test. The number of expected defects is the same for each day of the week. Because there was a total of 150 defects during the week, we would expect, if the null is true, to have 30 defects each day. Because the number of expected defects is greater than 5 for each day, the conditions for the chi-square test are present.
>
> **III.** $X^2 = \sum \dfrac{(O-E)^2}{E} = \dfrac{(36-30)^2}{30} + \cdots + \dfrac{(40-30)^2}{30} = 7.533$, df $= 4 \Rightarrow 0.10 <$
>
> P-value < 0.15 (from Table C). (χ^2 cdf (7.533.1000,4) = 0.11; or, you could use the χ^2 GOF-Test in the STAT TESTS menu of a TI-83/84 if you have it.)
>
> **IV.** The P-value is larger than any commonly accepted significance level. We do not have strong evidence that there are more defects produced on Monday and Friday than on other days of the week.

8. Because we have a single population from which we drew our sample and we are asking if two variables are related *within* that population, this is a chi-square test for independence.

> **I.** H_0: Type of respondent and opinion toward the legalization of marijuana are independent.
>
> H_A: Type of respondent and opinion toward the legalization of marijuana are not independent.
>
> **II.** We will do a χ^2 test of independence. The following table gives the observed and expected values for this situation (expected values are the second row in each cell; the expected values were calculated by using the MATRIX function and the χ^2-Test item in the STAT TESTS menu):
>
> | | | |
> |---|---|---|
> | 17
11.21 | 9
11.62 | 6
9.17 |
> | 33
35.03 | 40
36.31 | 27
28.66 |
> | 5
8.76 | 8
9.08 | 12
7.17 |

Continued

> Since all expected values are at least 5, the conditions necessary for the chi-square test are present.
>
> **III.** $\chi^2 = \sum \dfrac{(O-E)^2}{E} = \dfrac{(17-11.21)^2}{11.21} + \cdots + \dfrac{(12-7.17)^2}{7.17} = 10.27$, df $= (3-1)(3-1) = 4 \Rightarrow 0.025 < P\text{-value} < 0.05$ (from Table C).
>
> (Using the TI-83/84, P-value $= \chi^2\texttt{cdf(10.27,1000,4)= 0.036.}$)
>
> **IV.** Because $P < 0.05$, reject H_0. We have evidence that the type of respondent is related to opinion concerning the legalization of marijuana.

Solutions to Cumulative Review Problems

1. The t-test statistic for the slope of the regression line (under H_0: $\beta = 0$) is 6.09. This translates into a P-value of 0.002, as shown in the printout. This low P-value allows us to reject the null hypothesis. We conclude that we have strong evidence of a linear relationship between X and Y.

2. For $C = 0.99$ (a 99% confidence interval) at df $= n - 2 = 6$, $t^* = 3.707$. The required interval is $3.9795 \pm 3.707(.6529) = (1.56, 6.40)$. We are 99% confident that the true slope of the population regression line lies between 1.56 and 6.40.

3. a. This is a geometric distribution problem. The probability that the first success occurs on the 10th roll is given by $G(10) = (0.083)(1 - 0.083)^9 = 0.038$. (On the TI-83/84, $G(10) = \texttt{geometpdf(0.083,10)=0.038}$.)

   b. $\dfrac{1}{0.083} = 12.05$. On average, it will take about 12 rolls before you get a 10.

   (Actually, it's *exactly* 12 rolls on average since the *exact* probability of rolling a 10 is $\dfrac{1}{12}$, not 0.083, and $\dfrac{1}{1/12} = 12$.)

4. This is an example of a *matched-pairs design*. The identical twins form a block for comparison of the treatment. Statistics are based on the differences between the scores of the pairs of twins. Hence this is a one-variable situation—one population of values, each constructed as a difference of the matched pairs.

5. a. Residual $=$ actual $-$ predicted $= 3.8 - [-0.826 + 0.193(25.5)] = 3.8 - 4.096 = -0.296$.
   b. There does not seem to be any consistent pattern in the residual plot, indicating that a line is probably a good model for the data.

CHAPTER 14

Free Response Question 6: The Investigative Task

Question 6 in the Free Response section is called the **Investigative Task**, and it is different from the other free-response questions in three ways. First, it is expected to take longer. The exam recommends about 25 minutes on that question alone! Second, it is worth more than the other free-response questions. The Investigative Task is worth one-fourth, or 25% of the free-response grade. Because the free-response part of the exam is half of your overall score the Investigative Task ends up being one-eighth of the total test score! Third, it will address topics in a way that you may not have seen before.

That is because the Investigative Task is intended to test your ability to apply what you've learned in a new and different situation. It might involve working with or creating a statistic you've never seen, or you may be asked to read a different kind of plot from any you've seen, or use a familiar one in a different way. No worries! You have been doing and thinking about statistics all year, so trust yourself a bit. Follow where the prompts lead.

Because there is no way to predict what topic will be addressed, you cannot really study for it in the traditional sense. What you can and should do is make sure to have a solid understanding of the topics you've learned and then employ some strategies to help ensure your success:

First, realize that parts of the Investigative Task, usually the first parts, will probably be familiar. They almost always start on familiar ground and lead you to the investigative portion, which is the new stuff. It is this new part where you often have to think critically or apply existing statistical knowledge in a new way.

This is the idea of Question 6—to get you to think for yourself in a new situation. So keep calm, and go about doing what the question asks you to do—you have been thinking statistically for almost a year; the concepts in this question are likely to be well within your reach. While only the members of the test development committee can ever predict with any certainty what an Investigative Task will actually look like, here's a stab we took to help you:

Taylor and Harper are twins who want to buy a used car to drive when they go to college. Together, they collect information on the age (in years), miles driven (in thousands of miles), and selling price (in thousands of dollars) of a group of 20 of the same model of a popular car currently offered for sale in their area. This information is provided in the table below:

AGE (YEARS)	MILES (1000's)	$PRICE (1000's)	AGE (YEARS)	MILES (1000's)	$PRICE (1000's)
3	20	16.6	1	11	18.2
3	78	11.0	3	79	13.3
3	47	14.6	3	45	14.3
10	111	8.6	1	5	18.0
8	123	8.0	6	112	10.1
3	59	15.0	5	69	12.1
2	39	17.1	5	40	12.9
2	61	14.2	4	44	14.5
7	91	9.6	9	65	10.6
7	118	9.0	2	19	17.0

Taylor wants to use *Age* to predict the selling price of this car. Here are the summary statistics and appropriate plots for Taylor's linear regression:

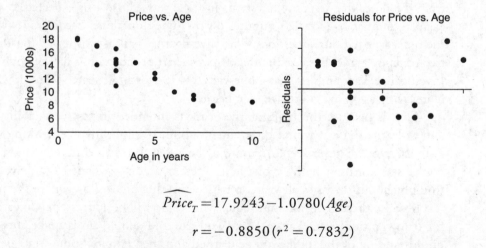

$$\widehat{Price}_T = 17.9243 - 1.0780(Age)$$

$$r = -0.8850 \, (r^2 = 0.7832)$$

Harper wants to use *Miles Driven* to predict the selling price of this car. Here are the summary statistics and appropriate plots for Harper's linear regression:

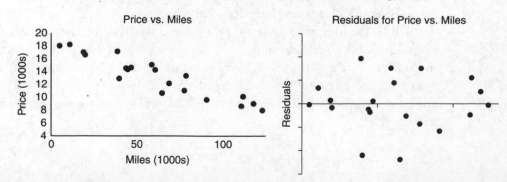

$$\widehat{Price}_H = 18.4598 - 0.0845(Miles)$$

$$r = -0.9370 \, (r^2 = 0.8780)$$

(a) Interpret the value of r^2 for Taylor's regression results in the context of this problem.

(b) Based on the information provided above, which linear regression do you think does a better job predicting the price of this car? Explain your answer.

Taylor believes that if they were to include BOTH Age and Miles, they could improve their ability to predict the price of this type of car. To that end, Taylor produces the equation:

$$\widehat{Price} = 18.8674 - 0.4903(Age) - 0.0566(Miles)$$

(c) Use this new regression equation to predict the price of a car that is 4 years old and has 32,000 miles. Show your work.

But Harper points out that to evaluate whether or not this new equation actually is more effective will require some new way to measure how well the model predicts the price of this model of car. So Harper consulted some sources and found a statistic that has the same interpretation as r^2, but can be used with more than one independent variable.

Let r_T = correlation for the linear relationship between *Age* and *Price*, r_H = correlation for the linear relationship between *Miles* and *Price*, and $r_{A\&M}$ = correlation for the linear relationship between *Age* and *Miles*. Harper finds the following formula for the coefficient of determination when you have **two** independent variables:

$$R^2 = \frac{(r_T)^2 + (r_H)^2 - 2(r_T)(r_H)(r_{A\&M})}{1 - (r_{A\&M})^2}$$

Taylor and Harper find that the relationship between *Age* and *Miles* appears to be a positive linear relationship with a correlation of 0.7687.

(d) Use this new formula as well as the information provided in the regression output at the beginning of the problem to find the value of R^2 for the twins' sample of cars. Then interpret this value in the context of the problem.

Harper consults a large national database of all cars of this type currently offered for sale with information as to price, age, and miles driven. Harper is able to ask for 200 random samples of 20 cars each and calculate R^2 for predicting *Price* using both *Age* and *Miles*. The dotplot shown below represents the 200 values of R^2 that Harper collected from these samples.

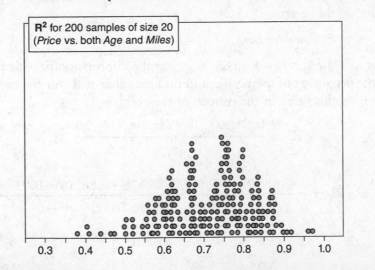

R^2 for 200 samples of size 20
(*Price* vs. both *Age* and *Miles*)

(e) Using Harper's dotplot and the value of R^2 you obtained in part (d), comment on whether you believe the sample of 20 cars the twins used in their analysis is likely a representative sample of all cars of this model currently for sale nationwide.

Solution

First, notice that while the raw data is included in the table, you do not need to type these values into your calculator—all relevant statistical summaries have been provided to help you answer this question.

Part (a): Interpret the value of r^2 for Taylor's regression results in the context of this problem.

First, make sure that you grab the appropriate value. $r^2 = 0.7832$. This should be interpreted in context as follows:

78.32% of the variation in the price of the cars in this sample can be explained by the linear relationship with the miles driven.

Note: This part is not new. You are familiar with interpreting r².

Part (b): Based on the information provided above, which linear regression do you think does a better job predicting the price of this car? Explain your answer.

Based on the scatterplots and results, Harper's model seems to do a better job at predicting the price of this car. Harper's model has a stronger correlation ($r = -0.9370$ is closer to negative one than is Taylor's $r = -0.8850$). And it would appear that the residual plot for Harper's model is much more randomly scattered and with less variation in that scatter.

Note: This part should also be familiar.

Part (c): Use this new regression equation to predict the price of a model of this car that is 4 years old and has 32,000 miles. Show your work.

(Hint: Be sure to check back with your graphs and use the appropriate units when you substitute. Note that the mileage is given in 1000's of miles, so you need to use 32 in the appropriate place as well as report the answer in thousands of dollars.)

$$\widehat{PRICE} = 18.8674 - 0.4903(Age) - 0.0566(Miles)$$

$$\widehat{PRICE} = 18.8674 - 0.4903(4) - 0.0566(32)$$

$$\widehat{PRICE} = 15.215$$

We would predict the price for a model of this car that is 4 years old and has 32,000 miles would be about $15,215.

Note: Still nothing new.

Part (d): Use this new formula as well as the information provided in the regression output at the beginning of the problem to find the value of R^2 for the twin's sample of cars. Then interpret this value in the context of the problem.

$$R^2 = \frac{(r_T)^2 + (r_H)^2 - 2(r_T)(r_H)(r_{A\&M})}{1 - (r_{A\&M})^2}$$

$$R^2 = \frac{(-0.8850)^2 + (-0.9370)^2 - 2(-0.8850)(-0.9370)(0.7687)}{1 - (0.7687)^2}$$

$$R^2 = 0.9443$$

In the context of this problem, 94.43% of the variation in the price of the cars in this sample can be explained by the relationship with both age and miles driven.

Note: OK, this looks new. You've probably never seen this statistic before. But relax! You were given a formula, and all you need to do is substitute and evaluate. One wrinkle, the sentence **"Taylor and Harper find that the relationship between Age and Miles appears to be a positive linear relationship with a correlation of 0.7687"** *tells you that the value of* $r_{A\&M}$ *is 0.7687. You need to recognize this based on the phrasing in this sentence. The other leap you need to make is in how this calculation of* R^2 *is interpreted. You were given a key piece of information at the beginning of part (e):* **"So Harper consulted some sources and found a statistic that has the same interpretation as** r^2**, but can be used with more than one independent variable."** *By reading the problem carefully you should be able to nail the interpretation of* R^2.

Part (e): Using Harper's dotplot and the value of R^2 you obtained in part (d), comment on whether you believe the sample of 20 cars the twins used in their analysis is likely a representative sample of all cars of this model.

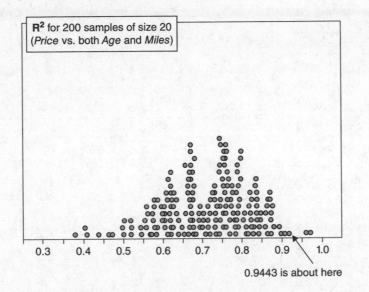

It would appear that only 2 of the 200 samples produced values of R^2 larger than 0.9443, or only 1% of the samples. This is a very small probability, and thus makes it unlikely that the sample the twins used in their analysis is a representative sample of all the cars of this model that are currently for sale nationwide.

Note: This **looks** *new, too. But it's really not. You've dealt with simulated sampling distributions in the past. We are simulating a different statistic, but sampling distributions all work the same way. Just find the location of your value among the simulated values and decide if what you got fits with that distribution or not. In other words, use the simulation to get a simulated* P-value.

General Tips for the Investigative Task

- Relax and remember that the first part of the question is usually familiar.
- As you work through the problem, note that the parts often tie together. You may see a direction that states "based on your answer to part (b)." If you get stuck on part (b), make up a **reasonable** answer (and say that you have done so!). Then when needed, go ahead and use this answer in the parts that follow.

- While each of us is unique in how we approach a test—some of us work the problems in order, others of us skip around until all the problems are completed—many teachers believe that you should not leave the Investigative Task for the very last. Consider then doing a couple of the first five free-response questions and then starting the Investigative Task. Get a couple of parts of the Investigative Task completed and then go back and finish the rest of the other free-response questions. Finally, use your remaining time to work on the rest of the parts of the Investigative Task.

STEP 5

Build Your Test-Taking Confidence

AP Statistics Practice Test 1
AP Statistics Practice Test 2

AP Statistics Practice Test 1

ANSWER SHEET FOR SECTION I

1 Ⓐ Ⓑ Ⓒ Ⓓ Ⓔ
2 Ⓐ Ⓑ Ⓒ Ⓓ Ⓔ
3 Ⓐ Ⓑ Ⓒ Ⓓ Ⓔ
4 Ⓐ Ⓑ Ⓒ Ⓓ Ⓔ
5 Ⓐ Ⓑ Ⓒ Ⓓ Ⓔ
6 Ⓐ Ⓑ Ⓒ Ⓓ Ⓔ
7 Ⓐ Ⓑ Ⓒ Ⓓ Ⓔ
8 Ⓐ Ⓑ Ⓒ Ⓓ Ⓔ
9 Ⓐ Ⓑ Ⓒ Ⓓ Ⓔ
10 Ⓐ Ⓑ Ⓒ Ⓓ Ⓔ
11 Ⓐ Ⓑ Ⓒ Ⓓ Ⓔ
12 Ⓐ Ⓑ Ⓒ Ⓓ Ⓔ
13 Ⓐ Ⓑ Ⓒ Ⓓ Ⓔ
14 Ⓐ Ⓑ Ⓒ Ⓓ Ⓔ
15 Ⓐ Ⓑ Ⓒ Ⓓ Ⓔ

16 Ⓐ Ⓑ Ⓒ Ⓓ Ⓔ
17 Ⓐ Ⓑ Ⓒ Ⓓ Ⓔ
18 Ⓐ Ⓑ Ⓒ Ⓓ Ⓔ
19 Ⓐ Ⓑ Ⓒ Ⓓ Ⓔ
20 Ⓐ Ⓑ Ⓒ Ⓓ Ⓔ
21 Ⓐ Ⓑ Ⓒ Ⓓ Ⓔ
22 Ⓐ Ⓑ Ⓒ Ⓓ Ⓔ
23 Ⓐ Ⓑ Ⓒ Ⓓ Ⓔ
24 Ⓐ Ⓑ Ⓒ Ⓓ Ⓔ
25 Ⓐ Ⓑ Ⓒ Ⓓ Ⓔ
26 Ⓐ Ⓑ Ⓒ Ⓓ Ⓔ
27 Ⓐ Ⓑ Ⓒ Ⓓ Ⓔ
28 Ⓐ Ⓑ Ⓒ Ⓓ Ⓔ
29 Ⓐ Ⓑ Ⓒ Ⓓ Ⓔ
30 Ⓐ Ⓑ Ⓒ Ⓓ Ⓔ

31 Ⓐ Ⓑ Ⓒ Ⓓ Ⓔ
32 Ⓐ Ⓑ Ⓒ Ⓓ Ⓔ
33 Ⓐ Ⓑ Ⓒ Ⓓ Ⓔ
34 Ⓐ Ⓑ Ⓒ Ⓓ Ⓔ
35 Ⓐ Ⓑ Ⓒ Ⓓ Ⓔ
36 Ⓐ Ⓑ Ⓒ Ⓓ Ⓔ
37 Ⓐ Ⓑ Ⓒ Ⓓ Ⓔ
38 Ⓐ Ⓑ Ⓒ Ⓓ Ⓔ
39 Ⓐ Ⓑ Ⓒ Ⓓ Ⓔ
40 Ⓐ Ⓑ Ⓒ Ⓓ Ⓔ

AP Statistics Practice Test 1

SECTION I

Time: 1 hour and 30 minutes

Number of questions: 40

Percentage of total grade: 50

Directions: Solve each of the following problems. Decide which is the best of the choices given and answer in the appropriate place on the answer sheet. No credit will be given for anything written on the exam. Do not spend too much time on any one problem.

1. A set of test scores has the following 5-number summary:

Minimum	Q1	Median	Q3	Maximum
2	18	22.5	28.5	33

Which statement about outliers must be true?

A. There is exactly one outlier on the lower end.
B. There is at least one outlier on the lower end.
C. There is exactly one outlier on the higher end.
D. There is at least one outlier on the higher end.
E. There are no outliers.

Predictor	Coef	SE Coef	t-ratio	p
Constant	147.1427	20.6957	7.110	0.0000
Sodium	0.2457	0.0210	11.673	0.0000

s = 136.69   r-sq = 0.542313   r-sq (adj) = 0.538333

2. The table above shows the regression output for predicting the number of calories from the number of milligrams of sodium for items at a fast-food chain. What percent of the variation in calories is explained by the regression model using sodium as a predictor?

A. 7.36%
B. 54.2%
C. 53.8%
D. 73.4%
E. 73.6%

3. A drug company will conduct a randomized controlled study on the effectiveness of a new heart disease medication called Heartaid. Heartaid is more expensive than the currently used medication. The analysis will include a significance test with H_0: Heartaid and the current medication are equally effective at preventing heart disease and H_a: Heartaid is more effective than the current medication at preventing heart disease. Which of these would be a potential consequence of a Type II error?

A. Patients will spend more money on Heartaid, even though it is actually not any more effective than the current medication.
B. Doctors will begin to prescribe Heartaid to patients, even though it is actually not any more effective than the current medication.
C. Patients will continue to use the current medication, even though Heartaid is actually more effective.
D. Researchers will calculate the wrong P-value, making their advice to doctors invalid.
E. Researchers will calculate the correct P-value, but will misinterpret it in their conclusion.

GO ON TO THE NEXT PAGE

4. Researchers in the Southwest are studying tortoises—a species of animal that is affected by habitat loss due to human development of the desert. A total of 78 tortoises are being studied by researchers at several sites. The data on gender and species of all these tortoises is organized in the table shown below.

Species	Male	Female	Total
Agassiz	18	33	**51**
Morafka	12	15	27
Total	**30**	**48**	**78**

If a tortoise from this study is to be selected at random, let A = *the tortoise is female* and B = *the tortoise is a Morafka*. Which of the following appropriately interprets the value of P(B|A)?

A. 31.3% is the probability that a randomly selected Morafka tortoise is a female tortoise.

B. 31.3% is the probability that a randomly selected female tortoise is a Morafka tortoise.

C. 19.2% is the probability that a randomly selected Morafka tortoise is a female tortoise.

D. 19.2% is the probability that a randomly selected female tortoise is a Morafka tortoise.

E. 31.3% is the probability that a randomly selected tortoise is a Morafka tortoise.

5. The main purpose of blocking in an experiment is to:

A. reduce bias.
B. reduce confounding.
C. reduce variation within treatments.
D. reduce variation between treatments.
E. reduce the probability of a Type I error.

6. The midterm scores for a statistics course were approximately normally distributed with a mean of 52 points and a standard deviation of 4 points. The final exam scores were approximately normally distributed with a mean of 112 points and a standard deviation of 10 points. One student had a score of 58 points on the midterm. If she had the same standardized score (*z*-score) on the final exam, what must her score have been?

A. 15 points
B. 58 points

C. 118 points
D. 122 points
E. 127 points

7. Which of the following is *not* an advantage of stratified random sampling over simple random sampling?

A. When done correctly, a stratified random sample is less biased than a simple random sample.

B. When done correctly, a stratified random sampling process has less variability from sample to sample than a simple random sample.

C. When done correctly, a stratified random sample can provide, with a smaller sample size, an estimate that is just as reliable as that of a simple random sample with a larger sample size.

D. A stratified random sample provides information about each stratum in the population as well as an estimate for the population as a whole, and a simple random sample does not.

E. When done correctly, a stratified random sample can provide a more reliable estimate than a simple random sample using the same sample size.

8. The correlation between height in inches and weight in pounds for a particular class is 0.65. If the heights are converted from inches to centimeters, what will the correlation be? (1 in. = 2.54 cm)

A. −0.65
B. −0.26
C. 0.10
D. 0.26
E. 0.65

GO ON TO THE NEXT PAGE

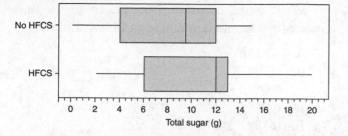

9. Breakfast cereals have a wide range of sugar content. Some cereals contain High Fructose Corn Syrup (HFCS) as a source of sugar and some do not. The boxplots above show the total sugar content of different types of cereal for those containing HFCS and for those that do not. Which statement is true based on the boxplots?

   A. The number of cereals with HFCS is about the same as the number of cereals without HFCS.
   B. The cereals with HFCS have a greater interquartile range than the cereals without HFCS.
   C. The cereals without HFCS have a greater range than the cereals with HFCS.
   D About half the cereals without HFCS have less sugar than about three-fourths of the cereals with HFCS.
   E. About half the cereals with HFCS have more sugar than about three-fourths of the cereals without HFCS.

10. Semir rolls a six-sided die every morning to determine what he will have for breakfast. If he rolls a 1 or 2, he takes time to cook himself a big breakfast. If he rolls a 3 or larger he grabs a quick lighter breakfast. When he cooks himself a big breakfast, there is a 15% chance he will be late for school. If he has a lighter breakfast, there is a 6% chance he will be late for school. What is the probability Semir will be on time for school any given day?

   A. 0.09
   B. 0.21
   C. 0.80
   D. 0.91
   E. 0.94

11. Researchers were gathering data on alligators in an attempt to estimate an alligator's weight from its length. They captured 29 alligators and measured their length and weight. They created three regression models. Each model, along with its residual plot, is shown below. y represents the weight in pounds and x represents the length in inches.

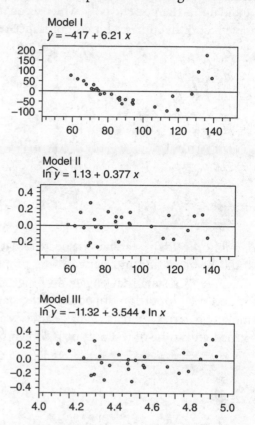

Which statement is true?

   A. There is a linear relationship between weight and length, and model I is most appropriate.
   B. There is a linear relationship between weight and length, and model II is most appropriate.
   C. There is a nonlinear relationship between weight and length, and model I is most appropriate.
   D. There is a nonlinear relationship between weight and length, and model II is most appropriate.
   E. There is a nonlinear relationship between weight and length, and model III is most appropriate.

GO ON TO THE NEXT PAGE

12. A polling company has been asked to do a poll to see if the proportion of people that would vote for their candidate has *improved* since the previous poll, as their analyst suspects. Using p_p = the proportion of all voters that would have supported their candidate in the previous poll, and p_c = the proportion of all voters that would support their candidate in the current poll, which is an appropriate pair of hypotheses for a significance test?

A. $H_0: p_c > p_p$
   $H_a: p_c \le p_p$
B. $H_0: p_c < p_p$
   $H_a: p_c > p_p$
C. $H_0: p_c > p_p$
   $H_a: p_c = p_p$
D. $H_0: p_c = p_p$
   $H_a: p_c > p_p$
E. $H_0: p_c = p_p$
   $H_a: p_c \ne p_p$

13. A public health researcher suspected that there would be a relationship between the proportion of people in a state that engage in binge alcohol use and the proportion of people in the state that smoke cigarettes. A regression analysis was performed and some of the output is shown below.

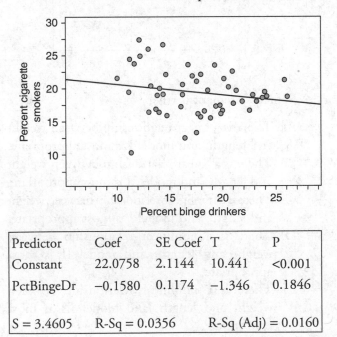

Predictor	Coef	SE Coef	T	P
Constant	22.0758	2.1144	10.441	<0.001
PctBingeDr	−0.1580	0.1174	−1.346	0.1846
S = 3.4605	R-Sq = 0.0356		R-Sq (Adj) = 0.0160	

What is the correlation coefficient for percent cigarette smokers and percent binge drinkers?

A. −0.1887
B. −0.0356
C. 0.0160
D. 0.0356
E. 0.1887

14. Marine biologists want to determine whether a small increase in the acidity of sea water would adversely affect marine life. They randomly assigned several *Entacmaea quadricolor*, one species of sea anemone, to one of eight aquariums that had been prepared with environments as similar as possible to the natural habitat. Then they randomly selected four tanks, and gradually increased the acidity of the water. They monitored the health of the anemones for the duration of the study. Which of these statements is NOT true about this study?

A. An advantage of using only one variety of sea anemones is that there should be less variability in the response to each treatment.
B. A disadvantage of using only one variety of sea anemones is that the scope of inference is limited only to that type of anemone.
C. An advantage of using aquariums is that it is easier to maintain control to avoid confounding factors.
D. A disadvantage of using aquariums is that the anemones might respond differently in the ocean than they do in aquariums.
E. If the anemones in the aquariums with increased acidity are less healthy than those without increased acidity, it cannot be determined that the increased acidity *caused* the response.

15. In a large high school, the student council wants to select a sample to survey the student body. They will first select 25 freshmen, then select 25 sophomores, then 25 juniors, and then finally 25 seniors. This is an example of a

A. Cluster sample
B. Stratified sample
C. Simple random sample
D. Convenience sample
E. Systematic sample

GO ON TO THE NEXT PAGE

Variable	N	Mean	Median	TrMean	StDev	SE Mean
High Temp	30	74.9	75	74.7	7.29	1.33

Variable	Minimum	Maximum	Q1	Q3
High Temp	60	90	69	79

16. Listed above are the summary statistics for the daily high temperatures (in degrees F) for the month of September in a Midwestern U.S. city. One particular day has a z-score of 0.70. What was the likely temperature that day?

    A. 68
    B. 70
    C. 74
    D. 76
    E. 80

17. In American football, the running back carries the football forward in a quest to score points. Many people create fantasy football teams by selecting players as their "team." These fans analyze data to judge which players are outstanding. The following regression output analyzes the linear relationship between the number of times a running back is given the football (carries) and the total number of yards that player gains in a season for a random sample of 18 running backs.

    The regression equation is
    Total yards gained = 30.947 + 4.3441(Carries)

Predictor	Coef	SE Coef
Constant	30.947	7.414
Carries	4.3441	1.811

    S = 56.647   R-sq = 45.9%

    The appropriate calculation for a 99% confidence interval for the slope of the least squares regression line for the total yards gained versus number of carries in a season for all running backs is

    A. $4.3441 \pm 2.576 \cdot (1.811)/\sqrt{18}$
    B. $4.3441 \pm 2.921 \cdot (1.811)$
    C. $4.3441 \pm 2.921 \cdot (1.811)/\sqrt{18}$
    D. $4.3441 \pm 2.898 \cdot (1.811)$
    E. $4.3441 \pm 2.898 \cdot (1.811)/\sqrt{18}$

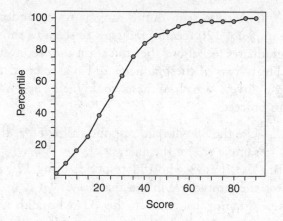

18. Above is a cumulative relative frequency plot for the scores of a large university class on a 90-point statistics exam. Which of the following observations is correct?

    A. The median score is at least 60 points.
    B. The distribution of scores is skewed to the right.
    C. The distribution of scores is skewed to the left.
    D. The distribution is roughly symmetric.
    E. If a passing score is 60, most students passed the test.

19. Self-efficacy (the belief that one has control over one's situation) as it related to job satisfaction was studied. When a group of teachers rated their ability to control their situation and their satisfaction with their job, the two variables had a correlation of 0.30. Which statement follows from this correlation?

    A. If you want teachers to be happy with their job, give them more control over their situation.
    B. If you want teachers to take more control over their situation, make them happier at their jobs.
    C. Teachers in the study who were more satisfied with their job were less confident in their ability to control their situation.
    D. There is evidence of a causal relationship.
    E. 9% of the variability in job satisfaction can be explained by the linear model with self-efficacy as a predictor.

GO ON TO THE NEXT PAGE

20. A university sent out a survey to a random sample of 120 recent graduates to see how those graduates feel about the education they received. Thirty-two of the graduates did not respond to the survey. Which of these would be the best way to proceed?

   A. Use the 88 who did respond, using 88 as the sample size in the analysis.
   B. Use the 88 who did respond, using 120 as the sample size in the analysis.
   C. Attempt to contact the 32 who did not respond and get the information from them.
   D. Select 32 more graduates at random and use them to replace the nonresponders.
   E. Start over, sending the survey to more graduates to increase the number of responses.

21. Some health professionals suspect that doctors are more likely to order cardiac tests for men than women, even when women describe exactly the same symptoms. Young doctors, training to work in the emergency room, were randomly assigned to two groups—Group A and Group B—and presented with the exact same description of common symptoms of heart disease. However, those in Group A were told that the patient was a 55-year-old female and those in Group B were told that the patient was a 55-year-old male. All other patient characteristics were exactly the same.

Group	Number of doctors in group	Number of doctors who ordered standard cardiac tests for the patient	Percentage of doctors who ordered standard cardiac tests for the patient
A	60	42	70%
B	40	34	85%

Which of the following is the appropriate statistic to test whether doctors are less likely to order cardiac tests when the patient is female?

A. $z = \dfrac{(0.70)-(0.85)}{\sqrt{(0.70)(0.85)}\sqrt{\dfrac{1}{60}+\dfrac{1}{40}}}$

B. $z = \dfrac{(0.70)-(0.85)}{\sqrt{(0.76)(0.24)}\sqrt{\dfrac{1}{60}+\dfrac{1}{40}}}$

C. $z = \dfrac{(0.70)-(0.85)}{\sqrt{(0.76)(0.24)}\sqrt{\dfrac{1}{42}+\dfrac{1}{34}}}$

D. $z = \dfrac{(0.70)-(0.85)}{\sqrt{\dfrac{(0.70)(0.30)}{60}+\dfrac{(0.85)(0.15)}{40}}}$

E. $z = \dfrac{(0.70)-(0.85)}{\sqrt{\dfrac{(0.70)(0.30)}{42}+\dfrac{(0.85)(0.15)}{34}}}$

22. A study looked at medical records of about 23,000 patients, mostly in Asia. They found that patients who drank at least three cups of green tea per day had a much lower incidence of depression. In an American newspaper article about this study, which of the following statements should not be made?

   A. It is possible that people who drink green tea also tend to have healthier lifestyles than those who don't.
   B. It is possible that people who drink green tea also tend to have a more positive outlook on life than those who don't.
   C. Because this is observational data, the association between drinking green tea and a lower incidence of depression is not an indication of a causal relationship.
   D. Because most of these patients lived in Asia, where tea drinking is much more common than in the United States, a similar association may not exist in the United States.
   E. People who want to avoid depression should consider drinking green tea on a daily basis.

23. A polling company wants to estimate, using a 95% confidence interval, the proportion of likely voters that would vote for a particular candidate in the upcoming presidential election. *Of the following choices*, which is the smallest sample size that will ensure a margin of sampling error less than or equal to 2 percentage points?

   A. 250
   B. 950
   C. 1,150
   D. 2,250
   E. 2,450

GO ON TO THE NEXT PAGE

24. The director of a local food bank asks for data on all donations given during the month of November. Of the 100 checks received, the average donation is $155 with a standard deviation of $32. Which of the following is the most appropriate statement?

   A. This November, the average donation is $155.
   B. 50% of all donations this November are more than $155.
   C. We are 95% confident that the average donation in November is between about $91 and $219.
   D. We are 95% confident that the average donation in November is between about $149 and $161.
   E. This November, about 95% of all donations are between $91 and $219.

25. A multiple-choice test has 30 questions, and each question has five options. A student has not studied and will guess on every question. Which calculation shows how to find the probability of getting between 15 and 20 (inclusive) questions correct?

   A. $\binom{30}{15} 0.2^{15} 0.8^{15} + \binom{30}{16} 0.2^{16} 0.8^{14} + \cdots$
      $+ \binom{30}{20} 0.2^{20} 0.8^{10}$

   B. $\binom{30}{20} 0.2^{20} 0.8^{10} - \binom{30}{15} 0.2^{15} 0.8^{15}$

   C. $0.2^{15} 0.8^{15} + 0.2^{16} 0.8^{14} + \cdots + 0.2^{20} 0.8^{10}$

   D. Use the normal approximation to find
      $$P\left(\frac{15-6}{\sqrt{30 \cdot 0.2 \cdot 0.8}} < z < \frac{20-6}{\sqrt{30 \cdot 0.2 \cdot 0.8}} \right).$$

   E. Use the normal approximation to find
      $$P\left(\frac{0.5-0.2}{\sqrt{\frac{0.2 \cdot 0.8}{30}}} < z < \frac{0.67-0.2}{\sqrt{\frac{0.2 \cdot 0.8}{30}}} \right).$$

26. Claire and Max each created a simulated sampling distribution for a test statistic. They each used samples of size 4 from a normal population with the same mean m and standard deviation s. Claire simulated the sampling distribution of $\frac{\bar{x}-\mu}{\sigma}$ and Max simulated the sampling distribution of $\frac{\bar{x}-\mu}{s}$, where s is the sample standard deviation. Which statement makes a correct comparison of their simulated sampling distributions?

   A. Both simulated sampling distributions are approximately normal.
   B. Both simulated sampling distributions have the same spread.
   C. Claire's simulated sampling distribution is skewed to the right, and Max's is roughly symmetric.
   D. Claire's simulated sampling distribution is roughly symmetric, and Max's is skewed to the left.
   E. Claire's simulated sampling distribution is approximately normal, and Max's is approximately a t-distribution.

27. A spreadsheet contains a data set that consists of 100 values such that $IQR > variance$. However, these values must be converted using the following formula: $NewValue = 2(OldValue) + 10$. Which of the following statements is true about the data set consisting of the NewValues?

   A. The IQR has been increased by 10 units.
   B. The IQR is now less than the variance.
   C. The IQR is now double its previous size.
   D. The IQR has been doubled and then increased by 10 units.
   E. The IQR has been cut in half.

28. A test for heartworm in dogs shows a positive result in 96% of dogs that actually have heartworm, and shows a negative result in 98% of dogs with no heartworm. If heartworm actually occurs in 10% of dogs, what is the probability that a randomly selected dog that tested positive for heartworm actually has heartworm?

   A. 11%
   B. 18%
   C. 84%
   D. 88%
   E. 96%

GO ON TO THE NEXT PAGE

29. A recent poll reported that 43% of Americans approve of the job the president is doing, with a margin of sampling error of ±3.2% at a 95% level of confidence. Which of these correctly interprets that margin of error?

    A. There is a 3.2% percent chance that the proportion of all Americans who approve of the job the president is doing is not 43%.
    B. There is a 95% chance that the proportion of all Americans who approve of the job the president is doing is 43%.
    C. About 95% of polls conducted in this way will find that between 39.8% and 46.2% of those sampled approve of the job the president is doing.
    D. About 95% of polls conducted in this way will give a sample proportion within 3.2 percentage points of the actual proportion of all Americans who approve of the job the president is doing.
    E. About 3.2% of all Americans approve of 95% of what the president does.

30. Two events A and B each have a nonzero probability. If A and B are independent, which of the following statements is true?

    A. $P(A \text{ and } B) = P(A) \cdot P(B)$
    B. A and B may or may not be mutually exclusive.
    C. A and B must be mutually exclusive.
    D. $P(A|B) = P(B|A)$
    E. $P(A \text{ and } B) = 0$

31. Players in the National Football League weigh, on average, about 248 pounds with a standard deviation of about 47 pounds. If four players are to be selected at random, the expected value of the random variable W, the total combined weight of the four players, is 992 pounds. The standard deviation of W is approximately

    A. 47 pounds
    B. 67 pounds
    C. 94 pounds
    D. 141 pounds
    E. 188 pounds

32. When a patient complains to the doctor about a certain set of symptoms, the doctor diagnoses the patient with Condition A 15% of the time. If a patient with these symptoms is diagnosed with Condition A, he or she is diagnosed with Condition B 70% of the time. A patient with these symptoms that is not diagnosed with Condition A is diagnosed with Condition B 10% of the time. What is the probability that a patient with this set of symptoms will be diagnosed with at least one of these conditions?

    A. 0.235
    B. 0.250
    C. 0.765
    D. 0.850
    E. 0.950

33. A doctor hopes that a new surgery technique will shorten the recovery time compared to the standard technique. To test this, he designed an experiment in which patients who required this type of surgery were randomly assigned to the standard technique or the new technique. Then the mean recovery time for each treatment group was compared. Assuming conditions for inference were met, which analysis should be used?

    A. A t-test for a mean.
    B. A t-test for a difference in means.
    C. A z-test for a mean.
    D. A z-test for a difference in means.
    E. A z-test for a difference in proportions.

GO ON TO THE NEXT PAGE

34. For a class project, a student wants to see if boys and girls at their large high school differ in the number of contacts they have stored in their phone. The student conducts a survey of 50 randomly sampled boys and 40 randomly selected girls, and asks them to report the number of contacts. Which of the following is true about this situation?

A. Because the population standard deviations are not known and conditions are met, the student *should* use a two-sample *t*-test.

B. Because the sample sizes are different, the student should *not* use a two-sample *t*-test.

C. Because the sample sizes are both greater than 30, the student should *not* use a two-sample *t*-test.

D. Because the shape of the population distribution is not known, the student should *not* use a two-sample *t*-test.

E. Because np and $n(1-p)$ are both at least 10, the student *should* use a two-proportion *z*-test.

35. Which of these is the best description of a *P*-value?

A. The probability of making a Type I error.

B. The probability of making a Type II error.

C. The probability of rejecting the null hypothesis if it is, in fact, false.

D. The probability of getting a test statistic at least as extreme as the observed test statistic, if the null hypothesis is true.

E. The probability of getting a test statistic at least as extreme as the observed test statistic, if the null hypothesis is false.

36. A farmer who raises hens for egg production wants his eggs to have a mean mass of 56 grams. He is considering the purchase of a different type of hen, so he took a random sample of 18 eggs laid by this type of hen. The distribution of the masses is symmetric and mound-shaped with a mean of 54.1 grams and no outliers. The farmer conducted a *t*-test to see if there is evidence that the eggs from these hens have a mean mass that is different from 56 g and got a test statistic of $t = -1.973$. If he uses a 5% significance level, which is the correct conclusion and reason?

A. Because t is more extreme than ± 1.96, he should reject the null hypothesis. He has convincing evidence at the 5% significance level that the mean mass of eggs from these hens is different from 56 grams.

B. Because t is less extreme than the critical value of t for 17 degrees of freedom, he should not reject the null hypothesis. He does not have convincing evidence at the 5% significance level that the mean mass of eggs from these hens is different from 56 grams.

C. Because t is less extreme than the critical value of t for 18 degrees of freedom, he should not reject the null hypothesis. He does not have convincing evidence at the 5% significance level that the mean mass of eggs from these hens is different from 56 grams.

D. Because t is more extreme than the critical value of t for 18 degrees of freedom, he should reject the null hypothesis. He has convincing evidence at the 5% significance level that the mean mass of eggs from these hens is different from 56 grams.

E. Because the sample mean was less than 56, he should use a one-sided alternative hypothesis. Thus, t is more extreme than the critical value of t for 17 degrees of freedom and he should reject the null hypothesis. He has convincing evidence at the 5% significance level that the mean mass of eggs from these hens is different from 56 grams.

GO ON TO THE NEXT PAGE

37. An experiment is conducted in which the response variable is the average gain in participants' performance in the long jump. A two-sample t-test with a 5% level of significance will be used to analyze the results. If all else is kept the same, which of the following descriptions of a possible change in procedure is true?

    A. Change from equal size treatment groups to very different size treatment groups would increase the power of the test.
    B. Change from a 5% significance level to a 1% significance level would increase the power of the test.
    C. Taking more careful measurements to reduce variability in the response would increase the power of the test.
    D. Increasing the sample size would reduce the probability of a Type I error.
    E. Increasing the sample size would increase the probability of a Type I error.

38. Researchers wish to test how rangelands growth is affected by wildfire. A very large open area is divided into 28 equally sized plots and subjected to a controlled burn. One year later, the researchers calculate the percentage of grass returned in each plot after the fire damage. The average percentage return for these 28 plots was 58.5% with a standard deviation of 6.5%. Which of the following represents a 95% confidence interval to estimate the average proportion of grass return one year after a fire?

    A. $0.585 \pm 1.96(0.065/\sqrt{27})$
    B. $0.585 \pm 2.052(0.065/\sqrt{27})$
    C. $0.585 \pm 2.052(0.065/\sqrt{28})$
    D. $0.585 \pm 1.96(0.065/\sqrt{28})$
    E. $0.585 \pm 2.052\sqrt{(0.585)(0.415)/27}$

39. Which of the following is NOT true of the χ^2 probability distribution function?

    A. The area under the χ^2 curve is 1.
    B. χ^2 is defined only for nonnegative values of the variable.
    C. For small degrees of freedom, the curve displays strong right-skewness.
    D. For the same α, as the number of degrees of freedom increases, the critical value for the rejection region decreases.
    E. χ^2 is actually a family of probability distributions defined by its degrees of freedom.

40. Tucson, Arizona, hosts the world's largest gem and mineral show every February. A government official claims that more than 50% of city residents have been to the Gem Show at least once. You decide to test this theory, take a simple random sample of 200 Tucson residents, and discover that 108 have attended. Which of the following is the appropriate calculation of the test statistic for the hypothesis proposed by the government official?

    A. $z = \dfrac{0.54 - 0.50}{\sqrt{(0.54)(0.46)/200}}$

    B. $z = \dfrac{0.54 - 0.50}{\sqrt{(0.50)(0.50)/200}}$

    C. $z = \dfrac{0.54 - 0.50}{\sqrt{(0.54)(0.46)/108}}$

    D. $z = \dfrac{0.54 - 0.50}{\sqrt{(0.50)(0.50)/108}}$

    E. $z = \dfrac{0.54 - 0.50}{\sqrt{(0.54)(0.50)/200}}$

END OF SECTION I

AP Statistics Practice Test 1

SECTION II

Time: 1 hour and 30 minutes

Number of questions: 6

Percentage of total grade: 50

General Instructions

There are two parts to this section of the examination. Part A consists of five equally weighted problems that represent 75% of the total weight of this section. Spend about 65 minutes on this part of the exam. Part B consists of one longer problem that represents 25% of the total weight of this section. Spend about 25 minutes on this part of the exam. You are not necessarily expected to complete all parts of every question. Statistical tables and formulas are provided.

- Be sure to write clearly and legibly. If you make an error, you may save time by crossing it out rather than trying to erase it. Erased or crossed-out work will not be graded.
- Show all your work. Indicate clearly the methods you use because you will be graded on the correctness of your methods as well as the accuracy of your final answers. Correct answers without support work may not receive credit.

Statistics, Section II, Part A, Questions 1–5

Spend about 65 minutes on this part of the exam; percentage of Section II grade: 75.

Directions: Show all your work. Indicate clearly the methods you use because you will be graded on the correctness of your methods as well as on the accuracy of your results and explanation.

1. In response to drought, some plants react with reduced biological nitrogen fixation (BNF), hindering their ability to grow at a typical rate. Scientists are working to develop techniques to supplement crops so that the BNF rate is not as greatly reduced during drought. One promising treatment is to be compared to standard growing conditions on randomly selected acres of soybeans. The following summary data is for the bushels per acre of soybeans for each group:

Variable	N	Mean	StDev	Min	Q1	Med	Q3	Max
Treatment	30	43.555	3.937	36.46	40.52	42.78	46.46	51.50
Control	38	35.813	6.576	27.14	31.99	35.60	42.60	46.27

   a. The maximum yield of the treated acres was 51.50 bushels of soybeans. Is this an unusually high yield compared to the other treated acres? Justify your answer.
   b. Researchers indicate that the average yield for the treatment group is statistically significantly higher than that of the control group. Explain what this means in simple terms.
   c. Field workers report to the scientists that the machine to measure the yield in the treatment group was poorly calibrated—each of the bushel values was over-reported by 0.5 bushels. So each of the measures should be reduced (subtracted) by 0.5 bushels. Describe the impact doing so would have on the mean, median, standard deviation, and interquartile range of the treatment group.

GO ON TO THE NEXT PAGE

2. A particular bottle-filling machine is supposed to put 16 ounces of water into a bottle. The amount actually dispensed is approximately normal with mean 16.1 ounces and a standard deviation of 0.12 ounces. A bottle is rejected by quality control if it contains less than 15.7 ounces or more than 16.3 ounces.

   a. What is the probability that a randomly selected bottle gets rejected by quality control?
   b. What is the probability that, out of 20 randomly selected bottles, at least 2 are rejected by quality control?
   c. What is the probability that a set of 5 randomly selected bottles contains, on average, more than 16.2 ounces of water?

3. A student was curious about a news article that reported the percentage of Americans who believe that a "higher power" affects who wins sporting events. With the assumption that the percentage reported in the article is correct, she simulated selecting samples of 100 Americans and calculating the percentage of those that believe a higher power has a hand in who wins a sporting event. The results of 200 trials are shown below.

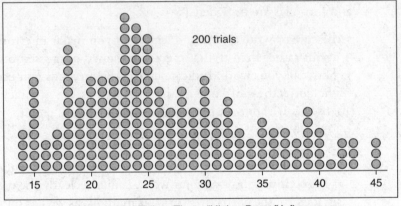

Percent Who Believe That a "Higher Power" Influences
the Results of a Sporting Event

   a. Based on the results of these trials, what appears to be the reported percentage of Americans that believes the results of a sporting event are influenced by a "higher power?" Explain.
   b. Another student believes that the proportion reported in the article is too low. He takes a random sample of 100 Americans and finds that 44 percent believe in the influence of a "higher power" on the outcome of a sporting event. Based on the dotplot above, does it appear the percentage in the article is correct? Explain why or why not.

4. Below are the residual plots generated for the linear regression on three separate sets of bivariate data.

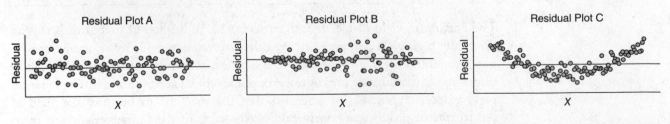

   a. Is a linear model appropriate for Data Set A? Justify your answer using Residual Plot A.

GO ON TO THE NEXT PAGE

b. What does Residual Plot B reveal about the reliability of the estimates from the regression line in Data Set B?

c. Explain what information is being conveyed about the linear regression for Data Set C in Residual Plot C.

5. New formulas of gel nail polish have recently been introduced. A local salon wished to test number of days of wear of the new formula against the old formula as a control. Identical bottles of both formulas are prepared and 14 clients have their nails polished on their first visit with one formula (chosen at random) and then the next visit with the other formula. The number of days of wear is then compared.

Client	A	B	C	D	E	F	G	H	I	J	K	L	M	N
New Gel Polish	18	19	21	19	17	14	16	19	12	15	18	16	14	14
Control—Old Version	15	21	19	16	13	17	12	18	12	13	12	18	15	14

Do these data provide evidence that the mean number of days of wear is greater for the new formula of gel nail polish than for the old? Use $\alpha = 0.05$.

Statistics, Section II, Part B, Question 6

Spend about 25 minutes on this part of the exam; percentage of Section II grade: 25.

Directions: Show all of your work. Indicate clearly the methods you use because you will be graded on the correctness of your methods as well as on the accuracy of your results and explanation.

6. Overuse of lawn chemicals causes environmental problems when runoff gets into waterways. Researchers tested three types of fertilizer spreaders typically used by home-owners: drop spreaders, hand-held broadcast rotary spreaders, and push rotary spreaders. The goal was to determine if the type of spreader influenced the ability of the user to apply the correct amount of fertilizer.

Fifteen volunteers who were homeowners with no particular training in the use of this equipment operated the spreaders. Each spreader was randomly assigned to a 1,000-square-foot plot of lawn. The researchers prepared the spreaders by correctly setting the dispensation amount and providing the same amount of fertilizer in each spreader. The spreader with fertilizer was weighed before and after the fertilizer was applied to determine the amount dispensed.

a. What were the experimental units? The treatments? The response variable?
   The plot below shows the amounts dispensed by each type of spreader in kg/100 m².

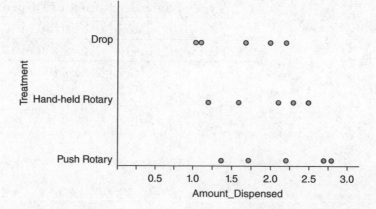

b. Based on the plot alone, comment on the strength of evidence of a difference in the amount of fertilizer dispensed by the three types of spreaders.

Volunteers with no particular training were used because regular homeowners are the population of interest. However, one researcher points out that there may be very different skill levels among the volunteers. He suggests using a randomized block design, with each volunteer as a block. Each volunteer will use all three spreaders in a randomized order. They ran the experiment with just five volunteers using this design. The results, along with the operator of the spreader, are displayed below.

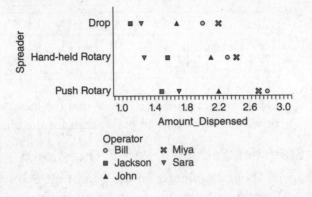

c. Based on the plot above, comment on the association between the operator of the spreader and the amount of fertilizer dispensed.

To remove the effect of the operators' skill levels on the response variable, the researchers calculated the mean amount dispensed for each operator. Miya's amounts dispensed are given in the table below.

Type of Spreader	Amount Dispensed
Drop	2.2 kg/100 m^2
Hand-held Rotary	2.4 kg/100 m^2
Push Rotary	2.7 kg/100 m^2

d. To adjust the amount dispensed for Miya, calculate the mean amount of fertilizer she dispensed in her three trials, and subtract that mean from each of her amounts dispensed. Show all computations.

Type of Spreader	Amount Dispensed	Adjusted Amount Dispensed
Drop	2.2 kg/100 m^2	
Hand-held Rotary	2.4 kg/100 m^2	
Push Rotary	2.7 kg/100 m^2	

The plot below shows the Adjusted Amount Dispensed for the other four volunteers.

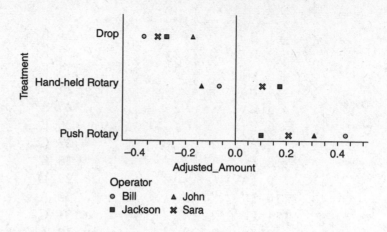

e. Add Miya's Adjusted Amount Dispensed for each type of spreader to the plot above. Then comment on what this new plot reveals about the differences in amounts dispensed between the types of spreaders.

END OF SECTION II

> Answers to Practice Test 1, Section I

1. B		21. B	
2. B		22. E	
3. C		23. E	
4. B		24. A	
5. C		25. A	
6. E		26. E	
7. A		27. C	
8. E		28. C	
9. E		29. D	
10. D		30. A	
11. D		31. C	
12. D		32. A	
13. A		33. B	
14. E		34. A	
15. B		35. D	
16. E		36. B	
17. B		37. C	
18. B		38. C	
19. E		39. D	
20. C		40. B	

› Solutions to Practice Test 1, Section I

1. (B) Using the outlier guideline, any value above $Q3 + 1.5 \cdot IQR = 28.5 + 1.5 \cdot 10.5 = 44.25$, or below $Q1 - 1.5 \cdot IQR = 18 - 1.5 \cdot 10.5 = 2.25$, is considered an outlier. Because the maximum is 33, there are no values above 44.25. The minimum of 2 is an outlier. There may be more than one value below 2.25, so all we can say is that there is at least one outlier on the lower end.

2. (B) You should recognize that this is the interpretation of r^2. According to the printout, that value is 0.542313, or about 54.2%.

3. (C) A Type II error means that the null hypothesis is false, but we fail to reject it. In this case, a false null hypothesis means that Heartaid actually does work better. Failing to reject that null hypothesis means they don't have convincing evidence that Heartaid is better, so they won't use it.

4. (B) This is asking for a conditional probability. $P(\text{Morafka}|\text{Female})$. Out of the 48 female tortoises, 15 are Morafka. $15/48 = 0.3125$, or about 31.3%.

5. (C) Randomization is what fights against confounding by making the groups as alike as possible. With blocking, similar units are grouped together so the effect of being in that group can be accounted for in the analysis.

6. (E) On the midterm, the z-score is $\frac{58-52}{4} = 1.5$. So she scored 1.5 standard deviations above the mean. To achieve the same standardized score on the final exam, she must score 1.5 standard deviations, or $1.5 \cdot 10 = 15$ points, above the mean on that exam. $112 + 15 = 127$ points.

7. (A) A simple random sample is already unbiased. A stratified random sample cannot improve on that.

8. (E) The correlation coefficient remains unchanged when converting units.

9. (E) The median of the HFCS group is the same as Q3 of the No HFCS group.

10. (D) $\frac{1}{3} \cdot 0.15 + \frac{2}{3} \cdot 0.06 = 0.09$. $1 - 0.09 = 0.91$.

11. (D) Because there is a curve in residual plot I, which is for weight vs. length, the relationship between those two variables is not linear. Because residual plot II shows a no curve and plot III does, plot II represents a better model.

12. (D) The null hypothesis always contains equality. Because they are interested in whether the candidate's support has improved, the alternative hypothesis is one-sided.

13. (A) The correlation, r, is the square root of r^2, which is given in the table. So r could be ±0.1887. Since the slope is negative, r is negative.

14. (E) Because this is a properly conducted randomized experiment, a causal relationship can be determined. E is the choice that is not true.

15. (B) The population is divided into homogenous subgroups, and a random sample is selected from each.

16. (E) $z = 0.70 = \frac{x-74.9}{7.29}$, $x \approx 80$.

17. (B) The formula for the confidence interval for the slope is $b_1 \pm t^* \cdot s_{b_1}$. From the printout $b_1 = 4.3441$ and $s_{b_1} = 1.811$. t^* for $18 - 2 = 16$ degrees of freedom and 99% confidence is 2.921.

18. (B) The steeper part of the graph corresponds to the higher bars on a histogram. So the higher bars would be on the left with a tail stretching toward the right.

19. (E) No causal relationship is established simply because of a correlation. But r^2 is 0.09.

20. (C) Nonresponse bias is a problem because the kind of people who don't respond might feel differently from those who do. Getting responses from more of the people that do respond won't fix the problem.

21. (B) Under the null hypothesis, the proportions are the same, so the standard error formula is based on the pooled proportion $\hat{p} = \dfrac{x_1 + x_2}{n_1 + n_2}$.

22. (E) Choice E implies a causal relationship, which is not justified because this is an observational study.

23. (E) The margin of error is largest for $\hat{p} = 0.5$, so if $\hat{p} \neq 0.5$, the margin of error will be even smaller. So use $0.02 = 1.96\sqrt{\dfrac{0.5 \cdot 0.5}{n}}$. Solve for n and round up.

24. (A) They have all the data for November, so there is no inference to do. 50% of the donations are above the median, not the mean. A is the only conclusion we can make.

25. (A) n is not large enough to use a normal approximation. This is a binomial situation, and choice A correctly applies the binomial calculation.

26. (E) Using s in the denominator makes Max's an approximate t-distribution. It has heavier tails and more spread than Claire's approximately normal distribution.

27. (C) Multiplying affects both measures of location and spread. Adding affects only measures of location.

28. (C) This is easiest to see in a two-way table.

29. (D) The confidence interval is based on one sample of many possible samples. The process has a 95% chance of capturing the true proportion of all Americans within the interval created. But there is nothing special about that one interval.

30. (A) That is the multiplication rule for independent events.

31. (C) Since these are four independent occurrences of W, the variances add. So the standard deviation is $\sqrt{47^2 + 47^2 + 47^2 + 47^2} = 94$.

32. (A) The bottom path in the tree diagram shows that the probability that a patient has neither condition is $0.85 \cdot 0.90 = 0.765$. $1 - 0.765 = 0.235$.

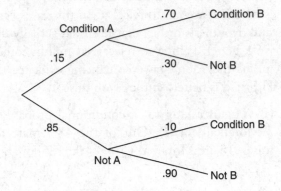

33. (B) They are comparing means without a known population standard deviation.

	Positive Test	Negative Test	Total
Heartworm present	$0.96 \cdot 0.1 = 0.096$	$0.1 - 0.096 = 0.004$	0.1
Heartworm not present	$0.9 - 0.882 = 0.018$	$0.98 \cdot 0.9 = 0.882$	$1 - 0.1 = 0.9$
Total	$0.096 + 0.018 = 0.114$	$0.004 + 0.882 = 0.886$	1

The required probability is $0.096/0.114 = 0.842$.

34. (A) They are comparing means, so a *t*-test would be the appropriate test. The students were randomly selected. With such large sample sizes, population shapes are not a concern. Conditions are met.

35. (D) That is the definition of a *P*-value.

36. (B) The critical value of t for $n - 1 = 17$ degrees of freedom is ±2.11. His test statistic was not outside those values.

37. (C) Reducing measurement variability would give a lower standard error of the mean and, therefore, a greater test statistic (in absolute value). Thus, the probability of rejecting the null hypothesis increases.

38. (C) This is a confidence interval for a mean with unknown σ, so *t* for 27 degrees of freedom is used rather than *z*. The standard error of the

mean is $\frac{s}{\sqrt{n}} = \frac{0.065}{\sqrt{28}}$.

39. (D) As the number of degrees of freedom increases, the entire distribution is moved to the right, so the critical value for the rejection region increases.

40. (B) The official should use the hypothesized value of *p*, which is 0.5, in the calculation. Also, $n = 200$, not 108.

> Solutions to Practice Test 1, Section II, Part A

1. a. The 1.5*IQR* guideline can be used. $IQR = 46.46 - 40.52 = 5.94$. $1.5IQR = 8.91$ bushels. The largest possible nonoutlier would be $46.46 + 8.91 = 55.37$ bushels. The maximum of 51.5 bushels is below this value, so it is not an unusually high soybean yield.

   (It would be acceptable to use the mean + 2 standard deviations, which is $43.555 + 2 \cdot 3.937 = 51.429$. Using this guideline, the maximum is more than 2 standard deviations from the mean, so one might consider it unusual.)

   b. Even if the treatment has no effect on soybean yield, the means of the groups will probably be different due to random variability. If the treatment group's yield was statistically significantly higher than that of the control group, that means the difference is too large to be reasonably attributed to chance.

   c. The mean and median would both decrease by 0.5 bushels. The standard deviation and interquartile range would be unchanged.

2. a. Using a normal model with $\mu = 16.1$ and $\sigma = 0.12$, the z-score for $15.7 = \dfrac{15.7 - 16.1}{0.12} = -3.33$. The z-score for $16.3 = \dfrac{16.3 - 16.1}{0.12} = 1.67$. $P(15.7 < x < 16.3) = P(-3.33 < z < 1.67) = 0.048$.

   b. This is a binomial situation with $n = 20$ and $p = 0.048$ (from part a).

   $$P(x \geq 2) = 1 - P(x \leq 1) = 1 - \binom{20}{0} 0.048^0 \cdot 0.952^{20} - \binom{20}{1} 0.048^1 \cdot 0.952^{19} \approx 0.25.$$

   c. This is the sampling distribution of $\bar{x}$, which will be approximately normal with $\mu_{\bar{x}} = \mu = 16.1$, and

   $$\sigma_{\bar{x}} = \frac{\sigma}{\sqrt{n}} = \frac{0.12}{\sqrt{5}} = 0.0537.$$

   $$P(\bar{x} > 16.2) = P\left(z > \frac{16.2 - 16.1}{0.0537} = 1.86\right) = 0.03$$

3. a. The mean of the distribution appears to be about 26%. This is the best estimate of the proportion used in the simulation.

   b. A sample proportion of 44% or more appeared only 3 times out of 100. Because that is less than 5% of the trials, this is convincing evidence that the proportion that appears in the article is too low.

4. a. Yes. The residual plot shows no pattern and the points show random scatter. A linear model is appropriate.

   b. The reliability of the estimates decreases as the *x* variable increases. This is evident from the fan shape in the residual plot.

   c. There is a definite curve in the residual plot, which shows that a linear model is not appropriate for this relationship.

5. This is a matched pairs design.

Hypotheses: $H_0: \mu_{diff} = 0$

$H_a: \mu_{diff} \neq 0$

Check conditions:
Treatments are assigned to subjects in a random order.
The plot of the differences is reasonably symmetric with no outliers.

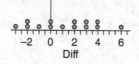

Conditions are met for a matched pairs *t*-test.
Computations:

$$t = \frac{\mu_{diff} - 0}{s_{diff}} = 1.705$$

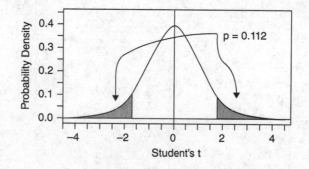

Comparing to a *t* distribution with 13 degrees of freedom, the *P*-value is 0.112.

Conclusion:

Because the *P*-value = 0.112 > α = 0.05, we fail to reject the null hypothesis. We do not have convincing evidence that there is a difference in the number of days of wear between the new formula and the old formula.

› Solutions to Practice Test 1, Section II, Part B

6. a. The experimental units are the plots of land. The treatments are the three types of spreaders. The response variable is the amount of fertilizer dispensed.

   b. There seems to be a slight difference in the amounts dispensed, but with the large amount of overlap between the groups it is not very convincing evidence of a difference.

   c. There seems to be a strong association between the operator and the amount dispensed. Jackson and Sara dispensed smaller amounts with all three spreaders than the other operators. Bill and Miya dispensed larger amounts.

   d. Miya's mean amount dispensed is $\dfrac{2.2 + 2.4 + 2.7}{3} = 2.43$.

Type of Spreader	Amount Dispensed	Adjusted Amount Dispensed
Drop	2.2 kg/100 m²	2.2 − 2.43 = −0.23
Hand-held Rotary	2.4 kg/100 m²	2.4 − 2.43 = −0.03
Push Rotary	2.7 kg/100 m²	2.7 − 2.43 = 0.27

e.

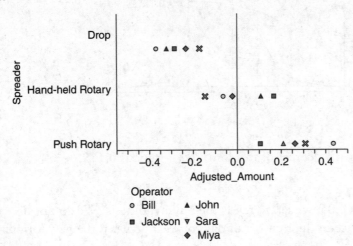

There is now much less overlap between the treatments. With the variability of the operator accounted for, the evidence of a difference is more convincing. With drop spreaders, users tend to dispense the least fertilizer, and with push rotary spreaders, they tend to dispense the most.

AP Statistics Practice Test 2

ANSWER SHEET FOR SECTION I

1 (A) (B) (C) (D) (E)   16 (A) (B) (C) (D) (E)   31 (A) (B) (C) (D) (E)
2 (A) (B) (C) (D) (E)   17 (A) (B) (C) (D) (E)   32 (A) (B) (C) (D) (E)
3 (A) (B) (C) (D) (E)   18 (A) (B) (C) (D) (E)   33 (A) (B) (C) (D) (E)
4 (A) (B) (C) (D) (E)   19 (A) (B) (C) (D) (E)   34 (A) (B) (C) (D) (E)
5 (A) (B) (C) (D) (E)   20 (A) (B) (C) (D) (E)   35 (A) (B) (C) (D) (E)
6 (A) (B) (C) (D) (E)   21 (A) (B) (C) (D) (E)   36 (A) (B) (C) (D) (E)
7 (A) (B) (C) (D) (E)   22 (A) (B) (C) (D) (E)   37 (A) (B) (C) (D) (E)
8 (A) (B) (C) (D) (E)   23 (A) (B) (C) (D) (E)   38 (A) (B) (C) (D) (E)
9 (A) (B) (C) (D) (E)   24 (A) (B) (C) (D) (E)   39 (A) (B) (C) (D) (E)
10 (A) (B) (C) (D) (E)   25 (A) (B) (C) (D) (E)   40 (A) (B) (C) (D) (E)
11 (A) (B) (C) (D) (E)   26 (A) (B) (C) (D) (E)
12 (A) (B) (C) (D) (E)   27 (A) (B) (C) (D) (E)
13 (A) (B) (C) (D) (E)   28 (A) (B) (C) (D) (E)
14 (A) (B) (C) (D) (E)   29 (A) (B) (C) (D) (E)
15 (A) (B) (C) (D) (E)   30 (A) (B) (C) (D) (E)

AP Statistics Practice Test 2

SECTION I

Time: 1 hour and 30 minutes

Number of questions: 40

Percentage of total grade: 50

Directions: Solve each of the following problems. Decide which is the best of the choices given and answer in the appropriate place on the answer sheet. No credit will be given for anything written on the exam. Do not spend too much time on any one problem.

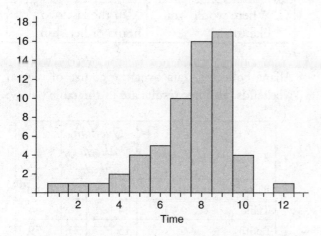

1. Sixty-two students were asked to complete a number puzzle. The number of minutes each student needed to complete the puzzle was recorded and the times are displayed in the histogram above. Which of the following describes the distribution?

   A. The distribution is approximately normal.
   B. The distribution is roughly symmetric.
   C. The distribution is bimodal.
   D. The distribution is skewed left.
   E. The distribution is skewed right.

Predictor	Coef	SE Coef	t-ratio	p
Constant	147.1427	20.6957	7.110	0.0000
Sodium	0.2457	0.0210	11.673	0.0000

s = 136.69   r-sq = 0.542313   r-sq (adj) = 0.538333

2. The table above shows the regression output for predicting the number of calories from the number of milligrams of sodium for items at a fast-food chain. Which is a correct interpretation of the slope of the regression line?

   A. Every milligram of sodium contains 0.2457 calories.
   B. A food item with one more milligram of sodium than another tends to have 0.2457 more calories.
   C. A food item with one more milligram of sodium than another tends to have 136.69 more calories.
   D. A food item with one more milligram of sodium than another tends to have 147.1427 more calories.
   E. A food item with one more calorie than another tends to have 147.1427 more milligrams of sodium.

GO ON TO THE NEXT PAGE

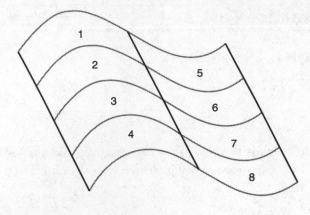

3. A new fertilizer is to be tested on tomato plants to determine if it increases the production of the plants. The available land is divided into eight plots. Four of the plots are on a hill and four are in a valley as shown. The plots in the valley are expected to get more water due to the influence of gravity, and those on the hill will get more direct sunlight. A randomized block design will be used, using two blocks of four plots each. Which of the following blocking schemes would be most appropriate?

A. Plots 1, 2, 3, and 4 in one block and plots 5, 6, 7, and 8 in the other
B. Plots 1, 2, 5, and 6 in one block and plots 3, 4, 7, and 8 in the other
C. Plots 1, 4, 5, and 8 in one block and plots 2, 3, 6, and 7 in the other
D. Plots 1, 3, 6, and 8 in one block and plots 2, 4, 5, and 7 in the other
E. The plots should be randomly assigned to blocks.

4. At a large hospital, the durations of emergency room visits, from check-in to discharge, are approximately normally distributed with a mean of 167 minutes and a standard deviation of 76 minutes. Which interval, to the nearest minute, contains the middle 80% of durations?

A. 15 minutes to 319 minutes
B. 53 minutes to 281 minutes
C. 70 minutes to 264 minutes
D. 91 minutes to 243 minutes
E. 103 minutes to 231 minutes

5. Aliana works 6-hour weekend shifts at an area chain restaurant as a hostess, and she wants to explore whether customer preferences are influenced by how you ask them about seating. Aliana's questioning strategy is as follows:

Question	Asked of all customers
A. Would you prefer a booth or a table?	...in the first two hours of her shift
B. Would you prefer a table or a booth?	...in the middle two hours of her shift
C. Where would you like to sit?	...in the last two hours of her shift

Aliana collected data over the course of several weekends, and her results are in the table below:

	Question Aliana asked:			
Party requested to sit at:	*A*	*B*	*C*	Totals
Table	4	22	18	44
Booth	15	23	21	59
Totals	**19**	**45**	**39**	**103**

Which of the following is a true statement?

A. The number of degrees of freedom for the χ^2 is 102.
B. The sample size is a large enough size because it is bigger than 30.
C. We do not meet the conditions for a χ^2 test because the Table/Question A cell is less than 5.
D. We do not meet the conditions for a χ^2 because Aliana did not randomly assign treatments to subjects.
E. We do not meet the conditions for a χ^2 test because Aliana did not ask the same number of customers each of the three questions.

GO ON TO THE NEXT PAGE

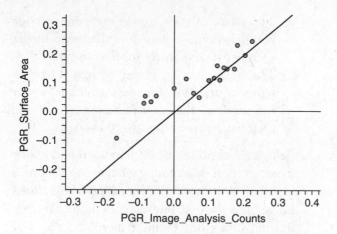

6. Biologists have two different methods for estimating the growth rate of a population of animals. One is based on counting the number of animals in a photographic image, and the other is based on a computer-calculated surface area of the animals in the image. (Larger animals are generally older and contribute more to reproduction.) The scatterplot above shows the estimates of the Population Growth Rate (PGR) for each method along with the line $y = x$. Which statement is true based on the scatterplot?

A. The *surface area* method overestimates the Population Growth Rate.
B. The *count* method underestimates the Population Growth Rate.
C. Both methods overestimate the Population Growth Rate.
D. Both methods give accurate estimates of the Population Growth Rate.
E. There is not enough information to tell how well either method estimates the actual Population Growth Rate.

7. An infrared thermometer measures the surface temperature of an object from a distance. The measurement errors for one particular model are approximately normally distributed with a mean of 0 degrees C and a standard deviation of 1.3 degrees C. If the temperature of an object is 20 degrees C, what is the probability that the thermometer will report the temperature as less than 18 degrees C?

A. 0.023
B. 0.062
C. 0.097
D. 0.903
E. 0.938

8. A study was conducted to test a new style of keyboard in preventing repetitive stress disorders. Volunteers who have had problems with such injuries were randomly assigned to use either a traditional keyboard or the new design. A significance test was conducted with the alternative hypothesis that a smaller proportion of those using the new keyboard will suffer injuries than those using the traditional keyboard. The resulting P-value was 0.07. Which is a correct interpretation of this P-value?

A. The null hypothesis should be rejected.
B. The null hypothesis should be accepted.
C. There is a 7% chance that the null hypothesis is correct.
D. There is a 7% chance of getting a difference between the two groups at least as large as the observed difference if the new keyboard is really no better at preventing injuries.
E. There is a 7% chance that the observed difference is due to chance.

9. For a class project, Charlotte recorded the heights of all 28 students in her class and calculated several statistics. She then realized she made an error recording the height of the tallest person in the class. She correctly had him listed as the tallest, but needed to add two inches to his recorded height to correct it. Which of these measures of spread must remain unchanged?

A. Mean absolute deviation
B. Standard deviation
C. Variance
D. Interquartile range
E. Range

GO ON TO THE NEXT PAGE

10. A large company grants its hourly employees pay raises after their six-month performance review. The amounts of the raises average $0.25 per hour, with a standard deviation of $0.05 per hour. Let $\hat{y}$ represent the average pay raise of 50 randomly selected hourly employees. Which calculation would give the approximate probability that $\hat{y}$ is at least $0.30 per hour?

A. $P(z \geq 0.30)$

B. $P\left(z \geq \dfrac{0.30 - 0.25}{0.05 / \sqrt{50}}\right)$

C. $P\left(z \geq \dfrac{0.30 - 0.25}{0.05 / 50}\right)$

D. $P\left(z \geq \dfrac{0.30 - 0.25}{0.05}\right)$

E. This probability cannot be calculated because no information is given about the shape of the distribution of all pay raises.

11. The table below shows the daily high temperatures for the month of October for two cities in different regions of the country.

Hendersonville		Sheboygan
	4	68
2	5	00234
76	5	55777779
411110	6	011122444
988765	6	699
332211	7	123
8777766	7	9
430	8	

Which of the following statements is true?

A. The range of the October temperatures for Hendersonville is greater than the range of October temperatures for Sheboygan.
B. The median October temperature for Hendersonville is less than the median October temperature for Sheboygan.
C. The median of the October temperatures for Hendersonville is greater than the third quartile of the October temperatures for Sheboygan.

D. The mean of the October temperatures for Hendersonville is less than the mean of the October temperatures for Sheboygan.
E. The interquartile range of the October temperatures for Hendersonville is about the same as the interquartile range of the October temperatures for Sheboygan.

12. Biologists around the world have increased efforts at conservation. Monitoring wildlife populations is important so that appropriate management efforts can be implemented, but such monitoring is often difficult. One study found a cheap and easy way to estimate the number of nesting sites of terns (a type of seabird) by monitoring the number of calls heard per minute. More calls happen when there are more birds, as one would expect. In fact, it turned out that the number of calls explained 71% of the variation in the abundance of nests between breeding sites. Which of the following statements is correct about the correlation between the number of calls and the abundance of nests?

A. The correlation coefficient is −0.71.
B. The correlation coefficient is 0.71.
C. The correlation coefficient is −0.84.
D. The correlation coefficient is 0.84.
E. The slope of the regression line is needed to determine the correlation.

13. A woman who works for a department store wants to sue the company for discrimination based on gender. Her reasoning is that women seem less likely to get promoted than men. She presented this table showing the employees eligible for promotion divided by both gender and whether they were promoted or not. Does the table provide some evidence to support her claim?

	Men	Women	Total
Promoted	9	12	21
Not Promoted	15	31	46
Total	24	43	67

A. No, because more women than men were promoted.
B. No, because the proportion of women who were promoted is the same as the proportion of men who were promoted.
C. No, because the proportion of women who were promoted is less than the proportion of men who were promoted.

GO ON TO THE NEXT PAGE

D. Yes, because the proportion of women who were promoted is less than the proportion of men who were promoted.

E. Yes, because the proportion of women who were promoted is less than the proportion of women who were not promoted.

14. Students in a statistics class, as part of an activity, measured the length of their hair in inches. Summary statistics are shown below.

Variable	N	Mean	Median	TrMean	StDev	SE Mean
Hair Length	41	9.18	3.5	8.75	8.91	1.39

Variable	Minimum	Maximum	Q1	Q3
Hair Length	0	29	2	18.5

Which of the following statements is true?

A. Seventy-five percent of the students had hair at least 18.5 inches long.

B. Fifty percent of the students had hair between 9.18 and 29 inches long.

C. Fifty percent of the students had hair between 2 and 18.5 inches long.

D. The distribution of hair lengths can be appropriately modeled by a normal distribution.

E. The distribution of hair lengths has at least one outlier.

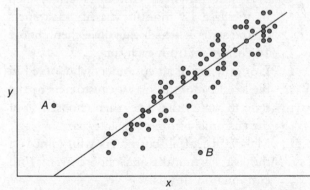

15. The scatterplot above shows data points and the regression line for predicting *y* from *x*. Which statement is true about the effect of removing point *A* or *B* on the regression model?

A. Removing *A* would decrease the slope but removing *B* would increase the slope.

B. Removing *A* would increase the slope but removing *B* would have little effect on the slope.

C. Removing *A* would decrease the correlation but removing *B* would increase the correlation.

D. Removing point *A* would increase the correlation and removing point *B* would not change the correlation.

E. Removing point *A* would increase the *y*-intercept and removing point *B* would decrease the *y*-intercept.

16. Which of these statements correctly explains bias?

A. It describes a process that creates estimates that are too high on average or too low on average.

B. It describes a process that results in a sample that does not represent the population.

C. It describes a lack of randomization in the sampling process.

D. It describes a sample that is not representative of the population.

E. It describes a sample that was selected by convenience.

17. After graduation, you have the opportunity to open a local pizza restaurant. But to cover all your expenses as well as pay yourself a salary, you need an average revenue of about $9,000 per month. Anything less and you risk losing your business. You take a random sample of 30 different pizza restaurants similar to the one you hope to open and get data on their monthly sales. For the hypotheses

$$H_0: \mu = \$9{,}000 \quad H_a: \mu > \$9{,}000$$

describe a Type I error and its consequence.

A. The consequences of a Type I error cannot be determined without an α-level.

B. You believe the average sales of pizza restaurants like yours is more than $9,000, so you open a pizza restaurant and have a high risk of losing your business.

C. You believe the average sales of pizza restaurants like yours is more than $9,000, so you open a pizza restaurant and have a high probability of a successful business.

D. You believe the average sales of pizza restaurants like yours is $9,000 or less, so you do not open a pizza restaurant yourself and will miss out on an opportunity to own a successful business.

E. You believe the average sales of pizza restaurants like yours is $9,000 or less, so you do not open a pizza restaurant because other pizza places are not successful.

GO ON TO THE NEXT PAGE

18. A particular crop of one variety of onion has weights that are approximately normally distributed with mean 9.8 oz. and standard deviation 2.1 oz. How does an onion in the 28th percentile for weight compare to the mean?

    A. 1.22 ounces below the mean
    B. 0.59 ounces below the mean
    C. 0.59 ounces above the mean
    D. 1.22 ounces above the mean
    E. 2.26 ounces above the mean

19. A pharmaceutical company wants to test a new cholesterol-reducing drug against the previous drug. It does not anticipate much association between cholesterol level and gender, but it does anticipate an association between cholesterol level and the amount of exercise a person gets. For a randomized block design, it should:

    A. Block on gender because it is not associated with cholesterol level.
    B. Block on gender because males and females are different.
    C. Block on the type of drug because it may be associated with cholesterol level.
    D. Block on exercise level because it is associated with cholesterol level.
    E. Block on both gender and exercise because blocking on more variables gives more reliable estimates.

20. A high school sent a survey to a randomly selected sample of 75 of last year's graduating class. 27 of those selected did not return the survey. The best plan of action would be to:

    A. Use the surveys that were returned and change the sample size to 48 for the analysis.
    B. Use the surveys that were returned and leave the sample size at 75 for the analysis.
    C. Randomly select 27 additional class members and send the survey to them.
    D. Start over with a new sample but use a larger sample size.
    E. Follow up with those that did not return the survey to encourage them to respond.

21. Which of these is a correct description of the term?

    A. A *factor* is a response variable.
    B. *Replication* means the experiment should be repeated several times.
    C. *Levels* are the same as *treatments*.
    D. *Experimental units* are the same as *subjects*.
    E. The *response variable* is the variable directly controlled by the researchers.

22. The president of an online music streaming service whose customers pay a fee wants to gather additional information about customers who have joined in the past 12 months. The company plans to send out an e-mail survey to a sample of current customers with a link that gives participants a month of streaming service for free once the survey has been completed. They know that musical tastes vary by geographical region. Which of the following sample plans would produce the most representative sample of its customers?

    A. Choose all of the customers who joined in the last month.
    B. Make a list of all the customers who joined in the last 12 months and choose a random sample of customers on this list.
    C. From the list of all customers who joined in the last 12 months, classify customers by the state in which they live, then choose 10 customers from each state.
    D. From the list of all customers who joined in the last 12 months, classify customers by the state in which they live, then choose 3% of the customers from each state.
    E. Make a list of all customers who joined in the last 12 months and choose every 10th customer on this list.

23. The reason we check that np and $n(1 - p)$ are at least 10 when calculating probabilities in a binomial situation is to:

    A. Ensure that $\hat{p}$ has a binomial distribution.
    B. Be sure the sample is random enough.
    C. Make sure the distribution of $\hat{p}$ is symmetric enough to use a normal approximation.
    D. Get an unbiased estimate for p.
    E. Be able to use $\sqrt{\dfrac{p(1-p)}{n}}$ as the standard deviation of $\hat{p}$.

GO ON TO THE NEXT PAGE

24. A maker of breakfast cereals plans to run a promotion in which an action figure is included in each box of cereal. The more popular figures will appear in fewer boxes. The probability of getting a box with each kind of toy is shown in the table below.

Toy	Serenity	Blackstar	Spartan	Kropp
Probability	0.12	0.18	0.30	0.40

What is the probability that the first two randomly selected boxes contain Serenity and Blackstar?

A. 0.0216
B. 0.0432
C. 0.0468
D. 0.2784
E. 0.3000

25. A department store has a promotion in which it hands out a "scratchcard" at the checkout register, with a percent discount concealed by an opaque covering. The customer scratches off the covering and reveals the amount of the discount. The table below shows the probability that a randomly selected card contains each percent discount.

Percentage of discount	5%	10%	15%	25%	50%
Probability	0.45	0.25	0.15	0.10	0.05

What is the expected value of the percentage of discount?

A. 5%
B. 12%
C. 15%
D. 21%
E. 40%

26. In basketball, an offensive rebound occurs when a player shoots and misses, and a player from the same team recovers the ball. For the 176 players on the roster for one season of professional men's basketball, the third quartile for the total number of offensive rebounds for one season was 143. If five players are selected at random (with replacement) from that season, what is the approximate probability that at least three of them had more than 143 rebounds that season?

A. 0.0127
B. 0.0879
C. 0.1035
D. 0.8965
E. 0.9121

27. A new smartwatch is manufactured in one part of a factory, then secured for shipping in another, independent part of the factory. The weight of the smartwatch has a mean of 62 grams and a standard deviation of 1.0 grams. The weight of the packaging (box, user's guide, bubble wrap, etc.) has a mean of 456 grams and a standard deviation of 6 grams. Together, the distribution of the weight of the smartwatch and its packaging would have the following mean and standard deviation:

A. Mean 518 grams; standard deviation 7.0 grams
B. Mean 518 grams; standard deviation 3.5 grams
C. Mean 518 grams; standard deviation 6.1 grams
D. Mean 394 grams; standard deviation 6.1 grams
E. Mean 394 grams; standard deviation 3.5 grams

28. Students in a small high school were asked to vote for a meal for a school event. The results of the survey and the grades of the students are summarized below.

	Hot dogs	Pizza	Tacos	**Total**
Freshmen	45	54	52	**151**
Sophomores	41	42	45	**138**
Juniors	37	48	59	**144**
Seniors	41	51	60	**152**
Total	**174**	**195**	**216**	**585**

Are the events *Student is a junior* and *Student chose pizza* independent?

A. Yes, because the proportion of juniors who chose pizza is the same as the proportion of all students who chose pizza.
B. Yes, because the proportion of students that preferred pizza who are juniors is different from the proportion of all students who are juniors.
C. No, because the proportion of juniors who chose pizza is the same as the proportion of all students who chose pizza.
D. No, because the proportion of students who are juniors that preferred pizza is different from the proportion of all students who are juniors.
E. No, because the proportion of juniors who chose pizza is different from the proportions of other grades who chose pizza.

GO ON TO THE NEXT PAGE

29. The total cholesterol level in a large population of people is strongly skewed right with a mean of 210 mg/dL and a standard deviation of 15 mg/dL. If random samples of size 16 are repeatedly drawn from this population, which of the following appropriately describes the sampling distribution of these sample means?

    A. The shape is unknown with a mean of 210 and a standard deviation of 15.
    B. The shape is somewhat skewed right with a mean of 210 and a standard deviation of 3.75.
    C. The shape is approximately normal with a mean of 210 and a standard deviation of 15.
    D. The shape is approximately normal with a mean of 210 and a standard deviation of 3.75.
    E. The shape is now skewed left with a mean of 210 and a standard deviation of 3.75.

30. In one metropolitan region, technical writers have an annual salary that is approximately normally distributed with a mean of $55,800. The first quartile of salaries is $48,815. What is the standard deviation?

    A. $6,984
    B. $10,356
    C. $10,476
    D. $13,968
    E. $20,709

31. Which of these explains why t should be used instead of z for inference procedures for means.

    A. The Central Limit Theorem applies to proportions but not means.
    B. We use z for proportions because proportions are approximately normal. Means are not.
    C. We use z for proportions because the sampling distribution of sample proportions is approximately normal, but that is not the case for the sampling distribution of sample means.

    D. The sampling distribution of sample means is a t-distribution, not a z-distribution.
    E. When using the sample standard deviation to estimate the population standard deviation, more variability is introduced into the sampling distribution of the statistic.

32. Which of the following is *least* likely to reduce bias in a sample survey?

    A. Following up with those who did not respond to the survey the first time
    B. Asking questions in a neutral manner to avoid influencing the responses
    C. Using stratified random sampling rather than simple random sampling
    D. Selecting samples randomly
    E. Allowing responses to be anonymous

33. In a study to determine whether there is a difference in color preferences for young children based on their gender, researchers had a random sample of children choose from a selection of toys that were identical except for their color. The table below shows the number of each color that was chosen by boys and by girls.

	Boys	Girls
Red	15	22
Blue	28	8
Green	20	14
Pink	3	17
Brown	9	2

How many degrees of freedom should be used in a test of independence of gender and color preference?

A. 2
B. 4
C. 9
D. 10
E. 137

GO ON TO THE NEXT PAGE

34. A local library has a scanner to detect library materials that have not been checked out. Each item has a chip somewhere inside. Upon checkout, the chip is deactivated so the scanner will not set off the alarm. The scanner has a 98% chance of detecting an active chip (meaning the material has not been checked out) and setting off the alarm. The scanner also has a 3% chance of sounding the alarm when someone passes through without an active chip. It is estimated that 0.5% of library customers actually try to leave the library with an active chip. What is the probability that, if the alarm sounds, the patron leaving the library has an item with an active chip?

    A. 0.0049
    B. 0.0348
    C. 0.1410
    D. 0.9700
    E. 0.9800

35. Which of the following is the best description of the power of a significance test?

    A. The probability that the null hypothesis is true.
    B. The probability of getting a Type I error.
    C. The probability of getting a Type II error.
    D. The probability of getting a test statistic at least as extreme as the actual test statistic, if the null hypothesis is true.
    E. The probability of rejecting the null hypothesis if it is, in fact, false.

36. A school board of a large school district is proposing a new dress code for students. Some students feel that this dress code unfairly targets female students. To see if there is a difference between boys and girls in their opposition to the new dress code, they conduct a poll of 60 randomly selected male and 70 randomly selected female high school students in the district. They find that 66 females oppose the dress code and 50 males oppose the dress code. Which of the following explains why a two-proportion z-test is not appropriate?

    A. The sample sizes are different.
    B. The sample sizes are too large.
    C. The number of successes and the number of failures for the two groups are not all large enough.

D. The shapes of the population distributions are not known.
E. The population standard deviations are not known.

37. Researchers are conducting an experiment using a significance level of 0.05. The null hypothesis is, in fact, false. If they modify their experiment to use twice as many experimental units for each treatment, which of the following would be true?

    A. The probability of a Type I error and the probability of a Type II error would both decrease.
    B. The probability of a Type I error and the power would both increase.
    C. The probability of a Type II error and the power would both increase.
    D. The probability of a Type I error would stay the same and the power would increase.
    E. The probability of a Type II error would stay the same and the power would increase.

38. A producer of skin care products has created a new formula for its cream to cure acne. To compare the effectiveness of the new cream to that of the old cream, it conducted a double-blind randomized experiment. Volunteers with acne tried the old formula on one side of their face and the new formula on the other, and which side got which formula was determined randomly. The response variable was the difference in the number of pimples (old formula – new formula). Which is the correct significance test to perform?

    A. A two-proportion z-test
    B. A two-sample t-test
    C. A matched pairs t-test
    D. A chi-square test of independence
    E. A chi-square goodness of fit test

39. Smartphones with larger screens tend to be more expensive than smartphones with smaller screens. A random sample of currently available smartphones was selected. A 95% confidence interval for the slope of the regression line to predict price from screen size is (61, 542). Which of the following statements concerning this interval is correct?

   A. If many samples of this size were taken, about 95% of confidence intervals for the slope based on those samples would contain the slope of the line for all smartphones that are currently available.

   B. If many samples of this size were taken, about 95% of those samples would have a regression line with a slope between 61 and 542.

   C. There is convincing evidence that the size of the screen is the reason for the difference in price.

   D. Because this interval is so large, there is not convincing evidence of a relationship between screen size and price.

   E. There is not enough information to make a statement about the relationship between screen size and price because memory capacity and other features must also be taken into account.

40. In the 50 states of the United States, the financial climate for people living in retirement varies. Some states are very tax friendly to retirees. Other states are much less tax friendly to retirees. As a member of the Oregon state legislature, you wonder if Oregon's tax policies are driving older, retired people to move to other states. You have a cousin in Florida with the opposite idea—that the tax-friendly nature of Florida attracts more retirees. Each of you takes a random sample of the adults in your respective states and finds the statistics in the table below.

State	Sample Size	Mean Age	Standard Deviation
Oregon	31	40.63	19.78
Florida	41	44.65	20.60

The 99% confidence interval for the difference in the mean age of adults in Oregon and adults in Florida is (−16.73, 8.69). Does this interval provide convincing evidence that the mean ages of adults in the two states differ?

   A. Yes, because 0 is in the interval.
   B. No, because 0 is in the interval.
   C. Yes, because the mean ages are different in the two samples.
   D. No, because the mean ages of the samples are less than one standard deviation apart.
   E. Yes, because this is a t-interval for a difference in means.

END OF SECTION I

AP Statistics Practice Test 2

SECTION II

Time: 1 hour and 30 minutes

Number of questions: 6

Percentage of total grade: 50

General Instructions

There are two parts to this section of the examination. Part A consists of five equally weighted problems that represent 75% of the total weight of this section. Spend about 65 minutes on this part of the exam. Part B consists of one longer problem that represents 25% of the total weight of this section. Spend about 25 minutes on this part of the exam. You are not necessarily expected to complete all parts of every question. Statistical tables and formulas are provided.

- Be sure to write clearly and legibly. If you make an error, you may save time by crossing it out rather than trying to erase it. Erased or crossed-out work will not be graded.
- Show all your work. Indicate clearly the methods you use because you will be graded on the correctness of your methods as well as the accuracy of your final answers. Correct answers without support work may not receive credit.

Statistics, Section II, Part A, Questions 1–5

Spend about 65 minutes on this part of the exam; percentage of Section II grade: 75.

Directions: Show all your work. Indicate clearly the methods you use because you will be graded on the correctness of your methods as well as on the accuracy of your results and explanation.

1. Universities generally post the rosters of all their student athletes, including which sport they play and their heights. The data in the table represent the height data for members of one university's fencing teams.

University Fencing Teams: Heights (inches)			
Men		Women	
64	71	62	65
64	72	62	65
68	72	63	67
69	74	63	67
69	74	63	68
69	74	64	69
70	74	64	69
70	74	64	72
70	79		

a. Construct an appropriate comparative graph for the heights of the members of the fencing teams.

b. Using your plot in part (a), describe the differences and similarities in the distributions of the heights of the two teams.

GO ON TO THE NEXT PAGE

2. In the fictional story "The Legend of Sleepy Hollow," the Headless Horseman chases Ichabod Crane through the New England countryside. A video game based on this legend has Crane choosing his route randomly, with a probability of 15% that he stays on the road and 85% that he cuts through the forest. If he stays on the road, there is a 90% chance of the Horseman catching Ichabod. If he cuts through the forest, there is only a 64% change of being caught. Find the probability that the next game has Mr. Crane cut through the forest, given that the Horseman will indeed catch him.

3. A local farmers' market sells grapefruit and claims that the mass of these citrus fruits is approximately normally distributed with a mean of 428.0 grams and a standard deviation of 21.1 grams. You purchase 10 grapefruit, and the average mass of these 10 fruits is 446.2 grams.

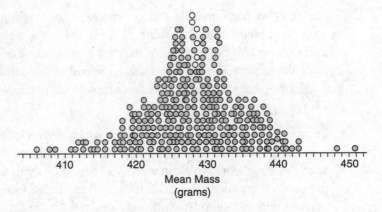

a. You wish to test the claim of the farmers' market, so you begin by using a computer model to simulate drawing 300 samples of size 10 from a population approximately N (428.0, 21.2) grams. Above is the dotplot of the means from these 300 samples.

Explain in simple language how the dotplot shows that the sample mean you are using is an unbiased estimator of the population mean.

b. Given your bag, with its average mass per grapefruit of 446.2 grams, do you think that the farmers' market is correct in its claim? Justify your answer using the dotplot above.

4. Cortisol is a hormone produced by the body to control inflammation. People with chronic inflammation and, hence, chronically elevated cortisol levels can develop problems with their immune system. To explore this, 100 adult volunteers with chronic inflammation will participate in a study to compare the effect of black tea versus coffee on cortisol levels. Each volunteer will be assigned at random to one of the two groups and provided with daily capsules that contain a concentrated form of either black tea or coffee. Each will also have his or her cortisol levels measured at the beginning of the study and then 20 weeks later.

a. Describe how you would assign the 100 volunteers to the two groups in such a way as to allow a statistically valid comparison of the two treatments.

b. Explain a *statistical* advantage to using capsules rather than having participants actually drink coffee or tea.

c. Is it reasonable to generalize the findings of this study to all adults with chronic inflammation? Explain.

GO ON TO THE NEXT PAGE

5. A large school district is planning an enrichment program for next summer. It is planning four course options for students: Music, Sports, Drama, or Academic Enrichment. For planning purposes, the district selected a random sample of 100 freshmen, a random sample of 100 sophomores, and a random sample of 100 juniors. (Since seniors will have graduated, they were not surveyed.) The selected students were asked which program they would choose to attend. The results for each class are shown in the graph below:

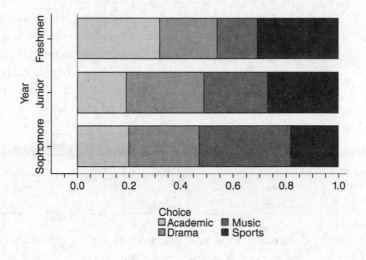

a. Describe any associations you see between year in school and choice of program.
b. District administrators want to determine if there is convincing evidence that students in different years have different preferences for programs throughout the whole district. Identify the hypothesis test they should use and state the degrees of freedom.
c. The *P*-value from the test is 0.0082. What conclusion should the administrators reach?

Statistics, Section II, Part B, Question 6

Spend about 25 minutes on this part of the exam; percentage of Section II grade: 25.

Directions: Show all of your work. Indicate clearly the methods you use because you will be graded on the correctness of your methods as well as on the accuracy of your results and explanation.

6. In 1937, the United States passed the Wagner-Steagall Act, also known as the National Housing Act. This piece of legislation was intended to remedy the unsafe housing conditions in which many low-income families were living. The National Housing Act contributed to today's common wisdom that when a household devotes more than 30 percent of its income to housing expenses that household is said to be burdened. A household that devotes more than 50 percent of its income to housing expenses is said to be severely burdened.

A graduate student interning at a nonprofit organization that addresses affordable housing has received a grant to study this issue. However, the grant will only cover in-depth investigation of eight households. The table below represents the percentage

GO ON TO THE NEXT PAGE

of income devoted to housing expenses (rent, utilities, etc.) for each of eight randomly selected low-income households.

Subject	A	B	C	D	E	F	G	H
%	46.3	59.1	70.4	52.2	54.5	47.0	51.9	53.4

a. Local officials say that for low-income households in this community the median percentage of income spent on housing is 48 percent. The graduate student would like to test the hypothesis that the median percentage is actually higher than 48. Explain why the graduate student should not use a t-test for this hypothesis.

b. Rather than a test using the numeric values, the graduate student decides to turn the data into categorical data by noting whether each subject's income is above or below the hypothesized median percentage of 48. Fill out the table appropriately to reflect the graduate student's change of the data:

Subject	A	B	C	D	E	F	G	H
%	46.3	59.1	70.4	52.2	54.5	57.0	51.9	53.4
Data Change								

c. Explain how the graduate student's decision to change the nature of the data addresses any issues raised in part (a).

d. If it were true that the median percentage of income devoted to housing expenses was 48 percent, then we would expect half of the population to spend less than 48 percent on housing expenses. Using the information in the table above, calculate the approximate probability that one or fewer clients would have a housing expense percentage of less than 48 percent.

e. Based on your answer to part (d), do you have convincing evidence that the graduate student's hypothesis is correct? Explain your answer.

END OF SECTION II

❯ Answers to Practice Test 2, Section I

1. D	21. D
2. B	22. D
3. A	23. C
4. C	24. B
5. D	25. B
6. E	26. C
7. B	27. C
8. D	28. A
9. D	29. B
10. B	30. B
11. C	31. E
12. D	32. C
13. D	33. B
14. C	34. C
15. B	35. E
16. A	36. C
17. B	37. D
18. A	38. C
19. D	39. A
20. E	40. D

› Solutions to Practice Test 2, Section I

1. (D) The location of the tail indicates skew. The tail is located toward the smaller values of time, therefore we call this distribution skewed left (since smaller values are to the left of larger values).

2. (B) Slope, traditionally written as $\Delta y/\Delta x$, becomes $\Delta CaloriesPredicted/\Delta Sodium = 0.2457/1$. So, every item with 1 additional mg. of sodium tends to have 0.2457 more calories.

3. (A) Since water and sunlight also influence plant growth, we block on those variables. Plots 1, 2, 3, and 4 all get more direct sunlight. Plots 5, 6, 7, and 8 all get increased water.

4. (C) The middle 80% of durations symmetrically straddles the mean of 167 minutes. That leaves 10% of the durations in each tail. Using Standard Normal Probabilities (Table A in the Appendix to this book), we find a z-score that corresponds to a 10% tail area that is ±1.28. Solving the equations $\pm 1.28 = (x - 167)/76$ gives us approximately 70 minutes and 264 minutes.

5. (D) Aliana's customers were not randomly selected. But neither did she randomly assign customers to her treatments—she spilt her shift into three parts and applied the treatment to all customers during the time period.

6. (E) You can tell that the surface area estimates tend to be higher than the count estimates, but you cannot tell how either of them compares to the correct value.

7. (B) Measurement errors are values calculated using the expression Thermometer Reading $-20°C$. If the reading is $18°C$, then the error would be $-2°C$. We then need to calculate the probability that the error is less than $-2°C$. Using Table A (see Appendix), we find $P\left(z < \dfrac{-2-0}{1.3}\right) = 0.062$.

8. (D) The definition of a P-value is the probability, in repeated sampling, of obtaining results at least as large/small as ours when the null hypothesis is actually true. Therefore, a P-value in this case says that there is a 7% chance of a difference at least as large if the new keyboard is no better than the old.

9. (D) Changing the largest value in a data set, in this case increasing it by two inches, would not affect either Q1 or Q3, therefore the IQR remains unchanged.

10. (B) The sampling distribution of the sample mean is approximately normal because $n = 50 > 30$. $\mu_{\bar{x}} = \mu = 0.25$, and $\sigma_{\bar{x}} = \dfrac{\sigma}{\sqrt{n}} = \dfrac{0.05}{\sqrt{50}}$.

11. (C) The median for Hendersonville is 71 degrees. The third quartile for Sheboygan is 64 degrees.

12. (D) The 71% refers to r^2. So $r = \pm\sqrt{0.71} \approx \pm 0.8$ Because the association is positive, $r = +0.84$.

13. (D) In this case $12/43 = 0.28$ women were promoted and $9/24 = 0.38$ men were promoted. The table does provide some evidence since these values are so different. It remains to be seen whether this difference is statistically significant.

14. (C) The values 2 and 18.5 represent Q1 and Q3 respectively. Approximately 50% of the values would be between Q1 and Q3.

15. (B) Point A "pulls up" on the left end of the line. Removing it would drop the left end, increasing the slope. Point B is pulling down near the mean value of x. Removing it would have little impact on the slope.

16. (A) Bias is defined as any process that systematically over- or underestimates. A process that creates estimates that are, on average, too high or too low is, by definition, biased.

17. (B) A Type I error means the null hypothesis is true, but you reject it. That means pizza places, on average, do not make over $9,000 per month, but you believe they do. You might open a business and do poorly.

18. (A) We need to find the z-score that corresponds to a lower-tail area/probability of 0.2800. Using Table A (see Appendix), that z-score is -0.58. $-0.58 \cdot 2.1$ ounces $= -1.22$ ounces, so this onion is 1.22 ounces below the mean.

19. (D) Because there is an association between exercise and cholesterol, we need to block on level of exercise.

20. (E) If we are to use the data that were returned, then we have to find a way to overcome nonresponse by getting survey results from those who did not return the survey.

21. (D) A factor is an explanatory variable. Replication means that, within an experiment, each treatment is applied to more than one experimental unit. Treatments are combinations of levels from different factors. The variable controlled by researchers is the explanatory variable, not the response variable. D is the only correct option.

22. (D) The key phrase in this question is "most representative sample of its customers." Choice (D) ensures that the sample selects customers from each state and that selection is proportional to the number of customers from each state. For example, if 25% of the customers are from California, then 25% of the sample will be from California.

23. (C) In order to use a normal approximation to a binomial model such as this, we should see if np and $n(1 - p)$ are both at least 10. (Note, some authors will check to see if both are at least 5 and others will check to see if both are at least 15.)

24. (B) We need to calculate P(Serenity in first box, Blackstar in second box) as well as P(Blackstar in first box, Serenity in second box) and then add them. This would be $(0.12)(0.18) + (0.18)(0.12) = 0.0432$.

25. (B) Using the expected value formula for a probability function we get $(0.05)(0.45) + (0.10)(0.25) + (0.15)(0.15) + (0.25)(0.10) + (0.50)(0.05) = 0.12$, or 12%.

26. (C) The probability of selecting a player above the third quartile is 0.25 because we are sampling with replacement, this situation meets the conditions of a binomial variable with $n = 5$ and probability of success $= 0.25$. Therefore, $P(x \geq 3) = P(3) + P(4) + P(5)$. Using the formula $P(x = k) = \binom{n}{k} p^k (1 - p)^{n-k}$ we get,

$P(3) + P(4) + P(5) = 0.1035$.

27. (C) The mean of the total weight is the same as the sum of the individual means, or 62 grams + 456 grams = 518 grams. We are told that the shipping is secured in an independent part of the factory, so therefore the variance of the total weight is given by $(1.0 \text{ grams})^2 + (6 \text{ grams})^2$. We find the standard deviation of the total weight by taking the square root of that variance, or $\sqrt{1^2 + 6^2} = 6.1$ grams.

28. (A) Recall that two events A and B are independent if $P(A|B) = P(A)$. The proportion of juniors in the group who vote for pizza is $48/144 = 0.33333$. The proportion of all students who vote for pizza is $195/585 = 0.3333$. Since these are the same value, *Student is a junior* and *Student choses pizza* are independent.

29. (B) Because $\bar{x}$ is an unbiased estimator of the mean, the sampling distribution of those sample means has the same value as the population. Therefore mean = 210. The standard deviation of the sampling distribution is given by $\sigma/\sqrt{n} = 15/\sqrt{4} = 3.75$. However, the population is substantially skewed right and the sample size is very small. Therefore we cannot say that the shape of the sampling distribution is approximately normal.

30. (B) The z-score for the first quartile ($p = 0.25$) is -0.674. $z = \dfrac{48{,}815 - 55{,}800}{s} = -0.67449$. So

$$s = \frac{48{,}815 - 55{,}800}{-0.67449} = 10{,}356.$$

31. (E) The correct option provides the explanation.

32. (C) The question asks which is least likely to reduce bias in a sample survey. Simple random sampling is unbiased, so using a stratified random sample would not improve on that. You can take a representative sample but still introduce bias unless you address the actions in choices (A), (B), (D), and (E).

33. (B) Degrees of freedom are calculated by (rows − 1) (columns − 1). In this case, $(5 - 1)(2 - 1) = 4$

34. (C) This can be done using a two-way table.

	Active chip	No active chip	Total
Alarm	$0.98 \cdot 0.005 = 0.0049$	$0.03 \cdot 0.995 = 0.02985$	$0.0049 + 0.02985 = 0.03475$
No alarm	$0.005 - 0.0049 = 0.0001$	$0.995 - 0.02985 = 0.96515$	$0.0001 + 0.96515 = 0.96525$
Total	0.005	0.995	1

The required probability is $0.0049/0.03475 = 0.1410$

35. (E) That is the definition of power.

36. (C) The number of successes and failures each needs to be greater than 10 for both males and females. Since the success/failure numbers for females are 66/4, we do not meet this condition.

37. (D) We did not change the significance level, which is the probability of making a Type I error. Increasing the sample size is one way to increase the power of the test.

38. (C) Each volunteer behaves as his or her own block. Therefore, a matched pairs test is appropriate.

39. (A) This choice correctly interprets the confidence level of 95%. When you take many, many, many samples of the same size, 95% of the confidence intervals you build around your sample results will contain the parameter you hope to estimate.

40. (B) Since zero is contained in the interval, zero (representing no difference in the means) is a plausible value. So, we do not have convincing evidence that there is a difference.

› Solutions to Practice Test 2, Section II

1. Fencing Team

   a. A back-to-back stem plot with split stems will do nicely:

Men		Women
	6	2 2 3 3 3
4 4	6	4 4 4 5 5
	6	7 7
9 9 9 8	6	8 9 9
1 0 0 0	7	
2 2	7	2
4 4 4 4 4	7	
	7	
9	7	

   Key: 6 | 2 = 62 inches

   b. The distribution of women's heights seems slightly skewed to the right, while that of the men seems somewhat symmetric. Both the center and spread of the distribution of men's heights is greater than that of the women's.

2. Using the tree diagram below: $P(\textit{Taking Forest given Caught}) = \dfrac{(0.85)(0.64)}{(0.15)(0.90)+(0.85)((0.64)} = 0.8012$

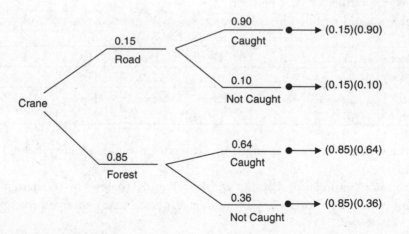

3. Grapefruit at the farmers' market.

   a. It appears that the center of this symmetric distribution is approximately 428 grams. In a symmetric distribution, the mean is approximately located at this center. Therefore, because the mean of this distribution is approximately equal to the population mean of 428 grams, we have evidence that we are using an unbiased estimator.

   b. There are only two simulated bags out of the 300 simulated samples that have a mean weight of more than 446.2 grams. This is a simulated probability of only 0.0067. Given how small a probability, it seems unlikely that the farmers' market is correct in its claim about the population of bags of grapefruit.

4. Cortisol
   a. Put all 100 volunteers' names on equally sized pieces of paper into a hat and mix thoroughly. Then draw the names one at a time out of the hat. The first 50 names drawn are the volunteers who will receive the black tea capsules. The remaining 50 names are the volunteers who will receive coffee capsules.
   b. Using capsules keeps the volunteers blind to which treatment they are receiving so that any effect can be attributed to the treatment (tea or coffee concentrate) rather than perhaps the placebo effect.
   c. It is not reasonable to generalize the results of this study to all adults. The sample was a voluntary one, not a random one.

5. Summer enrichment program
   a. There appears to be an association between course option choices and age as the distributions are so different. Specifically, the proportion of sophomores who chose music is higher than that of juniors and much higher than that of the freshmen. The proportion of juniors who chose academics is about the same as that of sophomores, but much lower than that of the freshmen.
   b. H_0: Choice of enrichment program is the same for the populations of freshmen, sophomores, and juniors.

      H_a: Choice of enrichment program is not the same for all three populations.
      (This would be best done with a Chi-square test that would have $(2 - 1)(4 - 1) = 3$ degrees of freedom).

   c. Since the P-value is so low, lower than any reasonable value of alpha, the district administrators should reject the null hypothesis. It appears that the choice of enrichment program is not the same for all three populations of students.

6. Affordable housing
   a. The graduate student should not use a t-test for this hypothesis because the sample is so small and contains one outlier of 70.4 percent.
   b. The table should be filled in as follows:

Subject	A	B	C	D	E	F	G	H
%	46.3	59.1	70.4	52.2	54.5	57.0	51.9	53.4
Data Change	Below	Above	Above	Above	Above	Above	Above	Above

   c. Once the variable has been changed to a categorical variable (above/below), the outlier ceases to be an issue.
   d. This is now a binomial probability with $n = 8$ and P(below the median) $= 0.50$. The approximate probability that 1 or fewer clients would have a housing expense percentage of less than 48 is found

      by $P(0) + P(1)$ calculated using the formula $P(x = k) = \binom{n}{k} p^k (1 - p)^{n-k}$ once for each probability.

      This is $0.00391 + 0.03125 = 0.03516$.
   e. Since the probability is greater than an alpha level of 0.01, I do not have convincing evidence that the graduate student's hypothesis is correct. (Alternatively: Since the probability is less than an alpha level of 0.05, I do have convincing evidence that the graduate student's hypothesis is correct.)

Appendixes

Formulas
Tables
Bibliography
Websites
Glossary

FORMULAS

I. Descriptive Statistics

$$\bar{x} = \frac{\sum x_i}{n}$$

$$s_x = \sqrt{\frac{1}{n-1}\sum(x_i - \bar{x})^2}$$

$$s_p = \sqrt{\frac{(n_1-1)s_1^2 + (n_2-1)s_2^2}{(n_1+1)+(n_2-1)}}$$

$$\hat{y} = b_0 + b_1 x$$

$$b_1 = \frac{\sum(x_i - \bar{x})(y_1 - \bar{y})}{\sum(x_i - \bar{x})^2}$$

$$b_0 = \bar{y} - b_1\bar{x}$$

$$r = \frac{1}{n-1}\sum\left(\frac{x_i - \bar{x}}{s_x}\right)\left(\frac{y_i - \bar{y}}{s_y}\right)$$

$$b_1 = r\frac{s_y}{s_x}$$

$$s_{b_1} = \frac{\sqrt{\dfrac{\sum(y_i - \hat{y})^2}{n-2}}}{\sqrt{\sum(x_i - \bar{x})^2}}$$

II. Probability

$$P(A \cup B) = P(A) + P(B) - P(A \cap B) \qquad P(A|B) = \frac{P(A \cap B)}{P(B)}$$

$$E(X) = \mu_x = \sum x_i p_i$$

$$Var(X) = \sigma_x^2 = \sum (x_i - \mu_x)^2 p_i$$

If X has a binomial distribution with parameters n and p, then

$$P(X = k) = \binom{n}{k} p^k (1-p)^{n-k} \qquad \mu_x = np \qquad \sigma_x = \sqrt{np(1-p)}$$

If X has a geometric distribution with parameter p, then:

$$P(X = k) = p(1 - p)^{k-1} \qquad \mu_x = \frac{1}{p} \qquad \sigma_x = \frac{\sqrt{1 - p}}{p}$$

If $\hat{p}$ is the proportion of a random sample with replacement of size n from a population with parameter p, then:

$$\mu_{\hat{p}} = p \qquad \sigma_{\hat{p}} = \sqrt{\frac{p(1 - p)}{n}}$$

If $\bar{x}$ is the mean of a random sample with replacement of size n from an infinite population with mean μ and standard deviation σ, then

$$\mu_{\bar{x}} = \mu \qquad \sigma_{\bar{x}} = \frac{\sigma}{\sqrt{n}}$$

III. Inferential Statistics

Standardized test statistic: $\dfrac{\text{statistic} - \text{parameter}}{\text{standard deviation of statistic}}$

Confidence interval: statistic $\pm$ (critical value) $\cdot$ (standard deviation of statistic)

Single-Sample

STATISTIC	STANDARD DEVIATION
Sample Mean	$\dfrac{\sigma}{\sqrt{n}}$
Sample Proportion	$\sqrt{\dfrac{p(1 - p)}{n}}$

Two-Sample

STATISTIC	STANDARD DEVIATION
Difference of sample means $(\sigma_1 \neq \sigma_2)$	$\sqrt{\dfrac{\sigma_1^2}{n_1} + \dfrac{\sigma_2^2}{n_2}}$
Difference of sample means $(\sigma_1 = \sigma_2)$	$\sigma\sqrt{\dfrac{1}{n_1} + \dfrac{1}{n_2}}$
Difference of sample proportions $(p_1 \neq p_2)$	$\sqrt{\dfrac{p_1(1 - p_1)}{n_1} + \dfrac{p_2(1 - p_2)}{n_2}}$
Difference of sample proportions $p_1 = p_2$	$\sqrt{p(1 - p)}\sqrt{\dfrac{1}{n_1} + \dfrac{1}{n_2}}$

Chi-square test statistic $= \displaystyle\sum \frac{(\text{observed} - \text{expected})^2}{\text{expected}}$

TABLES

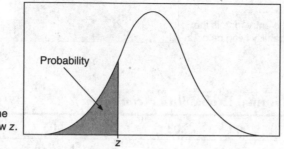

Table entry for z is the probability lying below z.

Probability

z

TABLE A   Standard Normal Probabilities

z	.00	.01	.02	.03	.04	.05	.06	.07	.08	.09
−3.4	.0003	.0003	.0003	.0003	.0003	.0003	.0003	.0003	.0003	.0002
−3.3	.0005	.0005	.0005	.0004	.0004	.0004	.0004	.0004	.0004	.0003
−3.2	.0007	.0007	.0006	.0006	.0006	.0006	.0006	.0005	.0005	.0005
−3.1	.0010	.0009	.0009	.0009	.0008	.0008	.0008	.0008	.0007	.0007
−3.0	.0013	.0013	.0013	.0012	.0012	.0011	.0011	.0011	.0010	.0010
−2.9	.0019	.0018	.0018	.0017	.0016	.0016	.0015	.0015	.0014	.0014
−2.8	.0026	.0025	.0024	.0023	.0023	.0022	.0021	.0021	.0020	.0019
−2.7	.0035	.0034	.0033	.0032	.0031	.0030	.0029	.0028	.0027	.0026
−2.6	.0047	.0045	.0044	.0043	.0041	.0040	.0039	.0038	.0037	.0036
−2.5	.0062	.0060	.0059	.0057	.0055	.0054	.0052	.0051	.0049	.0048
−2.4	.0082	.0080	.0078	.0075	.0073	.0071	.0069	.0068	.0066	.0064
−2.3	.0107	.0104	.0102	.0099	.0096	.0094	.0091	.0089	.0087	.0084
−2.2	.0139	.0136	.0132	.0129	.0125	.0122	.0119	.0116	.0113	.0110
−2.1	.0179	.0174	.0170	.0166	.0162	.0158	.0154	.0150	.0146	.0143
−2.0	.0228	.0222	.0217	.0212	.0207	.0202	.0197	.0192	.0188	.0183
−1.9	.0287	.0281	.0274	.0268	.0262	.0256	.0250	.0244	.0239	.0233
−1.8	.0359	.0351	.0344	.0336	.0329	.0322	.0314	.0307	.0301	.0294
−1.7	.0446	.0436	.0427	.0418	.0409	.0401	.0392	.0384	.0375	.0367
−1.6	.0548	.0537	.0526	.0516	.0505	.0495	.0485	.0475	.0465	.0455
−1.5	.0668	.0655	.0643	.0630	.0618	.0606	.0594	.0582	.0571	.0559
−1.4	.0808	.0793	.0778	.0764	.0749	.0735	.0721	.0708	.0694	.0681
−1.3	.0968	.0951	.0934	.0918	.0901	.0885	.0869	.0853	.0838	.0823
−1.2	.1151	.1131	.1112	.1093	.1075	.1056	.1038	.1020	.1003	.0985
−1.1	.1357	.1335	.1314	.1292	.1271	.1251	.1230	.1210	.1190	.1170
−1.0	.1587	.1562	.1539	.1515	.1492	.1469	.1446	.1423	.1401	.1379
−0.9	.1841	.1814	.1788	.1762	.1736	.1711	.1685	.1660	.1635	.1611
−0.8	.2119	.2090	.2061	.2033	.2005	.1977	.1949	.1922	.1894	.1867
−0.7	.2420	.2389	.2358	.2327	.2296	.2266	.2236	.2206	.2177	.2148
−0.6	.2743	.2709	.2676	.2643	.2611	.2578	.2546	.2514	.2483	.2451
−0.5	.3085	.3050	.3015	.2981	.2946	.2912	.2877	.2843	.2810	.2776
−0.4	.3446	.3409	.3372	.3336	.3300	.3264	.3228	.3192	.3156	.3121
−0.3	.3821	.3783	.3745	.3707	.3669	.3632	.3594	.3557	.3520	.3483
−0.2	.4207	.4168	.4129	.4090	.4052	.4013	.3974	.3936	.3897	.3859
−0.1	.4602	.4562	.4522	.4483	.4443	.4404	.4364	.4325	.4286	.4247
−0.0	.5000	.4960	.4920	.4880	.4840	.4801	.4761	.4721	.4681	.4641

Continued

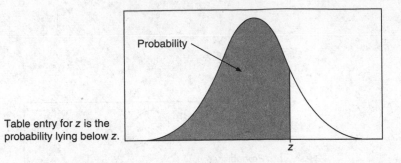

Table entry for z is the
probability lying below z.

TABLE A   Standard Normal Probabilities (*continued*)

z	.00	.01	.02	.03	.04	.05	.06	.07	.08	.09
0.0	.5000	.5040	.5080	.5120	.5160	.5199	.5239	.5279	.5319	.5359
0.1	.5398	.5438	.5478	.5517	.5557	.5596	.5636	.5675	.5714	.5753
0.2	.5793	.5832	.5871	.5910	.5948	.5987	.6026	.6064	.6103	.6141
0.3	.6179	.6217	.6255	.6293	.6331	.6368	.6406	.6443	.6480	.6517
0.4	.6554	.6591	.6628	.6664	.6700	.6736	.6772	.6808	.6844	.6879
0.5	.6915	.6950	.6985	.7019	.7054	.7088	.7123	.7157	.7190	.7224
0.6	.7257	.7291	.7324	.7357	.7389	.7422	.7454	.7486	.7517	.7549
0.7	.7580	.7611	.7642	.7673	.7704	.7734	.7764	.7794	.7823	.7852
0.8	.7881	.7910	.7939	.7967	.7995	.8023	.8051	.8078	.8106	.8133
0.9	.8159	.8186	.8212	.8238	.8264	.8289	.8315	.8340	.8365	.8389
1.0	.8413	.8438	.8461	.8485	.8508	.8531	.8554	.8577	.8599	.8621
1.1	.8643	.8665	.8686	.8708	.8729	.8749	.8770	.8790	.8810	.8830
1.2	.8849	.8869	.8888	.8907	.8925	.8944	.8962	.8980	.8997	.9015
1.3	.9032	.9049	.9066	.9082	.9099	.9115	.9131	.9147	.9162	.9177
1.4	.9192	.9207	.9222	.9236	.9251	.9265	.9279	.9292	.9306	.9319
1.5	.9332	.9345	.9357	.9370	.9382	.9394	.9406	.9418	.9429	.9441
1.6	.9452	.9463	.9474	.9484	.9495	.9505	.9515	.9525	.9535	.9545
1.7	.9554	.9564	.9573	.9582	.9591	.9599	.9608	.9616	.9625	.9633
1.8	.9641	.9649	.9656	.9664	.9671	.9678	.9686	.9693	.9699	.9706
1.9	.9713	.9719	.9726	.9732	.9738	.9744	.9750	.9756	.9761	.9767
2.0	.9772	.9778	.9783	.9788	.9793	.9798	.9803	.9808	.9812	.9817
2.1	.9821	.9826	.9830	.9834	.9838	.9842	.9846	.9850	.9854	.9857
2.2	.9861	.9864	.9868	.9871	.9875	.9878	.9881	.9884	.9887	.9890
2.3	.9893	.9896	.9898	.9901	.9904	.9906	.9909	.9911	.9913	.9916
2.4	.9918	.9920	.9922	.9925	.9927	.9929	.9931	.9932	.9934	.9936
2.5	.9938	.9940	.9941	.9943	.9945	.9946	.9948	.9949	.9951	.9952
2.6	.9953	.9955	.9956	.9957	.9959	.9960	.9961	.9962	.9963	.9964
2.7	.9965	.9966	.9967	.9968	.9969	.9970	.9971	.9972	.9973	.9974
2.8	.9974	.9975	.9976	.9977	.9977	.9978	.9979	.9979	.9980	.9981
2.9	.9981	.9982	.9982	.9983	.9984	.9984	.9985	.9985	.9986	.9986
3.0	.9987	.9987	.9987	.9988	.9988	.9989	.9989	.9989	.9990	.9990
3.1	.9990	.9991	.9991	.9991	.9992	.9992	.9992	.9992	.9993	.9993
3.2	.9993	.9993	.9994	.9994	.9994	.9994	.9994	.9995	.9995	.9995
3.3	.9995	.9995	.9995	.9996	.9996	.9996	.9996	.9996	.9996	.9997
3.4	.9997	.9997	.9997	.9997	.9997	.9997	.9997	.9997	.9997	.9998

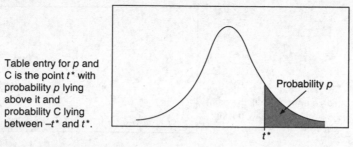

Table entry for *p* and *C* is the point *t** with probability *p* lying above it and probability *C* lying between −*t** and *t**.

Probability *p*

t*

TABLE B *t* Distribution Critical Values

df	.25	.20	.15	.10	.05	.025	.02	.01	.005	.0025	.001	.0005
					TAIL PROBABILITY *P*							
1	1.000	1.376	1.963	3.078	6.314	12.71	15.89	31.82	63.66	127.3	318.3	636.6
2	.816	1.061	1.386	1.886	2.920	4.303	4.849	6.965	9.925	14.09	22.33	31.60
3	.765	.978	1.250	1.638	2.353	3.182	3.482	4.541	5.841	7.453	10.21	12.92
4	.741	.941	1.190	1.533	2.132	2.776	2.999	3.747	4.604	5.598	7.173	8.610
5	.727	.920	1.156	1.476	2.015	2.571	2.757	3.365	4.032	4.773	5.893	6.869
6	.718	.906	1.134	1.440	1.943	2.447	2.612	3.143	3.707	4.317	5.208	5.959
7	.711	.896	1.119	1.415	1.895	2.365	2.517	2.998	3.499	4.029	4.785	5.408
8	.706	.889	1.108	1.397	1.860	2.306	2.449	2.896	3.355	3.833	4.501	5.041
9	.703	.883	1.100	1.383	1.833	2.262	2.398	2.821	3.250	3.690	4.297	4.781
10	.700	.879	1.093	1.372	1.812	2.228	2.359	2.764	3.169	3.581	4.144	4.587
11	.697	.876	1.088	1.363	1.796	2.201	2.328	2.718	3.106	3.497	4.025	4.437
12	.695	.873	1.083	1.356	1.782	2.179	2.303	2.681	3.055	3.428	3.930	4.318
13	.694	.870	1.079	1.350	1.771	2.160	2.282	2.650	3.012	3.372	3.852	4.221
14	.692	.868	1.076	1.345	1.761	2.145	2.264	2.624	2.977	3.326	3.787	4.140
15	.691	.866	1.074	1.341	1.753	2.131	2.249	2.602	2.947	3.286	3.733	4.073
16	.690	.865	1.071	1.337	1.746	2.120	2.235	2.583	2.921	3.252	3.686	4.015
17	.689	.863	1.069	1.333	1.740	2.110	2.224	2.567	2.898	3.222	3.646	3.965
18	.688	.862	1.067	1.330	1.734	2.101	2.214	2.552	2.878	3.197	3.611	3.922
19	.688	.861	1.066	1.328	1.729	2.093	2.205	2.539	2.861	3.174	3.579	3.883
20	.687	.860	1.064	1.325	1.725	2.086	2.197	2.528	2.845	3.153	3.552	3.850
21	.686	.859	1.063	1.323	1.721	2.080	2.189	2.518	2.831	3.135	3.527	3.819
22	.686	.858	1.061	1.321	1.717	2.074	2.183	2.508	2.819	3.119	3.505	3.792
23	.685	.858	1.060	1.319	1.714	2.069	2.177	2.500	2.807	3.104	3.485	3.768
24	.685	.857	1.059	1.318	1.711	2.064	2.172	2.492	2.797	3.091	3.467	3.745
25	.684	.856	1.058	1.316	1.708	2.060	2.167	2.485	2.787	3.078	3.450	3.725
26	.684	.856	1.058	1.315	1.706	2.056	2.162	2.479	2.779	3.067	3.435	3.707
27	.684	.855	1.057	1.314	1.703	2.052	2.158	2.473	2.771	3.057	3.421	3.690
28	.683	.855	1.056	1.313	1.701	2.048	2.154	2.467	2.763	3.047	3.408	3.674
29	.683	.854	1.055	1.311	1.699	2.045	2.150	2.462	2.756	3.038	3.396	3.659
30	.683	.854	1.055	1.310	1.697	2.042	2.147	2.457	2.750	3.030	3.385	3.646
40	.681	.851	1.050	1.303	1.684	2.021	2.123	2.423	2.704	2.971	3.307	3.551
50	.679	.849	1.047	1.299	1.676	2.009	2.109	2.403	2.678	2.937	3.261	3.496
60	.679	.848	1.045	1.296	1.671	2.000	2.099	2.390	2.660	2.915	3.232	3.460
80	.678	.846	1.043	1.292	1.664	1.990	2.088	2.374	2.639	2.887	3.195	3.416
100	.677	.845	1.042	1.290	1.660	1.984	2.081	2.364	2.626	2.871	3.174	3.390
1000	.675	.842	1.037	1.282	1.646	1.962	2.056	2.330	2.581	2.813	3.098	3.300
∞	.674	.841	1.036	1.282	1.645	1.960	2.054	2.326	2.576	2.807	3.091	3.291
	50%	60%	70%	80%	90%	95%	96%	98%	99%	99.5%	99.8%	99.9%
					Confidence level *C*							

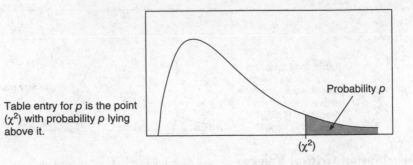

Table entry for p is the point (χ^2) with probability p lying above it.

TABLE C χ^2 Critical Values

df	TAIL PROBABILITY P											
	.25	.20	.15	.10	.05	.025	.02	.01	.005	.0025	.001	.0005
1	1.32	1.64	2.07	2.71	3.84	5.02	5.41	6.63	7.88	9.14	10.83	12.12
2	2.77	3.22	3.79	4.61	5.99	7.38	7.82	9.21	10.60	11.98	13.82	15.20
3	4.11	4.64	5.32	6.25	7.81	9.35	9.84	11.34	12.84	14.32	16.27	17.73
4	5.39	5.99	6.74	7.78	9.49	11.14	11.67	13.28	14.86	16.42	18.47	20.00
5	6.63	7.29	8.12	9.24	11.07	12.83	13.39	15.09	16.75	18.39	20.51	22.11
6	7.84	8.56	9.45	10.64	12.59	14.45	15.03	16.81	18.55	20.25	22.46	24.10
7	9.04	9.80	10.75	12.02	14.07	16.01	16.62	18.48	20.28	22.04	24.32	26.02
8	10.22	11.03	12.03	13.36	15.51	17.53	18.17	20.09	21.95	23.77	26.12	27.87
9	11.39	12.24	13.29	14.68	16.92	19.02	19.68	21.67	23.59	25.46	27.88	29.67
10	12.55	13.44	14.53	15.99	18.31	20.48	21.16	23.21	25.19	27.11	29.59	31.42
11	13.70	14.63	15.77	17.28	19.68	21.92	22.62	24.72	26.76	28.73	31.26	33.14
12	14.85	15.81	16.99	18.55	21.03	23.34	24.05	26.22	28.30	30.32	32.91	34.82
13	15.98	16.98	18.20	19.81	22.36	24.74	25.47	27.69	29.82	31.88	34.53	36.48
14	17.12	18.15	19.41	21.06	23.68	26.12	26.87	29.14	31.32	33.43	36.12	38.11
15	18.25	19.31	20.60	22.31	25.00	27.49	28.26	30.58	32.80	34.95	37.70	39.72
16	19.37	20.47	21.79	23.54	26.30	28.85	29.63	32.00	34.27	36.46	39.25	41.31
17	20.49	21.61	22.98	24.77	27.59	30.19	31.00	33.41	35.72	37.95	40.79	42.88
18	21.60	22.76	24.16	25.99	28.87	31.53	32.35	34.81	37.16	39.42	42.31	44.43
19	22.72	23.90	25.33	27.20	30.14	32.85	33.69	36.19	38.58	40.88	43.82	45.97
20	23.83	25.04	26.50	28.41	31.41	34.17	35.02	37.57	40.00	42.34	45.31	47.50
21	24.93	26.17	27.66	29.62	32.67	35.48	36.34	38.93	41.40	43.78	46.80	49.01
22	26.04	27.30	28.82	30.81	33.92	36.78	37.66	40.29	42.80	45.20	48.27	50.51
23	27.14	28.43	29.98	32.01	35.17	38.08	38.97	41.64	44.18	46.62	49.73	52.00
24	28.24	29.55	31.13	33.20	36.42	39.36	40.27	42.98	45.56	48.03	51.18	53.48
25	29.34	30.68	32.28	34.38	37.65	40.65	41.57	44.31	46.93	49.44	52.62	54.95
26	30.43	31.79	33.43	35.56	38.89	41.92	42.86	45.64	48.29	50.83	54.05	56.41
27	31.53	32.91	34.57	36.74	40.11	43.19	44.14	46.96	49.64	52.22	55.48	57.86
28	32.62	34.03	35.71	37.92	41.34	44.46	45.42	48.28	50.99	53.59	56.89	59.30
29	33.71	35.14	36.85	39.09	42.56	45.72	46.69	49.59	52.34	54.97	58.30	60.73
30	34.80	36.25	37.99	40.26	43.77	46.98	47.96	50.89	53.67	56.33	59.70	62.16
40	45.62	47.27	49.24	51.81	55.76	59.34	60.44	63.69	66.77	69.70	73.40	76.09
50	56.33	58.16	60.35	63.17	67.50	71.42	72.61	76.15	79.49	82.66	86.66	89.56
60	66.98	68.97	71.34	74.40	79.08	83.30	84.58	88.38	91.95	95.34	99.61	102.7
80	88.13	90.41	93.11	96.58	101.9	106.6	108.1	112.3	116.3	120.1	124.8	128.3
100	109.1	111.7	114.7	118.5	124.3	129.6	131.1	135.8	140.2	144.3	149.4	153.2

BIBLIOGRAPHY

Advanced Placement Program Course Description, New York: The College Board, 2010–2016.

Bock, D., Velleman, P., De Veaux, R. 2015. *Stats: Modeling the World*, 4th ed. Boston: Pearson.

Moore, David S. 2007. *Introduction to the Practice of Statistics*, 6th ed. New York: W.H. Freeman and Company.

Peck, R., Olsen, C., Devore, J. 2011. *Introduction to Statistics and Data Analysis*, 4th ed. Belmont, CA: Duxbury/Thomson Learning.

Starnes, D., Tabor, J., Yates, D., Moore, D. 2014. *The Practice of Statistics*, 5th ed. New York: W. H. Freeman and Company.

Utts, J., Heckard, R. 2011. *Mind on Statistics*. 4th ed. Belmont, CA: Duxbury/Thomson Learning.

Watkins, A., Scheaffer, R., Cobb, G. 2008. *Statistics in Action: Understanding a World of Data*, 2nd ed. Kendall-Hunt Press.

Yates, D., Starnes, D., Moore, D. 2009. *Statistics Through Applications*, 2nd ed. New York: W.H. Freeman and Company.

WEBSITES

Here is a list of websites that contain information and links that you might find useful in your preparation for the AP Statistics exam:

AP Central: http://apcentral.collegeboard.com/home

Bureau of the Census: http://www.census.gov/

Data and Story Library (DASL): http://lib.stat.cmu.edu/DASL/

Applets: http://wise.cgu.edu/

Statkey: http://lock5stat.com/statkey/

Art of Statistics: http://www.artofstat.com/webapps.html

Rossman/Chance applets: http://www.rossmanchance.com/applets/

GLOSSARY

Alternative hypothesis—the theory that the researcher hopes to confirm by rejecting the null hypothesis

Association—when some of the variability in one variable can be accounted for by the other

Bar chart—graph in which the frequencies or relative frequencies of categories are displayed with bars

Bimodal—distribution with two most common values; see **mode**

Binomial distribution—probability distribution for a random variable X in a binomial setting;

$$P(X = x) = \binom{n}{x}(p)^x(1-p)^{n-x},$$

where n is the number of independent trials, p is the probability of success on each trial, and x is the count of successes out of the n trials

Binomial setting (experiment)—when each of a fixed number, n, of observations either succeeds or fails, independently, with probability p

Bivariate data—having to do with two variables

Block—a group of experimental units thought to be homogenous with respect to the response variable

Block design—procedure by which experimental units are put into homogeneous groups in an attempt to reduce variability due to the group on the response variable

Blocking—see **block design**

Boxplot (box-and-whisker plot)—graphical representation of the five-number summary of a dataset. Each value in the five-number summary is located over its corresponding value on a number line. A box is drawn that ranges from Q1 to Q3 and "whiskers" extend to the maximum and minimum values from Q1 and Q3.

Categorical data—data whose values range over categories rather than values. See also **qualitative data**

Census—attempt to contact every member of a population

Center—the "middle" of a distribution; either the mean or the median

Central limit theorem—theorem that states that the sampling distribution of a sample mean becomes approximately normal when the sample size is large

Chi-square (χ^2) goodness-of-fit test—compares a set of observed categorical values to a set of expected values under a set of hypothesized proportions for the categories;

$$\chi^2 = \sum \frac{(O-E)^2}{E}$$

Cluster sample—The population is first divided into sections or "clusters." Then we randomly select an entire cluster, or clusters, and include all of the members of the cluster(s) in the sample.

Coefficient of determination (r^2)—measures the proportion of variation in the response variable explained by regression on the explanatory variable

Complement of an event—set of all outcomes in the sample space that are not in the event

Completely randomized design—when all subjects (or experimental units) are randomly assigned to treatments in an experiment

Conditional probability—the probability of one event succeeding given that some other event has already occurred

Confidence interval—an interval that, with a given level of confidence, is likely to contain a population value; (estimate) ± (margin of error)

Confidence level—the probability that the procedure used to construct an interval will generate an interval that does contain the population value

Confounding variable—has an effect on the outcomes of the study but whose effects cannot be separated from those of the treatment variable

Contingency table—see **two-way table**

Continuous data—data that can be measured, or take on values in an interval; the set of possible values cannot be counted

Continuous random variable—a random variable whose values are continuous data; takes all values in an interval

Control—see **statistical control**

Convenience sample—sample chosen without any random mechanism; chooses individuals based on ease of selection

Correlation coefficient (r)—measures the strength of the linear relationship between two quantitative variables;

$$r = \frac{1}{n-1}\sum_{i=1}^{n}\left(\frac{x_i - \bar{x}}{s_x}\right)\left(\frac{y_i - \bar{y}}{s_y}\right)$$

Correlation is not causation—just because two variables correlate strongly does not mean that one caused the other

Critical value—values in a distribution that identify certain specified areas of the distribution

Degrees of freedom—number of independent datapoints in a distribution

Density function—a function that is everywhere non-negative and has a total area equal to 1 underneath it and above the horizontal axis

Descriptive statistics—process of examining data analytically and graphically

Dimension—size of a two-way table; $r \times c$

Discrete data—data that can be counted (possibly infinite) or placed in order

Discrete random variable—random variable whose values are discrete data

Dotplot—graph in which data values are identified as dots placed above their corresponding values on a number line

Double blind—experimental design in which neither the subjects nor the study administrators know what treatment a subject has received

Empirical Rule (68-95-99.7 Rule)—states that, in a normal distribution, about 68% of the terms are within one standard deviation of the mean, about 95% are within two standard deviations, and about 99.7% are within three standard deviations

Estimate—sample value used to approximate a value of a parameter

Event—in probability, a subset of a sample space; a set of one or more simple outcomes

Expected value—mean value of a discrete random variable

Experiment—study in which a researcher measures the responses to a treatment variable, or variables, imposed and controlled by the researcher

Experimental units—individuals on which experiments are conducted

Explanatory variable—explains changes in response variable; treatment variable; independent variable

Extrapolation—predictions about the value of a variable based on the value of another variable outside the range of measured values

First quartile—the value which identifies the 25th percentile; the value that has at least 25% of the data at or below it and at least 75% of the data at or above it.

Five-number summary—for a dataset, [minimum value, Q1, median, Q3, maximum value]

Geometric setting—independent observations, each of which succeeds or fails with the same probability p; number of trials needed until first success is variable of interest

Histogram—graph in which numerical data are grouped into intervals and the frequencies or relative frequencies within each interval are displayed with bars

Homogeneity of proportions—chi-square hypothesis in which proportions of a categorical variable are tested for homogeneity across two or more populations

Independent events—knowing one event occurs does not change the probability that the other occurs; $P(A) = P(A|B)$

Independent variable—see **explanatory variable**

Inferential statistics—use of sample data to make inferences about populations

Influential observation—observation, usually in the x direction, whose removal would have a marked impact on the slope of the regression line

Interpolation—predictions about the value of a variable based on the value of another variable within the range of measured values

Interquartile range—value of the third quartile minus the value of the first quartile; contains middle 50% of the data

Least-squares regression line—of all possible lines, the line that minimizes the sum of squared errors (residuals) from the line

Line of best fit—see **least-squares regression line**

Lurking variable—one that has an effect on the outcomes of the study but whose influence was not part of the investigation

Margin of error—measure of uncertainty in the estimate of a parameter; (critical value) · (standard error)

Marginal totals—row and column totals in a two-way table

Matched pairs—experimental units paired by a researcher based on some common characteristic or characteristic

Matched pairs design—experimental design that utilizes each pair as a block; one unit receives one treatment, and the other unit receives the other treatment

Mean—sum of all the values in a dataset divided by the number of values

Median—halfway through an ordered dataset, below and above which lies an equal number of data values; 50th percentile

Mode—most common value in a distribution

Mound-shaped (bell-shaped)—distribution in which data values tend to cluster about the center of the distribution; characteristic of a normal distribution

Mutually exclusive events—events that cannot occur simultaneously; if one occurs, the other doesn't

Negatively associated—larger values of one variable are associated with smaller values of the other; see **association**

Nonresponse bias—occurs when subjects selected for a sample do not respond

Normal curve—familiar bell-shaped density curve; symmetric about its mean; defined in terms of its mean and standard deviation;

$$f(x) = \frac{1}{\sigma\sqrt{2\pi}} e^{-\frac{1}{2}\left(\frac{x-\mu}{\sigma}\right)^2}$$

Normal distribution—distribution of a random variable X so that $P(a < X < b)$ is the area under the normal curve between a and b

Null hypothesis—hypothesis being tested—usually a statement that there is no effect or difference between treatments; what a researcher wants to disprove to support his/her alternative

Numerical data—see **quantitative data**

Observational study—when variables of interest are observed and measured but no treatment is imposed in an attempt to influence the response

Observed values—counts of outcomes in an experiment or study; compared with expected values in a chi-square analysis

One-sided alternative—alternative hypothesis that varies from the null in only one direction

One-sided test—used when an alternative hypothesis states that the true value is either less than or greater than the hypothesized value but not both

Outcome—simple events in a probability experiment

Outlier—a data value that is far removed from the general pattern of the data

P(A and B)—probability that *both* A and B occur; $P(A \text{ and } B) = P(A) \cdot P(A|B)$

P(A or B)—probability that *either* A or B occurs; $P(A \text{ or } B) = P(A) + P(B) - P(A \text{ and } B)$

P-value—probability of getting a sample value at least as extreme as that obtained by chance alone assuming the null hypothesis is true

Parameter—measure that describes a population

Percentile rank—proportion of terms in the distributions less than the value being considered

Placebo—an inactive procedure or treatment

Placebo effect—effect, often positive, attributable to the patient's expectation that the treatment will have an effect

Point estimate—value based on sample data that represents a likely value for a population parameter

Positively associated—larger values of one variable are associated with larger values of the other; see **association**

Power of a test—probability of rejecting a null hypothesis against a specific alternative

Probability distribution—identification of the outcomes of a random variable together with the probabilities associated with those outcomes

Probability histogram—histogram for a probability distribution; horizontal axis shows the outcomes, vertical axis shows the probabilities of those outcomes

Probability of an event—relative frequency of the number of ways an event can succeed to the total number of ways it can succeed or fail

Probability sample—sampling technique that uses a random mechanism to select the members of the sample

Proportion—ratio of the count of a particular outcome to the total number of outcomes

Qualitative data—data whose values range over categories rather than values. See also **categorical data**

Quantitative data—data whose values are numerical

Quartiles—25th, 50th, and 75th percentiles of a dataset

Random phenomenon—unclear how any one trial will turn out, but there is a regular distribution of outcomes in a large number of trials

Random sample—sample in which each member of the sample is chosen by chance and each member of the population has an equal chance to be in the sample

Random variable—numerical outcome of a random phenomenon (random experiment)

Randomization—random assignment of experimental units to treatments

Range—difference between the maximum and minimum values of a dataset

Replication—repetition of each treatment enough times to help control for chance variation

Representative sample—sample that possesses the essential characteristics of the population from which it was taken

Residual—in a regression, the actual value minus the predicted value

Resistant statistic—one whose numerical value is not influenced by extreme values in the dataset

Response bias—bias that stems from respondents' inaccurate or untruthful response

Response variable—measures the outcome of a study

Robust—when a procedure may still be useful even if the conditions needed to justify it are not completely satisfied

Robust procedure—procedure that still works reasonably well even if the assumptions needed for it are violated; the *t*-procedures are robust against the assumption of normality as long as there are no outliers or severe skewness.

Sample space—set of all possible mutually exclusive outcomes of a probability experiment

Sample survey—using a sample from a population to obtain responses to questions from individuals

Sampling distribution of a statistic—distribution of all possible values of a statistic for samples of a given size

Sampling frame—list of experimental units from which the sample is selected

Scatterplot—graphical representation of a set of ordered pairs; horizontal axis is first element in the pair, vertical axis is the second

Shape—geometric description of a dataset: mound-shaped; symmetric, uniform; skewed; etc.

Significance level (α)—probability value that, when compared to the *P*-value, determines whether a finding is statistically significant

Simple random sample (SRS)—sample in which all possible samples of the same size are equally likely to be the sample chosen

Simulation—random imitation of a probabilistic situation

Skewed—distribution that is asymmetrical with data bunched at one end and a tail stretching out in the other

Skewed left (right)—asymmetrical with more of a tail on the left (right) than on the right (left)

Spread—variability of a distribution

Standard deviation—square root of the variance;

$$s = \sqrt{\frac{\sum (x - \bar{x})^2}{n-1}}$$

Standard error—estimate of population standard deviation based on sample data

Standard normal distribution—normal distribution with a mean of 0 and a standard deviation of 1

Standard normal probability—normal probability calculated from the standard normal distribution

Statistic—measure that describes a sample (e.g., sample mean)

Statistical control—holding constant variables in an experiment that might affect the response but are not one of the treatment variables

Statistically significant—a finding that is unlikely to have occurred by chance

Statistics—science of data

Stemplot (stem-and-leaf plot)—graph in which ordinal data are broken into "stems" and "leaves"; visually similar to a histogram except that all the data are retained

Stratified random sample—groups of interest (strata) chosen in such a way that they appear in approximately the same proportions in the sample as in the population

Subjects—human experimental units

Survey—obtaining responses to questions from individuals

Symmetric—data values distributed equally above and below the center of the distribution

Systematic bias—the mean of the sampling distribution of a statistic does not equal the mean of the population; see **unbiased estimate**

Systematic sample—probability sample in which one of the first *n* subjects is chosen at random for the sample and then each *n*th person after that is chosen for the sample

t-distribution—the distribution with $n - 1$ degrees of freedom for the *t* **statistic**

t statistic—

$$t = \frac{\bar{x} - \mu}{s/\sqrt{n}}$$

Test statistic—

$$\frac{\text{estimator} - \text{hypothesized value}}{\text{standard error}}$$

Third quartile—the value which identifies the 75th percentile; the value that has at least 75% of the data at or below it and at least 25% of the data at or above it.

Treatment variable—see **explanatory variable**

Tree diagram—graphical technique for showing all possible outcomes in a probability experiment

Two-sided alternative—alternative hypothesis that can vary from the null in either direction; values much greater than or much less than the null provide evidence against the null

Two-sided test—a hypothesis test with a **two-sided alternative**

Two-way table—table that lists the outcomes of two categorical variables; the values of one category are given as the row variable, and the values of the other category are given as the column variable; also called a contingency table

Type I error—the error made when a true null hypothesis is rejected

Type II error—the error made when a false null hypothesis is not rejected

Unbiased estimate—mean of the sampling distribution of the estimate equals the parameter being estimated

Undercoverage—some groups in a population are not included in a sample from that population

Uniform—distribution in which all data values have the same frequency of occurrence

Univariate data—having to do with a single variable

Variance—average of the squared deviations from their mean of a set of observations;

$$s^2 = \frac{\sum (x - \bar{x})^2}{n-1}$$

Voluntary response bias—bias inherent when people choose to respond to a survey or poll; bias is typically toward opinions of those who feel most strongly

Voluntary response sample—sample in which participants are free to respond or not to a survey or a poll

Wording bias—creation of response bias attributable to the phrasing of a question

z-score—number of standard deviations a term is above or below the mean;

$$z = \frac{x - \bar{x}}{s}$$

NOTES

NOTES

The Cross-Platform Prep Course

McGraw-Hill Education's multi-platform course gives you a variety of tools to help you raise your test scores. Whether you're studying at home, in the library, or on-the-go, you can find practice content in the format you need—print, online, or mobile.

Print Book

This print book gives you the tools you need to ace the test. In its pages you'll find smart test-taking strategies, in-depth reviews of key topics, and ample practice questions and tests. See the Welcome section of your book for a step-by-step guide to its features.

Online Platform

The Cross-Platform Prep Course gives you additional study and practice content that you can access *anytime, anywhere*. You can create a personalized study plan based on your test date that sets daily goals to keep you on track. Integrated lessons provide important review of key topics. Practice questions, exams, and flashcards give you the practice you need to build test-taking confidence. The game center is filled with challenging games that allow you to practice your new skills in a fun and engaging way. And, you can even interact with other test-takers in the discussion section and gain valuable peer support.

Getting Started

To get started, open your account on the online platform:

Go to www.xplatform.mhprofessional.com

↓

Enter your access code, which you can find on the inside back cover of your book

↓

Provide your name and e-mail address to open your account and create a password

↓

Click "Start Studying" to enter the platform

It's as simple as that. You're ready to start studying online.

Your Personalized Study Plan

First, select your test date on the calendar, and you'll be on your way to creating your personalized study plan. Your study plan will help you stay organized and on track and will guide you through the course in the most efficient way. It is tailored to *your* schedule and features daily tasks that are broken down into manageable goals. You can adjust your test date at any time and your daily tasks will be reorganized into an updated plan.

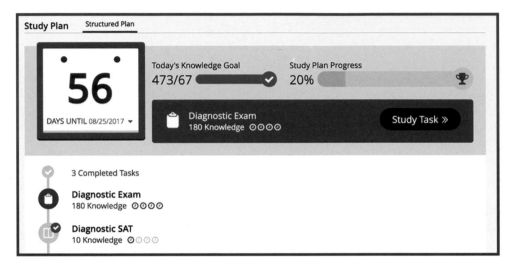

You can track your progress in real time on the Study Plan Dashboard. The "Today's Knowledge Goal" progress bar gives you up-to-the minute feedback on your daily goal. Fulfilling this every time you log on is the most efficient way to work through the entire course. You always get an instant view of where you stand in the entire course with the Study Plan Progress bar.

If you need to exit the program before completing a task, you can return to the Study Plan Dashboard at any time. Just click the Study Task icon and you can automatically pick up where you left off.

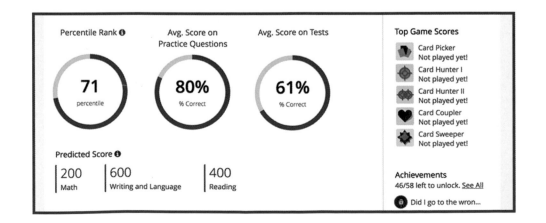

Practice Tests

One of the first tasks in your personalized study plan is to take the Diagnostic Test. At the end of the test, a detailed evaluation of your strengths and weaknesses shows the areas where you need the most focus. You can review your practice test results either by the question category to see broad trends or question-by-question for a more in-depth look.

The full-length tests are designed to simulate the real thing. Try to simulate actual testing conditions and be sure you set aside enough time to complete the full-length test. You'll learn to pace yourself so that you can work toward the best possible score on test day.

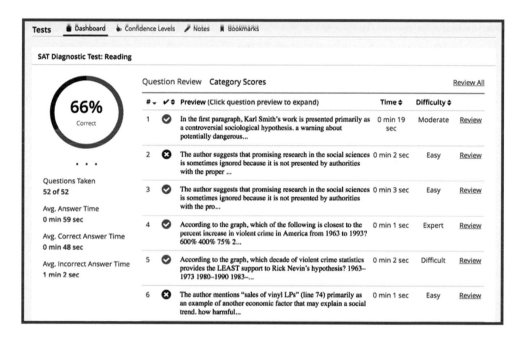

Lessons

The lessons in the online platform are divided into manageable pieces that let you build knowledge and confidence in a progressive way. They cover the full range of topics that you're likely to see on your test.

After you complete a lesson, mark your confidence level. (You must indicate a confidence level in order to count your progress and move on to the next task.) You can also filter the lessons by confidence levels to see the areas you have mastered and those that you might need to revisit.

> *Use the bookmark feature to easily refer back to a concept or leave a note to remember your thoughts or questions about a particular topic.*

♠ Home

▤ Study Plan

⊞ Game Center

🖥 Discussions

☑ Historical Reports

▥ Lessons

🗂 Flashcards

▤ Practice

🖥 Tests

Lessons   Table of Contents   Confidence Levels   Notes   Bookmarks

Browse

◀ ▶   What's new in the redesigned SAT?   🔖

What's new in the redesigned SAT?

Beginning in spring 2016, the redesigned SAT features ten major changes.

1. More time per question

The redesigned SAT gives you more time per question, making it less likely that you will underperform due to time restrictions.

Section	Old SAT Time Per Question	New SAT Time Per Question
Writing and Language	43 seconds (49 questions, 35 mins)	48 seconds (44 questions, 35 mins)
Mathematics	78 seconds (54 questions, 70 mins)	84 seconds (57 questions, 80 mins)
Reading	63 seconds (67 questions in 70 mins)	75 seconds (52 questions, 65 mins)

Rate Your Confidence

High

Medium

Low

Continue »

Category :
Attacking the New SAT: Twelve FAQs

Practice Questions

All of the practice questions are reflective of actual exams and simulate the test-taking experience. The "Review Answer" button gives you immediate feedback on your answer. Each question includes a rationale that explains why the correct answer is right and the others are wrong. To explore any topic further, you can find detailed explanations by clicking the "Help me learn about this topic" link.

Quantitative-Data Sufficiency

🕐 This question: 00:17   🕐 Total: 00:17   Done practicing

Question 1 of 24

Review Answer **ON**

Direction :The following data sufficiency problem consist of a question and two statements, labeled (1) and (2), in which certain data are given. You have to decide whether the data given in the statements are *sufficient* for answering the question.
A. Statement (1) ALONE is sufficient, but statement (2) alone is not sufficient.
B. Statement (2) ALONE is sufficient, but statement (1) alone is not sufficient.
C. BOTH statements TOGETHER are sufficient, but NEITHER statement ALONE is sufficient.
D. EACH statement ALONE is sufficient.
E. Statements (1) and (2) TOGETHER are NOT sufficient.

Q. Is x an integer?

(1) x is evenly divisible by $\frac{1}{3}$

(2) x is evenly divisible by 2.

Solution:
The correct answer is B.
A reminder: The question stem starts with *is*, so this is a *yes* or *no* problem. The first step is to determine whether statement (1) is sufficient, which we can do by trying out some real numbers. If x is 2, which is evenly divisible by $\frac{1}{3}$, then the answer would be *yes*. But if x were 1.5, which is also evenly divisible by $\frac{1}{3}$, then the answer would be *no*. Since you can get either *yes* or *no*, statement (1) is not sufficient and you should cross off A and D. Now let's move on to statement (2). There are no non-integer numbers that are divisible by 2, so *yes* is the only possibility. Statement (2) is sufficient, so we can eliminate C and E. Thus the answer is B.

❌ Incorrect

Help me learn about this topic

Correct answer   Your answer

A

B

C

D   ✕

E

Continue

Difficulty Level: **Unrated**

You can go to the Practice Dashboard to find an overview of your performance in the different categories and sub-categories.

| Practice | Dashboard | Confidence Levels | Notes | Bookmarks |

| Dashboard | | | Reset All Questions |

77% Correct

Questions Taken
159 of 491

Avg. Answer Time
2 min 52 sec

Avg. Correct Answer Time
2 min 20 sec

Avg. Incorrect Answer Time
3 min 6 sec

Avg. Session Duration
18 min 41 sec

Question Categories

Category Name	Completion	% Correct
Quantitative-Problem Solving	94 of 297	72% »
Arithmetic	75 of 75	84%
Elementary Algebra	35 of 189	60%
Geometry	0 of 49	--
Quantitative-Data Sufficiency	49 of 49	81%
Arithmetic	6 of 6	75%
Elementary Algebra	30 of 30	86%
Geometry	13 of 13	82%
Verbal	0 of 129	--
Reading Comprehension	0 of 46	--
Critical Reasoning	0 of 64	--
Sentence Correction	0 of 19	--

Dashboard

The dashboard is constantly updating to reflect your progress and performance. The Percentile Rank icon shows your position relative to all the other students enrolled in the course. You can also find information on your average scores in practice questions and exams.

A detailed overview of your strengths and weaknesses shows your proficiency in a category based on your answers and difficulty of the questions. By viewing your strengths and weaknesses, you can focus your study on areas where you need the most help.

Strengths & Weaknesses ⓘ

Writing and Language

| Beginner | Basic | Intermediate | Proficient | Advanced | Expert |

Reading

| Beginner | Basic | Intermediate | Proficient | Advanced | Expert |

Math

| Beginner | Basic | Intermediate | Proficient | Advanced | Expert |

The Heart of Algebra	N/A	Beginner	Basic	Intermediate	Proficient	Advanced	Expert
Advanced Mathematics	N/A	Beginner	Basic	Intermediate	Proficient	Advanced	Expert
Additional Topics	N/A	Beginner	Basic	Intermediate	Proficient	Advanced	Expert

🔒 Riding dolphins
🔒 What doesn't kill yo...
🔒 Good Karma
🔒 Stenographer

Top Performers
👤 David Flores
👤 Jenny
👤 Elle
👤 Shadaj L
👤 Grant Foray
👤 Coco Ma
👤 Angela Jeon
👤 Sarah Kim
👤 Madison McClane
👤 Gabi Kreuscher
...
👤 Tenant Admin

Flashcards

The hundreds of flashcards are perfect for learning key terms quickly, and the interactive format gives you immediate feedback. You can filter the cards by category and confidence level for a more organized approach. Or, you can shuffle them up for a more general challenge.

1 of 14

All the

includ

all fra

Real Numbers

Did you get it right?

| No | Kinda | Yes |

Another way to customize the flashcards is to create your own sets. You can either keep these private or share or them with the public. Subscribe to Community Sets to access sets from other students preparing for the same exam.

Flashcards   Study   Community Sets   My Sets

Create Set

Title

Math review

Description

Topic – Select one of our topics from the dropdown or add your own

Math equations

Access

● Public
○ Private

Create Set   Cancel

Game Center

Play a game in the Game Center to test your knowledge of key concepts in a challenging but fun environment. Increase the difficulty level and complete the games quickly to build your highest score. Be sure to check the leaderboard to see who's on top!

| Game Center | ⊞ Dashboard | 🂠 Card Picker | ◎ Card Hunter I | ◎ Card Hunter II | ♡ Card Coupler | ⬚ Card Sweeper |

Card Picker

Your highest score: 0
Your rank: Not yet determined

[Start playing]

Leaderboard
Top 1: Alyssa - 232001
Top 2: Fatima - 188598
Top 3: Sheena Lu - 80160

Card Hunter I

Your highest score: 0
Your rank: Not yet determined

[Start playing]

Leaderboard
Top 1: Siddh - 1997
Top 2: Siddh - 1997
Top 3: Eileen - 1915

Card Hunter II

Your highest score: 0
Your rank: Not yet determined

[Start playing]

Leaderboard
Top 1: Siddh - 23990
Top 2: Eileen - 9037
Top 3: Ira - 8927

Card Coupler

Your highest score: 0
Your rank: Not yet determined

[Start playing]

Leaderboard
Top 1: Toni - 1398
Top 2: Siddh - 1062
Top 3: Ri - 1051

Social Community

Interact with other students who are preparing for the same test. Start a discussion, reply to a post, or even upload files to share. You can search the archives for common topics or start your own private discussion with friends.

Discussions   Public   Private

Public Discussions ⓘ   [Filter ▾]

What do you want to discuss...

PA Hey, I have to take the SAT Exam in 55 days and I'm just starting to study now. Is it possible or should I take the exam at a later date?

Created 4 days ago by Prosperity Antoine   0 replies | Last activity: 4 days ago
Reply to this Comment

RJ Is it possible to practice a full day exam as i would when walking into to take the SAT???

Created 2 months ago by River Jones   0 replies | Last activity: 2 months ago
Reply to this Comment

Mobile App

The companion mobile app lets you toggle between the online platform and your mobile device without missing a beat. Whether you access the course online or from your smartphone or tablet, you'll pick up exactly where you left off.

Go to the iTunes or Google Play stores and search "McGraw-Hill Education Cross-Platform App" to download the companion iOS or Android app. Enter your e-mail address and the same password you created for the online platform to open your account.

Now, let's get started!